Magnetic Shape Memory Alloys

Xuexi Zhang · Mingfang Qian

Magnetic Shape Memory Alloys

Preparation, Martensitic Transformation and Properties

哈爾濱工業大學出版社
HARBIN INSTITUTE OF TECHNOLOGY PRESS

Springer

Xuexi Zhang
School of Materials Science
and Engineering
Harbin Institute of Technology
Harbin, China

Mingfang Qian
School of Materials Science
and Engineering
Harbin Institute of Technology
Harbin, China

ISBN 978-981-16-6335-2 ISBN 978-981-16-6336-9 (eBook)
https://doi.org/10.1007/978-981-16-6336-9

Jointly published with Harbin Institute of Technology Press
The print edition is not for sale in China (Mainland). Customers from China (Mainland) please order the print book from: Harbin Institute of Technology Press.

This Springer imprint is published by the registered company Springer Nature Singapore Pte Ltd.
The registered company address is: 152 Beach Road, #21-01/04 Gateway East, Singapore 189721, Singapore

Preface

Shape memory alloys, polymers and ceramics are important intelligent materials for various applications. Traditional shape memory alloys (SMAs) such as Ti-Ni have been widely used in engineering and biomedical fields. These SMAs are driven by heat or stress. In the past few decades, a new type of SMAs, i.e., magnetic shape memory alloys (MSMAs), attracted much attentions since they can be driven by an external magnetic field, and thus can work at much higher frequencies than traditional Ti-Ni alloy owing to the rapid generation of reversible strain under magnetic field. MSMAs can respond to temperature, stress and magnetic fields and output stress, strain and heat effects. So, they are promising for actuators, sensors and refrigeration applications.

In 1996, Ullakko obtained 0.19% recoverable magnetic field-induced strain (MFIS) in Ni_2MnGa single crystals. Soon O'Handerly proposed that the high MIFS was attributed to the reorientation of twin variants in the martensite that has high magnetic anisotropy energy. In the following years, the obtained MFIS in Ni-Mn-Ga single-crystal alloys increased rapidly and reached the theoretical value of 9.4% at 2003. This makes the MSMAs represented by Ni-Mn-Ga gain extensive attention for scientist and engineers. In 2006, Kainuma reported the magnetic-induced phase transformation in a Co-doped Ni-Mn-In alloy that showed extremely high work output during the metamagnetic transition process. Then, Ni-Mn-In(Sn,Sb) and related quaternary alloys set off another research upsurge in the field of MSMAs, especially for the room-temperature refrigeration.

Ni-Mn-Ga single crystals exhibit excellent MFIS; but the growth rate of single crystals is slow and the composition is prone to serious segregation. The preparation cost of single crystals is also high. Polycrystalline alloys are feasible to prepare with low cost; however, the MFIS in polycrystalline alloys is low due to the constraints of twin boundary motion induced by grain boundaries. By introducing texture and pores (size of pores is the same order of grains) in polycrystalline bulk alloys or preparing reduced sized microwires to reduce the restriction of grain boundaries, the MFIS can be enhanced. These reduced sized materials with porous or wire structure has a large specific surface area and thus some advantages over bulk counterpart, such as reduced thermal/magnetic hysteresis and eddy current loss, increased heat transfer efficiency and accelerated demagnetization time. These characters make small-sized Ni-Mn-X (X = Ga, In, Sn, Sb) alloy especially attractive for high-efficient solid-state refrigeration, which is superior than conventional vapor compression technique.

For more than ten years, the preparation, processing, phase transformation and properties of MSMAs, especially their small-sized materials, have been systematically studied. This book tried to summarize the latest development in this field, aiming at accelerating the communication of the researchers in this field. In this book, the development history and characters of MSMAs (Chap. 1) are briefly reviewed. The preparation and properties of bulk alloys, especially single-crystal alloys are given in Chap. 2. Then, the preparation techniques, phase transformations and functional properties of foams (Chap. 3), microwires (Chaps. 4, 5) and particles (Chap. 6) are introduced. Finally, the possible application fields are discussed in Chap. 7. This book is mainly composed of the research work published by the authors and their supervised graduate and postgraduate students; the latest research results of other researcher are also included. This book focuses on the size effect of MSMAs and describes the preparation, phase transformation and functional characteristics under multiple conditions. This book can be used as a reference book for researchers, engineers and postgraduates in this field.

During writing this book, Prof. Lin Geng (Harbin Institute of Technology) gave warm support and help. The postgraduate students Longsha Wei, Hehe Zhang, Ruochen Zhang, Siyao Ma, Muhammad Imran and Xuejie Zhu participated in the material organization and draft writing. The authors express his/her gratitude to them. Financial supports from National Science Foundation of China (NSFC), Heilongjiang Touyan Team and China Postdoctoral Science Foundation for the research work are also appreciated.

Harbin, China
2021

Xuexi Zhang
Mingfang Qian

Contents

Chapter 1
An Overview on Magnetic Shape Memory Alloys

1.1 Introduction

Smart materials can respond actively to input stimuli in a controllable and reversible way. The external stimulus may be mechanical stress, temperature, light, heat, sound, electricity, magnetic field, pH change and much more. As a result, smart materials may act as actuators and sensors.

Smart materials include alloys, polymers and ceramics. Shape memory materials are a kind of smart materials that can recover their original shape from plastic deformation under an external stimulus, which is known as the shape memory effect (SME). They also exhibit superelasticity (in alloys) or visco-elasticity (in polymers) under certain conditions. Nowadays, SME has been found in alloys, polymers and ceramics.

Studies on shape memory alloys (SMAs) dated back to 1932, when Holland and Sweden observed the shape change and recovery in a gold–cadmium alloy. This is the first demonstration of SME. Later in 1963, Buehler, the United States Naval Ordnance Research Institute, found the SME in nickel titanium alloys and opened the door for their worldwide applications. The functional properties of shape memory materials are mainly derived from thermoelastic martensitic transformation driven by heat and stress. However, in the 1990s, a new class of SMAs combining ferromagnetic to paramagnetic (or antiferromagnetic) change in parent and martensite phases was discovered. They are later named magnetic SMAs (MSMAs) since they can be driven actively by a magnetic field. As a result, the working frequency of MSMAs may be very high (depending on the external magnetic field change frequency).

MSMAs consist of two categories: (1) ferromagnetic MSMAs (FMSMAs), including Ni-based (e.g., Ni–Mn–Ga, Ni–Fe–Ga, Ni–Mn–Al), Co-based (e.g., Co–Ni–Ga and Co–Ni–Al) and Fe-based (e.g., Fe–Pd and Fe–Pt); (2) meta-magnetic MSMAs (MMSMAS), including Ni–Mn–X (X = Sn, In, Sb).

MSMAs have not only magnetic SME, but also traditional properties such as SME, superelasticity, magnetic-field-induced strain (MFIS), magnetoresistance and magnetocaloric effect (MCE). The MFIS of single-crystal Ni–Mn–Ga alloy is an

X. Zhang and M. Qian, *Magnetic Shape Memory Alloys*,
https://doi.org/10.1007/978-981-16-6336-9_1

order of magnitude larger than that of magnetostrictive materials such as Terfenol-D and is also larger than that of piezoelectric and magnetostrictive materials [1]. However, the output power of Ni–Ti alloy is higher than that of Ni–Mn–Ga alloys.

So far, MFIS of the single-crystal Ni–Mn–Ga alloys has reached 10% [2]. However, the single-crystal preparation is difficult, the cost is high, and the composition segregation is prone to occur, which seriously deteriorates its properties and limits its practical application. MFIS of polycrystalline Ni–Mn–Ga fine-grained alloys is small ($< 0.01\%$) due to the constraints caused by grain boundary that hinders the movement of twin boundaries and large orientation difference between adjacent grains that counteracts the overall strain. Obviously, increasing grain size, introducing texture or training can reduce the constraints of grain boundaries to the movement of twin boundaries. For example, about 1% MFIS [3] can be obtained in well-trained polycrystalline Ni–Mn–Ga alloys with strong texture.

The magnetocrystalline anisotropic energy (K_u) acts as the driving force for MFIS in FMSMAs to overcome the twin boundary motion resistance (e.g., $\sigma_{tw} =$ 2–6 MPa [4–6] for Ni–Mn–Ga single crystals). Ganor et al. [7] analyzed that K_u can be increased by reducing sample size, while σ_{tw} can be reduced [8] by adjusting composition, changing martensite structure and reducing grain boundary area.

Giant MCE has been found in Ni–Mn–X (X = Ga, In, Sn, Sb) alloys. Magnetic refrigeration materials have the characteristics of high magnetic entropy density, simple composition structure, low pollution, excellent refrigeration efficiency and low power consumption. Replacing the refrigeration materials such as Freon for refrigerators, freezers, air conditioners and other refrigeration equipment can effectively eliminate environmental pollution, which is of great significance and is expected to become a new refrigeration material [9].

In summary, FMSMAs have made great progress in preparation technology, magnetostructural transitions, physical properties and applications. In the past decades, research on small-sized materials has received great attention. The small-sized materials include particles, microwire, films, ribbons and porous alloys. The porous alloys are considered as small-sized materials since they are built up by micron-sized nodes and struts. So, the small-sized materials exhibit at least one characteristic length in micron size. These small-sized materials can be feasibly manipulated to form oligocrystals, bamboo grains or even single crystals, which renders them novel properties not possessed by bulk counterparts. The high specific surface area can enhance the heat exchange capacity and reduce the thermal hysteresis; the single-crystal microwires also exhibit enhanced output power. Furthermore, small size materials show reduced eddy current and small inertia, leading to increased response frequency and working hysteresis [10]. Representative works include the size effect theories of conventional SMAs [11] and Ni–Mn–Ga foam alloy with novel architectures [12]. Nowadays, small-sized materials exhibit high application potential in various minor devices.

1.2 Phase Diagram, Composition and Crystal Structure

MSMAs are ordered intermetallic compounds. The crystal structure belongs to Heusler alloy type X_2YZ, where X, Y are transition group elements and Z are III–V group elements. These alloys are subjected to significant magnetic property change during the magnetostructural transition, which renders them various properties.

1.2.1 Ni–Mn–Ga Alloys

1.2.1.1 Development Scenario

Ni_2MnGa alloys have both ferromagnetic and thermoelastic martensitic transformations. In 1978, Stager and Campbell [13] found that the magnetic moment of Ni_2MnSn alloy was mainly localized at Mn atomic sites. Later in 984, Webster et al. [14] studied the crystal structure and magnetic properties of the stoichiometric Ni_2MnGa alloy by neutron diffraction and found that the distance between Ni–Ni, Ga–Ga and Mn–Mn is 2.6 Å, 3 Å and 4.2 Å, respectively. The magnetic coupling between Mn and Mn is realized by the traveling electrons of Ni and Ga atoms: Mn–Mn interactions contribute to a local magnetic moment ~ 4.17 B along <111> direction, while the local magnetic moment of Ni atoms is < 0.3 B. When the temperature is higher than Curie point, the magnetic moment of Mn atom is constant, but the direction becomes disordered, leading to transition from ferromagnetic to paramagnetic state.

The breakthrough in Ni_2MnGa research began in 1996. Ullakko et al. [15] found 0.2% MFIS in Ni_2MnGa single crystals. In the following years, the MFIS increased rapidly: Murray et al. [6] obtained 6% MFIS in $Ni_{47.4}Mn_{32.1}Ga_{20.5}$ single crystals in 5M martensite in 2000; Sozinov et al. [2] obtained ~ 10% MFIS in Ni–Mn–Ga single crystals in 7M in 2003. Indeed, the lattice parameters show that the limit of MFIS in 5M and 7M martensite is 6.0% and 9.4%, respectively. So, the strains obtained by Murray and Sozinov have actually reached the theoretical limit. Unlike 5M and 7M that have lower twinning stress, the twinning stress of NM martensite is much higher. However, under a high enough magnetic field, MFIS of NM can reach 20%. Despite the superior MFIS of single crystals, the preparation of single crystals is difficult, expensive and prone to composition segregation. So, the research of polycrystalline Ni–Mn–Ga alloys has attracted more and more attention.

In 2007, Boonyongmaneerat and Dunand reported an MFIS of 0.24% in porous Ni–Mn–Ga alloys with single pores [16]. The obtained MFIS is similar to the widely used magnetostrictive Terfenol-D alloy. Subsequently, Chmielus and Zhang reported MFIS of 8.7% in porous Ni–Mn–Ga alloys [17, 18] with dual pores when the alloy was subjected to a magnetothermal cycle [17]. Gaitzsch et al. found ~ 1% MFIS in Ni–Mn–Ga microwires indirectly obtained from the EBSD plots [3]. After that, the unidirectional solidification was used to prepare the polycrystalline alloys with strong textures, which produced lowered constraints by grain boundaries and thus improved MFIS [19].

The study of Ni–Mn–X (X = In, Sn, Sb) alloy structures, and magnetic and transport properties began in the 1960s, but no thermoelastic MT was found until 2004 when Sutou et al. [20] observed the complete MT in Mn-rich Ni–Mn–X alloys and the excellent properties brought by the variable magnetic transitions.

1.2.1.2 Phase Diagram

From Ni–Mn–Ga alloy phase diagram (Fig. 1.1) [21], the crystal structure changes with temperature and composition. In the process of cooling, liquid–solid phase change occurs first at a very narrow temperature range. Then, the solid–solid transition temperature varies with Mn content: liquid to A2 phase (disordered structure) at < 15 at.%Mn, and liquid phase to B2′ phase (partial ordered cubic structure) at 15–40 at.% Mn content. The dotted line in Fig. 1.1 is the solid transition line from B2′ and fully ordered $L2_1$ phase; such ordered transition temperature changes greatly with composition. For $Ni_{48}Mn_{30}Ga_{22}$ alloy, the cubic $L2_1$ type ordered structure appears at temperatures 373–293 K [22]. The atom rearrangement during transition from B2′ to $L2_1$ can hardly be observed by X-ray scattering and electron diffraction but can be characterized by neutron diffraction since the ordered structure is sensitive to the latter.

The chemical formula of $L2_1$ ordered structure is X_2YZ. Webster et al. [14] studied the lattice structure of $L2_1$ in stoichiometric Ni_2MnGa in 1984. Considering a superlattice structure with eight cubes, Ni atoms occupy the body center of eight cubes, Mn atoms occupy the edge and body center of the cubic cell, and Ga atoms occupy

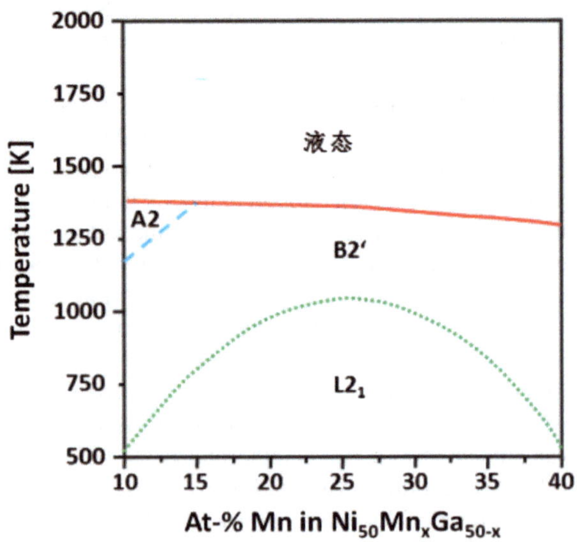

Fig. 1.1 Temperature composition phase diagram of $Ni_{50}Mn_xGa_{50-x}$ alloy [21]

the corner and face center. The contribution of Ga atoms to magnetic moments is negligible, while the contribution of Mn atoms is ten times that of Ni. It is well known that Mn is an antiferromagnetic metal. However, the distance between Mn–Mn atoms in Ni–Mn–Ga crystal structure is slightly larger than that in Mn metal, thus changing the exchange energy between the electrons in the outer layer of the Mn atom, and makes it transition from paramagnetic to ferromagnetic, which contributes to the ferromagnetism of the Ni–Mn–Ga alloys [23].

1.2.1.3 Crystal Structure

The solid–solid transition from B2′ to $L2_1$ occurs at a high temperature; thus, the resulting $L2_1$ phase is the parent austenite phase. As the temperature continues to decrease, a second-order transition from paramagnetic austenite to ferromagnetic austenite may occur first; the transition temperature is called Curie point (T_c). Then, the $L2_1$ ferromagnetic parent austenite with high crystal symmetry transforms to low symmetry paramagnetic martensite phase, which is called first-order martensitic transformation (MT).

The produced martensite has various crystal structures, such as five-layer modulated martensite with near-quartet structure (5M), seven-layer modulated martensite with near-orthogonal structure (7M) and non-modulated martensite with tetragonal structure (NM) [24]. The 5M and 7M martensites exhibit long-period stacking characteristics of the modulated structure and can be characterized using the orthogonal axis coordinate system composed of three spindle $a_M(1/2[\bar{1}10]_{(A)})$, $b_M([001]_{(A)})$ and $c_M(1/2[110][110]_{(A)})$ [25]. The lattice parameters Ni–Mn–Ga alloy parent and martensite phases are shown in Table 1.1.

(1) A 5M martensite. 5M martensite follows the parent phase $(110)[1\bar{1}0]_A$ periodic dislocation or a long-period stacking along with the parent phase $(110)_A$ surface, with five $(110)_A$ planes stacking as a cycle. 5M martensite is an approximate tetragonal structure in the parent cubic coordinate system. The *c*-axis is an easy magnetization axis and the theoretical MFIS is 6.0%. On the other

Table 1.1 Ni–Mn–Ga alloy based on $L2_1$, 5M, 7M and NM structure parameters in different coordinate systems [26, 27]

Phase	Structure	Lattice parameters (Å)			Angle (°)		
		a	b	c	α	β	γ
Austenite	–	5.82	5.82	5.82	90	90	90
Martensite	$5M_M$	4.24	5.66	20.5	90	90.5	90
	$7M_M$	4.23	5.51	29.4	90	93.5	90
	$5M_A$	5.94	5.94	5.59	$c/a = 0.94$		
	$7M_A$	6.12	5.80	5.50	$c/a = 0.90$		
	NM	5.46	5.46	6.58	$c/a = 1.21$		

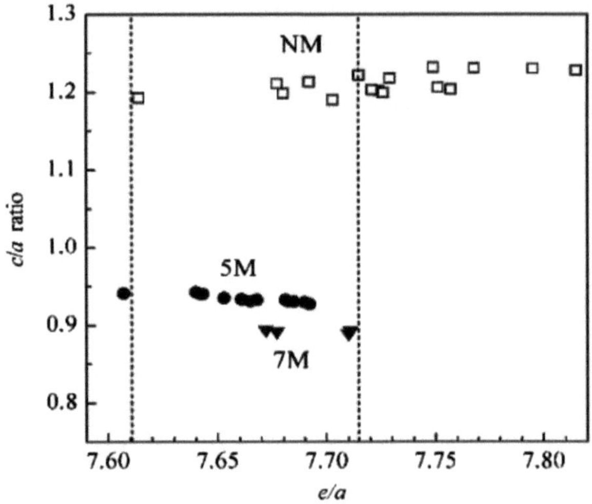

Fig. 1.2 Relationship between *c*/*a* value and *e*/*a* in 5M, 7M and NM martensite structures at a temperature 298 K [28]

hand, 5M martensite is monoclinic in the orthogonal coordinate system, and the lattice parameters and β angles change with the composition.

(2) A 7M martensite. 7M martensite modulation period is $7(110)_A$ planes. It is a monoclinic structure in the martensite coordinate system and an approximate orthogonal structure in the cubic coordinate system of the parent phase. The lattice constants are shown in Table 1.1, where the *c*-axis is also the easy magnetization axis. Its theoretical MFIS is 9.4%. Figure 1.2 shows the relationship between *c*/*a* values and valence electron concentration (*e*/*a*) under different structural martensites at a temperature 298 K. It is found that 5M and 7M structures may coexist within a certain range, but overall 5M appears in a wider range of *e*/*a* values.

(3) NM martensite. NM martensite is a non-modulated tetragonal structure. The lattice constant (Table 1.1) is different from 5 and 7M. It is $c/a > 1$ and has no easy magnetization axis. The surface composed of *a*- and *b*-axes is an easy magnetization plane. The theoretical MFIS is as high as 20.5%; however, the twinning stress is very high and thus is only accessible under a high external magnetic field. Therefore, only limited MFIS is obtainable under a magnetic field of 1 T, which is affordable for permanent magnets. The NM martensite generally appears in the region with higher *e*/*a* as shown in Fig. 1.2.

In addition, it is noted from Fig. 1.2 that the modulation structure mainly depends on the composition of the alloy (*e*/*a*). These alloys, with different modulation states, have different twin stresses and MFIS depending on *c*/*a* [29–31]. TEM diffraction is often adopted to identify 5M and 7M modulations: In 5M, four additional weak diffraction spots between each two main diffraction spots exist in the direction of the inversion space [110], and in 7M, six additional weak diffraction spots between every two main diffraction spots in the direction of the inversion space [110]. Other modulation conditions, such as 4M, have also been found. During MT, the stability of the 5M, 7M and NM decreases in turn, resulting in the successive appearance of 5M, 7M and NM martensite during temperature decrease. Accordingly, 5M martensite is directly transformed from the parent phase, while the NM martensite may be generated by a two-step transformation from 5M → NM or 7M → NM.

1.2.2 Ni–Mn–X (X = In, Sn, Sb) Alloy

In Ni–Mn–Ga alloys, the martensitic transformation is accompanied by the ferromagnetic transition from ferromagnetic parent phase to ferromagnetic martensite. However, the difference in magnetization between them is small. Ferromagnetic martensite which is sensitive to the external magnetic field can produce a greater response, which explains why most magnetostrictive studies focus on Ni–Mn–Ga systems. Nevertheless, Ni–Mn–X (X = In, Sn, Sb) system exhibits completely different magnetic properties. In 2006, Kainuma et al. [32] found the meta-magnetic MT in Ni–Co–Mn–In from paramagnetic or antiferromagnetic martensite to ferromagnetic austenite. The magnetic-field-induced MT produced a 3% strain under an 8 T magnetic field. More interestingly, the output stress reached as high as 100 MPa, which is much higher than Ni–Mn–Ga alloys.

Figure 1.3 shows the dependence of MT temperature and Curie point on the composition of $Ni_{50}Mn_{50-y}X_y$ (X = In, Sn, Sb) alloys. When $y < 16.5\%$, MT occurs in $Ni_{50}Mn_{50-y}X_y$ alloys; the MT temperature increases with the decrease of Mn content. The martensitic transformation of Ni–Mn–X (X = In, Sn, Sb) is more sensitive to the change of alloy composition than that of Ni–Mn–Ga, and ferromagnetic martensite only exists in narrow Mn-rich regions. The martensite phases in the Ni–Mn–X (X = In, Sn, Sb) alloys include 5M, 7M, NM, 4O, 6M structures.

Figure 1.4 shows the $Ni_{50}Mn_{50-y}Sn_y$ alloy longitudinal section phase diagram. It can be seen that different from the Ni–Mn–Ga alloy phase diagram, Ni–Mn–Sn can form a variety of second phases during cooling, such as T phase and Ni_3Sn_2 phase; this is also the reason why water cooling is usually applied after high temperature annealing heat treatment; otherwise, the second phase may affect the magnetic properties, MT temperature, SME, superelasticity and variable magnetic transitions.

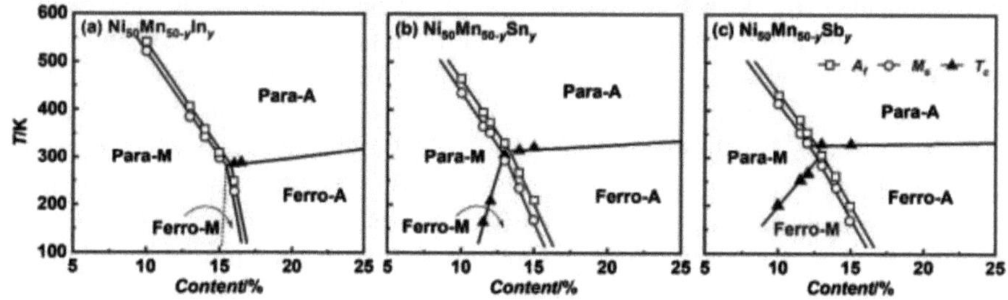

Fig. 1.3 Phase diagram of $Ni_{50}Mn_{50-y}X_y$ (X = In, Sn, Sb) alloys [20]

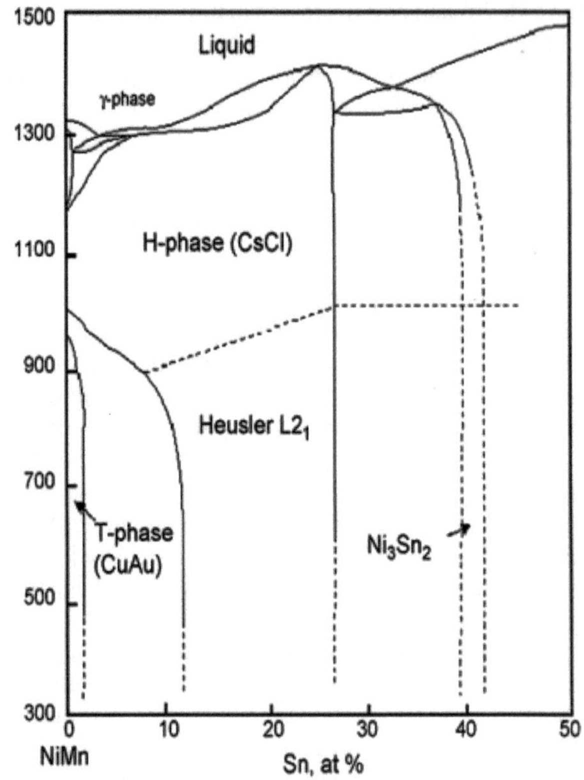

Fig. 1.4 $Ni_{50}Mn_{50-y}Sn_y$ alloy longitudinal section phase diagram [35]

Moreover, to obtain the alloy composition with L2$_1$ ordered structure, the content of Sn should be 12–26 at.%.

The Mn-rich Ni–Mn–X (X = In, Sn, Sb) parent phase exhibits paramagnetic or ferromagnetic states, and the martensite phase shows weak magnetism, such as antiferromagnetic, paramagnetic and superparamagnetic. During magnetic transition, the difference of magnetization between high-temperature austenite and low-temperature martensite is large, making the alloy prominent than Ni–Mn–Ga in magnetoresistive and magnetothermal properties. Ni–Mn–X (X = In, Sn, Sb) alloy magnetism also originates from magnetic coupling between Mn atoms. The distance between Mn and Mn is less than 2.83 Å as antiferromagnetic interaction and greater than or equal to 2.83 Å as ferromagnetic action [33]. Figure 1.5 shows a magnetic schematic diagram of different atomic occupancies in Ni-Mn-Sn alloy. When the alloy is a positive stoichiometric $Ni_{50}Mn_{25}Sn_{25}$, the distance between the nearest neighbor atoms 1 and 2 in Fig. 1.5b is relatively long. Mn–Mn ferromagnetic interactions are mainly accomplished indirectly by free electrons of Ni and X. When the alloy is Mn-rich, the excess Mn atoms occupy the crystal position of Sn atom 3 in Fig. 1.5b, and the atomic arrangement in Fig. 1.5c is obtained. The distance between Mn–Mn atoms and the Mn atoms occupying the position decreases, which coupling

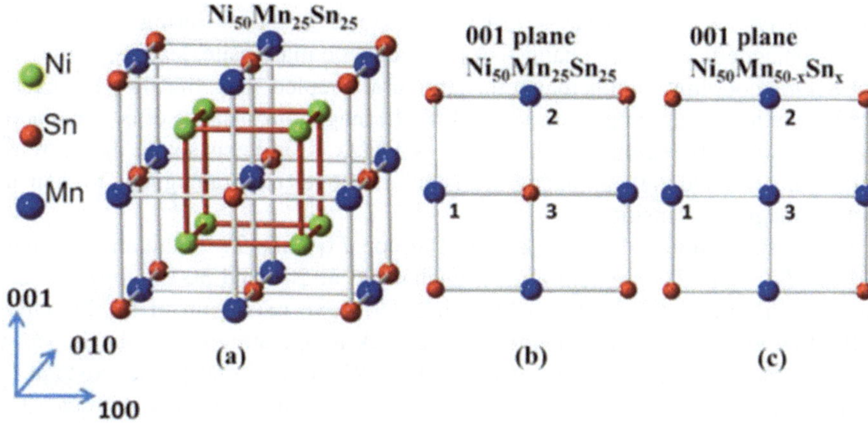

Fig. 1.5 Magnetic schematics of different atomic occupancy Mn in Ni–Mn–Sn alloys [36]

induces the antiferromagnetic action. However, ferromagnetism still originates from the magnetic coupling between the correct occupying Mn atoms in the lattice. After MT, the lattice shear further shortens the distance between the Mn and Mn and makes the martensite in the coexistence state of ferromagnetic and antiferromagnetic phases [34].

The magnetic structure phase diagram [37] of a Fe-doped $Ni_{50-x}Fe_xMn_{38}Sn_{12}$ alloy is shown in Fig. 1.6. The relationship between *e*/*a* of the alloy and M_P, $T_M{}^M$ and $T_C{}^A$ is revealed. M_P, $T_M{}^A$ and $T_C{}^A$ divide the magnetic structure phase diagram of $Ni_{50-x}Fe_xMn_{38}Sn_{12}$ alloy into three regions. PA, FA, PM and FM are paramagnetic austenite, ferromagnetic austenite, paramagnetic martensite and ferromagnetic martensite, respectively. Region I shows PA → FA → FM with the decrease of temperature ($x > 4.2$), and the MT occurs in ferromagnetic martensite region. In

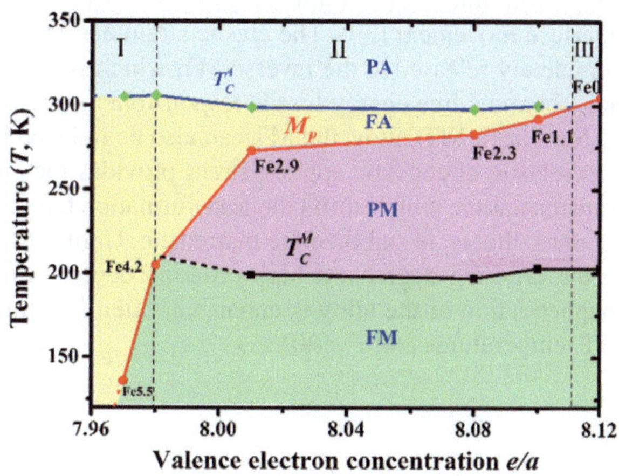

Fig. 1.6 Magnetostructural phase diagram in $Ni_{50-x}Fe_xMn_{38}Sn_{12}$ alloys [37]

region II, PA → FA → PM → FM ($4.2 \geq x > 0.5$) occurs during cooling, and the ferromagnetic austenite changes to paramagnetic martensite, leading to magnetostructural coupling. In this magnetic–structural coupled state, the alloy has excellent magnetic driving properties, such as MCE, magnetoresistance effect, magnetic SME and so on. In region III, PA → PM → FM ($x \leq 0.5$) exists with the decrease of temperature, and the paramagnetic austenite changes to paramagnetic martensite without magnetic transition.

For Fe substitution of Ni, the effect of *e/a* on $T_M{}^A$ and $T_C{}^A$ is relatively small, and the $T_M{}^A$ and $T_C{}^A$ remain almost unchanged at 200 and 300 K, respectively. The temperature corresponding to the first derivative maximum of the FC curve at the low magnetic field is defined as the peak temperature of martensite (M_P). M_P decreases with the increase of Fe content. When a second phase exists at Fe doping content > 2.9 at.%, the M_P increases approximately linearly with *e/a*. The composition-dependent M_P is: $M_P = 233.1$Mn(at.%)–54.4Ni(at.%)–141.9Fe(at.%)–459.4Sn(at.%). The strong composition-dependent M_P is related to the increase of Sn in the matrix after the second phase appears and the change of unit cell volume [38].

1.3 Magnetostructural Transition

1.3.1 Types of Martensitic Transformation in MSMAs

1.3.1.1 Martensitic Transformation

MSMAs have first-order MT and second-order magnetic transition. The driving force for the MT is also the Gibbs free energy similar to conventional MT in alloys such as irons and NiTi alloys. During MT, there is also reversible elastic strain energy and irreversible elastic strain energy loss, as well as friction loss induced by defects, interface movement [39]. The elastic strain energy is formed in forward MT and is completely released at the inverse MT, which is favorable for a reversible MT. The non-chemical free energy loss is responsible for the temperature hysteresis [39].

Similar to NiTi alloy, the MT can also be induced by external stress, creating the superelastic effect. The applied stress provides the energy needed for a part of the transformation, which shifts the transformation temperature to the high-temperature regimes, that is, to stabilize the martensite. Unlike NiTi alloy, the MT is also related to the ordering degree. At high ordering degrees in the $L2_1$ phase, the saturation magnetization of the alloy is enhanced, often accompanied by the increase of both MT temperatures and T_c [40].

1.3.1.2 Premartensitic Transition

The martensite can be divided into three categories according to the different temperature ranges of martensitic transformation: (1) type I: The M_s point is much lower than that room temperature and T_c (i.e., M_s < room temperature < T_c); (2) type II: The M_s is generally near-room temperature (i.e., $M_s \approx$ room temperature < T_c); (3) type III: The M_s is much higher than room temperature and T_c (i.e., $M_s > T_c >$ room temperature) [41]. Premartensitic transformation often occurs at low transformation temperature, i.e., in the first class of martensite. However, the appearance of the premartensite transition is also accompanied by changes in physical properties such as the magnetization and damping capacity, from which one can detect the occurrence of the premartensite transition [42]. The phenomenon of phonon mode softening with increasing temperature during premartensite transition is also observed [43].

1.3.1.3 Intermediate Martensitic Transformation

The stability of the three martensites, i.e., 5M, 7M and NM, in the Ni–Mn–Ga alloy is different. Unstable martensite can be changed to stable martensite under certain conditions, that is, intermediate MT. The intermediate MT has many ways. Chernenko et al. [44] found that, in a Ni–Mn–Ga alloy, 5M or 7M martensites firstly appeared during MT and then further transformed to NM martensite at even lower temperature. This obviously showed an intermediate MT. However, during heating, only a single-step transformation from NM to austenite was found.

Segui et al. [45] studied the effect of stress on the transformation of intermediate martensite and pointed out that NM martensite is the ground state of martensite, while 5M or 7M martensites can be transformed to NM martensite [46]. With the increase of the applied load, the austenite phase first changes to 5M $\rightarrow$ 7M $\rightarrow$ NM; or to 7M $\rightarrow$ NM; or directly to NM. The intermediate MT has a high microstructure and internal stress sensitivity. For example, Wang et al. [47] found that the 14 MPa internal stress caused by the internal lattice distortion of the sample can make the intermediate MT completely disappeared. Chernenko et al. [41] confirmed that the elastic strain energy in the non-chemical free energy has a great influence on the intermediate MT. Furthermore, in small-sized materials such as microwire and ribbon, the internal stress and fine grains produced by the rapid solidification process hinder intermediate MT.

1.3.2 Composition-Dependent Martensitic Transformation Temperature

1.3.2.1 Ni–Mn–Ga Alloys

Chernenko et al. [21] studied the relationship between stoichiometry of each element and MT temperature in 1995 (Fig. 1.7). Based on the MT temperature and enthalpy of phase change (ΔH), Ni–Mn–Ga alloys were divided into three categories, as shown in Table 1.2 [21]. Here, M_s is the martensitic starting temperature, ΔH is the enthalpy change, and T_c is the Curie point of ferromagnetic martensite and paramagnetic austenite transformation. In Table 1.2, *e*/*a* is the number of free electrons of the Ni–Mn–Ga alloy. The number of free electrons of Ni, Mn and Ga atoms is $10(3d^{9.9}4s^{0.1})$, $7(3d^54s^2)$ and $10(4s^24p^1)$ [48]. Therefore, *e*/*a* is:

$$e/a = \frac{10\mathrm{Ni}_{\mathrm{at.\%}} + 7\mathrm{Mn}_{\mathrm{at.\%}} + 3\mathrm{Ga}_{\mathrm{at.\%}}}{\mathrm{Ni}_{\mathrm{at.\%}} + \mathrm{Mn}_{\mathrm{at.\%}} + \mathrm{Ga}_{\mathrm{at.\%}}} \quad (1.1)$$

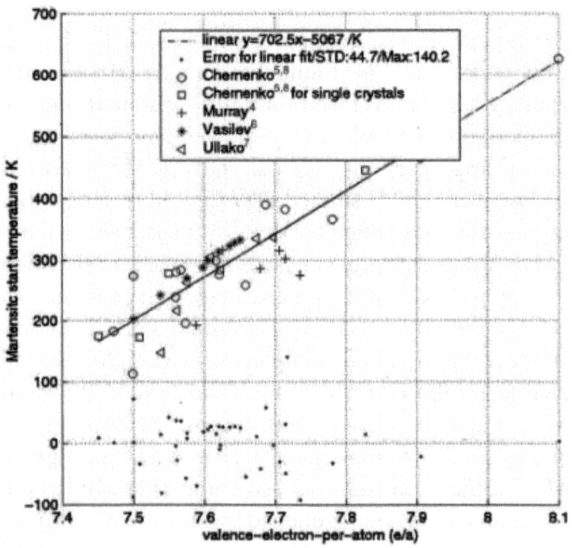

Fig. 1.7 Relationship between martensite transition temperature M_s and *e*/*a* of Ni–Mn–Ga alloy [21]

Table 1.2 Classification of Ni–Mn–Ga alloys according to M_s and enthalpy of phase change (ΔH) in Ni–Mn–Ga alloys [21]

Type	*e*/*a*	M_s (T)	ΔH (J/g)
I	< 7.55	M_s < RT < T_c	~ 1.6
II	7.55–7.7	$M_s \approx$ RT < T_c	~ 4.2
III	> 7.7	$M_s > T_c$	~ 8.5

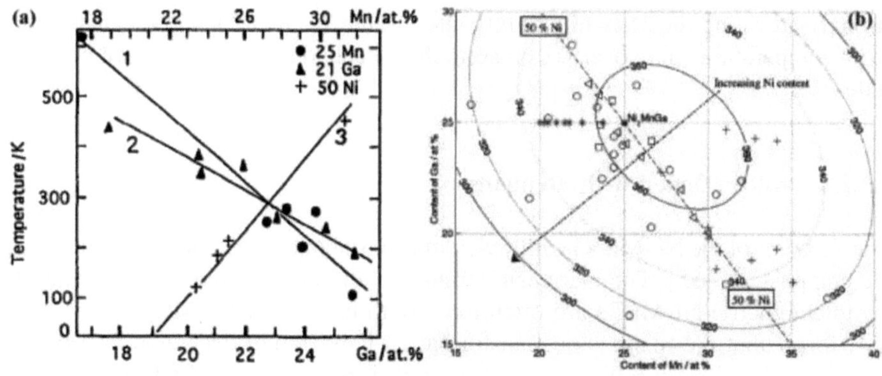

Fig. 1.8 Ni–Mn–Ga transformation temperatures as a function of composition. **a** MT temperature (M_s) versus composition [16], **b** ferromagnetic transition temperature (T_c) versus composition [50]

It is noticed that MT temperatures have a strong composition dependence. Chernenko et al. [21] found that when the content of Mn elements is constant, increasing the content of Ga elements makes the M_s decrease; when the content of Mn elements is constant, increasing the content of elements makes the M_s decrease. Wu and Yang [49] studied the relationship between the composition, enthalpy change and M_s of the alloy, and reached the following relations:

$$M_s(\text{K}) = 25.44\text{Ni(at.\%)} - 4.86\text{Mn(at.\%)} - 38.83\text{Ga(at.\%)} \tag{1.2}$$

$$\Delta H(\text{J/g}) = 0.72\text{Ni(at.\%)} - 0.16\text{Mn(at.\%)} - 1.23\text{Ga(at.\%)} \tag{1.3}$$

Jin et al. [50] summarized the relationship between M_s and *e*/*a* of Ni–Mn–Ga alloy as shown in Fig. 1.8. It should be noted that the measured results of a large number of alloys reach an average deviation of 44.7 K; but the maximum deviation reaches 140.2 K, which indicates that the MT temperature is also strongly related to other factors, such as internal stress and atomic order [51].

The relationship between the M_s temperature and *e*/*a* can be obtained by linear fitting according to Fig. 1.8: $M_s = 702.5(e/a) - 5067$.

The ordering state of the parent alloy affects the MT temperature as well. In stoichiometric Ni_2MnGa alloy with M_s temperature 202 K, the transformation temperature may be reduced as low as 100 K at a low chemical ordering state [52]. However, the T_c of the Ni–Mn–Ga alloy varies at a much less rate with composition change than M_s [53]. In stoichiometric Ni_2MnGa alloy with the largest Mn–Mn atomic spacing and the strongest ferromagnetism, the T_c value is the smallest (~ 376 K). Mn–Mn neighbor atomic distance decreases as the composition deviates from the stoichiometric ratio. As a result, the interaction between Mn–Mn atoms, i.e., the ferromagnetic interaction, weakens. So, the T_c value increases slightly [54]. By strongly adjusting the M_s while lightly altering the T_c, the MT and magnetic transition temperatures may overlap, creating the so-called magnetostructural coupling state.

1.3.2.2 Alloys Doped with Elements

The T_c point of the Ni–Mn–Ga alloy rich in Mn is about 80 °C and varies little with the composition [55]. The martensitic transformation temperatures of 5M and 7M in stoichiometric Ni_2MnGa are often too low to be applied for MFIS. Jiang et al. [54] found that the martensitic transformation temperature decreases at a rate of 42 K/at.%Cu with Cu replacement of Ni and rises at a rate of 65 K/at.%Cu with Cu replacement of Ga. On the other hand, T_c can increase by 3 K/at.%Cu with Cu substitution of Ni and decrease by 8 K/at.%. It is well known that the martensitic transformation temperature of MSMAs increases with increasing *e*/*a* and decreases with increasing cell volume [56]. Therefore, Cu ($3d^{10}4s^1$) substitution Ga ($4s^24p^1$) inevitably causes an increase in *e*/*a* and the volume of the Cu atoms is smaller than that of the Ga atoms, which makes the cell shrink, and thus, MT temperature increases. The reason why the Cu substitution Ga makes the T_c decrease is that the presence of Cu atoms affects the distance between Mn–Mn atoms. Cu substitution Mn also improves the alloy ordering degree [57].

Studies also showed that, when a large number of Cu are used to replace Ga (> 18%), a second phase appears and the plasticity of the alloy is significantly improved. As the Cu content decreases slightly ($8\% < x \leq 10\%$), the alloy becomes single phase, showing excellent plasticity and high-temperature SME [58, 59]. When the Cu content is further reduced ($3\% < x \leq 4\%$), the MT temperature increases and T_c decreases [60]. As a result, the suitable content of Cu can simultaneously adjust MT temperature, improve plasticity and reduce costs.

Transition family elements (e.g., Fe, Co, etc.), rare earth elements (e.g., Gd, Dy, etc.) and main family elements (In, Si, etc.) can also be used to modify MT temperatures in Ni–Mn–Ga alloys. Fe and Co are ferromagnetic metal elements in transition group; replacing Ni with Fe/Co for Ga leads to a decrease in *e*/*a* and thus a decrease in MT temperature. Similarly, based on *e*/*a* change, Co substitution for Ni leads to a decrease in MT temperature, while substitution for Mn or Ga causes a rise in transformation temperature [61]. By contrast, Fe substitution for Mn increases *e*/*a* value and MT temperature; however, Fe is an austenitic stabilizing element, and its addition makes martensite difficult to form. In addition, the addition of Fe increases the electron spin interaction, which makes the magnetic properties of the alloy increase and the T_c rise [62]. The solid solubility of Gd, Dy and other elements in Ni–Mn–Ga alloys is very low and therefore tends to precipitate at grain boundaries [63]. Studies showed that Nd-doped elements can improve the magnetization of the alloy in a high *e*/*a* range [63]. Additionally, the addition of Gd elements significantly increases the MT temperature [64]. Other doping elements such as Si, Sn, Ge and In in the main family are helpful to reduce the MT temperature, but the influence of elements such as C, Pb, Bi and C is not significant [65–67].

To sum up, without sacrificing the excellent properties of the Ni–Mn–Ga alloy itself, alloying is important for obtaining high-performance MSMAs with adjustable transition temperatures, high plasticity, excellent magnetocaloric properties and large MFIS.

1.4 Ferromagnetic-Related Properties in Ni–Mn–Ga Alloys

1.4.1 Shape Memory Effect and Superelasticity

Besides magnetic SME, the Ni–Mn–Ga alloy has excellent thermal driving SME and SE. The production of 5.2% SME [68] and 6% SE [44] has been reported in Ni–Mn–Ga single crystals. Zhang et al. [69] obtained 10.9% SE in microwires prepared by Taylor method in Ni–Mn–Ga polycrystalline alloys. Callaway et al. [68] obtained 3.8% two-way shape memory strain in $Ni_{53}Mn_{25}Ga_{22}$ single crystals of NM martensite structure after thermal–mechanical cyclic training. Chernenko et al. [70] obtained 9% two-way strain in $Ni_{57.5}Mn_{22.5}Ga_{20}$ single crystals. Furthermore, many scholars have improved the plasticity of polycrystalline alloys by adding the fourth component method, thus obtaining better TWSME [71].

$Ni_{52.5}Mn_{23.5}Ga_{24}$ single crystal can produce a two-way SME with a maximum total strain of 4.0% under a magnetic field [72]. CoNiGa single crystal can produce a two-way shape memory effect of 2.3% forward and 2.2% reverse under 20 kOe magnetic field [73]. Kainuma et al. [32] obtained a large strain via magnetic-field-induced martensite transition in Ni–Mn–In–Co alloys.

1.4.2 Magnetic-Field-Induced Strain (MFIS)

1.4.2.1 Magnetocrystalline Anisotropy

Magnetocrystalline anisotropy stems from different magnetization properties in a specific phase, such as martensite in Ni–Mn–Ga alloys. The magnetocrystalline anisotropy energy is K_u defined as the work required to shifting the magnetic moment from the easy magnetization axis to the difficult magnetization axis. Based on the magnetization curves of a single variant state Ni–Mn–Ga alloy in easy and difficult magnetization directions [74], the magnetic driving force of the applied magnetic field on the alloy leading to the movement of the twin variant is:

$$f_{\text{mag}}(H) = \int_0^H (m_a(H) - m_t(H))\text{d}H = -(g_a(H) - g_t(H)) \tag{1.4}$$

where H is the magnitude of the applied magnetic field, $f_{mag}(H)$ is the driving force of the magnetic field, and $m_a(H)$ and $m_t(H)$ are the magnetization in the direction of easy magnetization and difficult magnetization, respectively. This driving force is equal to the magnetocrystalline anisotropy energy K_u when the magnetic field intensity $H \to \infty$:

$$f_{mag}(H)\xrightarrow[H\to\infty]{}K_u = \int_0^{\infty} (m_a(H) - m_t(H))dH \tag{1.5}$$

The K_u values of martensite at room temperature for different structures in Ni–Mn–Ga alloy are shown in Table 1.3. At the twin boundary, the maximum stress σ_{max} is controlled by the magnetocrystalline anisotropy energy K_u $\sigma_{mag} < \sigma_{max} = \frac{K_u}{\varepsilon_{max}}$, where ε_{max} is the theoretical maximum strain value of martensite with different structures, which can be estimated by $\frac{K_u}{\varepsilon_{max}} > \sigma_{tw}$.

Moreover, the process of twin boundary motion does not occur under constant stress, but under starting σ^s and ending σ^f stresses. So, the maximum strain value of the theory is produced when the following condition is satisfied $\frac{K_u}{\varepsilon_{max}} > \sigma^f$.

The values of σ_{tw} and σ_{mag} of different types of martensite are summarized in Table 1.3. It can be seen that the critical stress of twin boundary motion in 5M and 7M martensites is low, generally in 1–2 MPa. In particular, 5M martensite in single crystals with II-type twins showed critical stress down to 0.1–0.3 MPa [76, 83]. However, in the case of multiple variant state, a rise in the critical stress may occur [77]. The critical stress of twin boundary motion in NM martensite is much larger (12–20 MPa) than 5M (2.5 MPa) and 7M (1.6 MPa).

1.4.2.2 Mechanism of MFIS

Martensite has different crystal structures and thus different twin boundary motion critical stresses. When the 10M $Ni_{50.0}Mn_{28.0}Ga_{22.0}$ alloy of tetragonal phase is transformed into monoclinic phase, type I and II twins can be produced [77, 84], with critical movement stress 0.8–1.2 MPa and 0.05–0.3 MPa, respectively.

The generation of MFIS in Ni–Mn–Ga alloys is due to the movement of twin boundaries under the action of magnetic field. Several models have been proposed to explain the mechanism: numerical micromagnetism model [70], analytical thermodynamic model [85], thermodynamic model [85], etc. Among them, O'Handley proposed a double variant model [86] that can explain the cause of not only twin boundary motion, but also the phase boundary motion, and the nonlinear behavior of magnetization and strain at saturation stage. For the magnetic-field-driven twin boundary motion in the case of different magnetic anisotropy energy under the dual variant models, it is assumed that there are only two variants in the single crystal in the absence of an external magnetic field. The magnetic moment direction of each variant has a certain angle difference. When the applied magnetic field acts on these two variants, the magnetic moment of each variant turns to the same direction as the

Table 1.3 Magnetic anisotropy energy (K_u), theoretical maximum strain (ε_{max}), magnetic force (σ_{mag}), critical stress (σ_{tw}) and measured MFIS in single-crystal alloys with different structures

Alloy		K_u (10^5 J/m^3)	ε_{max} (%)	σ_{mag} (MPa)	σ_{tw} (MPa)	MFIS (%)	References
5M	$Ni_{48}Mn_{30}Ga_{22}$	1.3	5.78	2.25	1.04–2.12	5	[74]
	$Ni_{49.7}Mn_{29.1}Ga_{21.2}$	1.75	5.8(cal)	3(cal)	< 2	–	[75]
	$Ni_{49.8}Mn_{28.5}Ga_{21.7}$	1.5	–	–	–	6%	[6]
	$Ni_{49.2}Mn_{29.6}Ga_{21.2}$	1.22	6.07	2.0	2.0	4.78%	[76]
	$Ni_{49.2}Mn_{29.6}Ga_{21.2}$	1.45	5.89	2.5	1–2.1	5.8%	[77]
	$Ni_{50.7}Mn_{28.4}Ga_{20.9}$	1.7	6.2	2.8	1–4	> 6%	[78]
7M	$Ni_{48.8}Mn_{29.7}Ga_{21.5}$	1.6/0.7	10.66	1.5	< 1.5	9.5%	[2, 79]
		1.6/0.7	10.66	1.5	1.1–1.9	9.4%	[77]
		1.6/0.7	10.6	1.5	1.2	~ 9.2%	[79]
	$Ni_{50.5}Mn_{29.4}Ga_{20.1}$	1.7	10.7	1.6	3–5	0.47	[78]
NM	$Ni_{52.1}Mn_{27.3}Ga_{20.6}$	−2.03	20.5	1.0	12–20	< 0.02%	[77]
		−2.03	20.5	1	15	< 0.02%	[80]
	$Ni_{53.1}Mn_{26.6}Ga_{20.3}$	–	18.02	–	–	0.17	[81]
	$Ni_{50.5}Mn_{30.4}Ga_{19.1}$	2.6–5	20.6	–	–	–	[82]
		1.9	19.3	~ 1	> 6	0	[78]

external magnetic field when the magnetic crystal anisotropy energy K_u is low, or the external magnetic field energy M_sH is relatively large ($K_u < M_sH$). In this case, the twin boundary does not move. On the other hand, when the two twin variants have high enough K_u, the force acting on the twin boundary (i.e., the Zeeman energy) exceeds the twinning stress. In this case, the external field drives the movement of twin boundary, which produces MFIS.

The single variant model proposed by O'Handley [86] is described in detail in martensite with a near tetragonal structure [87]. At zero magnetic field, the magnetic moments of different variants are arranged individually along the direction of the easy magnetization axis. If a magnetic field is applied along the easily magnetized direction of the intermediate variant, the magnetic moment of the two variants also tends to be arranged along the external field direction, which makes the energy on both sides of the variant different. This energy difference applies as a driving force σ_{mag} for twin boundary motion. According to the formula (1–5), the magnetic driving force of the initial twin boundary motion is proportional to the size of the external field. Therefore, when the external field is very small (H_1), the magnetic force is not high enough to drive the twin boundary motion. In this case, the magnetization direction rotates in these twin variants to re-orientate along the external field by steering the magnetic moment. With the increasing (H_2) of the applied magnetic field, the magnetic force also increases high enough to drive the twin boundary motion by favorably oriented variants expanding at the sacrifice of the unfavorable variants. Eventually, all the easy magnetization axes grow through the intermediate variant and are arranged along the external field direction. In this process, the MFIS is produced accordingly. The detailed MFIS of some alloys is shown in Table 1.4.

The strains achieved in Ni–Mn–Ga alloys can be reversible and irreversible, which are called magnetoelasticity and magnetoplasticity, respectively [88]. Since the movement of twin boundary can also be actuated by applying stress, the previous irreversible strain can be recovered by applying stress in the direction perpendicular to the magnetic field [89].

1.4.2.3 MFIS in Bulk and Small-Sized Alloys

(1) Single-crystal alloy

Since 1996 Ullakko [85] found 0.19% in Ni_2MnGa single-crystal MFIS more than 20 years before, the research on Ni–Mn–Ga alloy is mainly focused on single crystals. Murray et al. [6] obtained 6% MFIS in $Ni_{47.4}Mn_{32.1}Ga_{20.5}$ single crystals of 5M martensite at room temperature; in 2003, Sozinov et al. [2] obtained nearly 10% strain in Ni–Mn–Ga single crystals in 7M under a magnetic field 1 T. Strain in 5M and 7M has reached the theoretical limit. The maximum strain value in NM martensite in Ni–Mn–Ga alloy is 0.17% [81] at present because the critical stress of grain boundary movement of martensite is too high. If the critical stress of twin boundary motion is reduced to the level of magnetic force by some kind of training, it has great potential to produce 20% MFIS. Although the properties (MFIS) of

Table 1.4 Magnetic-field-induced strain (MFIS), crystal structure of parent and martensite phases, martensitic transformation temperature (T_M) and toughness (compared with Ni–Mn–Ga alloy) [90]

Alloy	MFIS	Crystal structure	Toughness
Ni–Mn–Ga	6–10% (single crystal) at room or low temperature	$L2_1$/5M/7M at room or low temperature	/
Ni–Mn–Ga–Fe	1–5.5% (single crystal) at room or low temperature [91]	$L2_1$/5M/7M at room or low temperature	Better than NiMmGa
Ni–Mn–Al	0.17% (single crystal), 0.01% (single crystal) at 253 K [92]	B2($L2_1$)/5M/6M/7M at room or low temperature	Improved from γ phase
Ni–Fe–Ga	0.02% (single crystal) at ~ 100 K [93]	$L2_1$/10M 6M 14M at room or low temperature	Improved from γ phase
Ni–Fe–Ga–Co	0.7% (single crystal) at 300 K [94]	/	Improved from γ phase
Co–Ni–Al	0.06% (single crystal) at 165 K [95]; 0.013% (single crystal) at 293 K [96]	B2/L10 below room temperature	Improved from γ phase
Co–Ni–Ga	0.011% (polycrystalline) ribbons; 0.003% (single crystal) at RT [97]	B2/L10 below room temperature	Improved from γ phase
Fe_3Pt	~ 0.6% (single crystal) at 4.2 K [98]	Ordered FCC/FCT ~ 100 K	Good intrinsic toughness
Fe–Pd	3.1% (single crystal) at 77 K [99]; 0.01–0.05% (polycrystalline) [100]; 0.06–0.07% (polycrystalline) ribbons [101]	Disorderly FCC/FCT structures at room or low temperature	Good intrinsic toughness

single crystals are superior, the preparation of single crystals is difficult and the cost is high, and the component segregation formed by the preparation process affects the properties of materials, so the study of polycrystalline alloys has attracted great attention.

(2) Polycrystalline alloys

Polycrystalline Ni–Mn–Ga alloys are simple to prepare, but the fine grains hinder the movement of twin boundary motion, which makes the MFIS value very small. A large MFIS can be obtained by forming texture in bulk polycrystalline to reduce the orientation difference of twin variants, inducing grain growth to reduce the resistance of twin boundary movement or making foams to eliminate some grain boundaries.

Directional solidification technology is used to produce a strong texture state, and thermal–mechanical training can be beneficial to MFIS acquisition. The introduction of texture can reduce the orientation difference between adjacent grains, which is beneficial to MFIS. Potschke et al. [102] used this method to introduce $<100>_A$ strong wire texture into polycrystalline Ni–Mn–Ga alloys, which showed 0.16% MFIS after grain growth and training. Gaitzsch et al. [3] used the same method to prepare 7M polycrystalline alloys and obtained 1% MFIS under compression stress.

In porous alloys, grain boundaries are replaced by pores, and martensite twins are extended to the whole pore edges by grain growth. Boonyongmaneera et al. studied the effect of Ni–Mn–Ga polycrystalline pores on the MFIS in 2007 [16]. The results showed that the pores could reduce the intergranular constraints during the twin boundary movement, thus increasing the MFIS, to obtain 0.115% reversible strain, which is close to the Terfenol-D. As a result, it overcomes the problem of the high cost of single-crystal preparation and opens up a new way for the MFIS study. Chmielus et al. [17] studied the foam alloy with double-pore structure, and the pore size is smaller or comparable to the grain size. After heat treatment, the bamboo grain was formed at the pore edges. The existence of the free surface at nodes and struts further reduced the re-orientation resistance of martensite twin, so the MFIS of 8.7% was obtained. Also, subsequent reports [18] confirmed that double-pore foam materials decreased less after multiple cycles compared to MFIS single pores.

Thermally deformed polycrystalline Ni–Mn–Ga alloys can be obtained by plastic deformation, including hot rolling [103], hot extrusion [104], isothermal forging [22] and high-pressure torsion [105]. After Ni–Mn–Ga polycrystalline alloy was subjected to isothermal forging to create a strong texture, the subsequent heat treatment changed the composition of texture, leading to different twin boundary motion characteristics [22].

Polycrystalline Ni–Mn–Ga alloy composites can be prepared by pre-aligning Ni–Mn–Ga microwires or particles in the resin under a directional magnetic field and subsequent curing. In this process, the easy magnetization axis of the particles or microwire was arranged along the external magnetic field. However, the un-trained composite materials showed little MFIS [106]. After magnetic–thermal training, 0.1% MFIS [107] was obtained.

The operating frequency of bulk polycrystalline materials is limited by eddy current and inertia [108]. In addition, for polycrystalline small size materials, when the grain size is equivalent to a certain characteristic size of the specimen, such as the thickness of the film, the diameter of the microwire, the width of the strip and the ball diameter of the particle, the surrounding free surface of the grain can reduce the resistance of the twin grain boundary movement. In addition, the decrease of material size is beneficial to the magnetocrystalline anisotropy energy [7].

In the process of fabricating ribbon or microwire, fast solidification leads to the formation of fine grains with strong texture. Gutierrez et al. [109] reported that a thin ribbon with a grain size of 1–3 μm and the thickness of 80 μm remains a strong texture after annealing, which induced the grain growth to 40 μm. Lazpita et al. [110] obtained 0.002% MFIS by applying a 1 T magnetic field to the as-prepared ultrafine grain ribbon. Guo et al. [111] prepared Ni–Mn–Ga ribbons by melt spinning and obtained 0.25% MFIS. For ribbons and films that have 2-D constraints, the grain growth resistance is usually high, hindering the formation of grains covering the whole thickness of the ribbon/film. However, it is more feasible to create bamboo grains in microwires by annealing [112] due to the 1-D constraints. Scheerbaum et al. [113] treated Ni–Mn–Ga microwires into bamboo-shaped grain structure by heat treatment and obtained 1% MFIS under a magnetic field of 2 T.

Also, there are many studies on the MFIS of thin plates and sheets, but with much lower MFIS. Pőtschke et al. [102] prepared Ni–Mn–Ga thin plate by directional solidification and obtained 0.16% MFIS. Bernard et al. [114] prepared the Ni–Mn–Ga film and obtained 0.025% MFIS annealed at 600 °C for 10 h.

1.5 Meta-Magnetic Behavior and Related Properties

1.5.1 Meta-Magnetic Field-Induced Phase Transition and Strain

The magnetic-field-induced phase transition originates from the Zeeman energy, which is equal to the difference between the external magnetic field and the magnetization difference of austenite/martensite ΔM. Because the Zeeman energy is not sensitive to the orientation of the grains, this effect can also be found in polycrystalline alloys. Table 1.5 lists the magnetic-field-induced strains of some well-studied Ni–Mn-based meta-magnetic alloys. Kainuma et al. [32] first reported the magnetotransformation strain of Ni–Co–Mn–In single crystals in 2006: The crystal was precompressed 3% in the low-temperature martensite phase and then applied a magnetic field to 7 T to reach a strain recovery of 2.9%. The strain recovery was achieved by field-induced martensite to austenitic transition since the austenite phase shows a ferromagnetic nature while the martensite para- or antiferromagnetic one. Unfortunately, the magnetic-induced phase transition strain is one way (does not recover when the magnetic field is removed). In addition, the output stress under magnetic field 7 T is estimated to be 108 MPa, which is much higher than that produced by the twin boundary motion. A similar magnetostrain effect was also found in Ni–Co–Mn–Sn polycrystal [115]. Later, Krenke et al. [116] obtained a reversible magnetostrain of 0.14% in a Ni–Mn–In alloy without prestress. Liu et al. [117] revealed that the reversible magnetostrain of Ni–Co–Mn–In alloy reached 0.25% along the columnar grain direction, while only −0.11% along the vertical magnetic field direction.

1.5.2 Magnetocaloric Effect

In recent years, more and more attention has been paid to environmental problems. The traditional refrigeration technology mostly uses the gas working fluid such as Freon and ammonia, which causes the environment to be polluted and destroyed greatly. There is a commitment to the development of non-fluorinated gas working fluids, but most still have greenhouse effects and energy conversion efficiency needs

Table 1.5 Magnetostrain of some Ni–Mn-based magnetic shape memory alloys

Material	Temperature/K	Magnetic field/T	Strain/reversibility	Mechanism
$Ni_{45}Co_5Mn_{36.7}In_{13.3}$ single crystal [32]	298	8	2.9%, irreversible	Magnetic field-driven phase change
$Ni_{43}Co_7Mn_{39}In_{11}$ polymorphic [115]	310	8	0.3%, reversible	Magnetic field-driven phase change
$Ni_{50}Mn_{34}In_{16}$ polymorphic [116]	195	5	0.14%, reversible	Magnetic field-driven phase change
$Ni_{45.2}Mn_{36.7}In_{13}Co_{5.1}$ polymorphic [117]	310	5	0.25%, reversible	Magnetic field-driven phase change
$Ni_{48.8}Mn_{29.7}Ga_{21.5}$ single crystal [79]	300	1	9.5%, irreversible	Magnetic field-driven twin boundary motion
Porous $Ni_{50.6}Mn_{28}Ga_{21.4}$ polymorphic [16]	300	1	0.115%, reversible	Magnetic field-driven twin boundary motion
Porous NiMnGa polymorphic [17]	300	1	8.7%, reversible	Magnetic field-driven twin boundary motion

to be improved. Magnetic refrigeration technology has been widely concerned by researchers because of its energy-saving and environment-free characteristics.

1.5.2.1 Principle of Magnetocaloric Effect

Magnetocaloric effect (MCE) is a phenomenon that uses the principle of entropy ΔS_m of electron spin system of magnetic material to achieve the purpose of endothermic and exothermic with the change of external magnetic field [118]. The entropy of magnetic materials is composed of spin electron magnetic entropy (S_m), lattice vibration entropy (S_l) and conduction electron entropy (S_e), where only ΔS_m can be controlled by a magnetic field and the S_e is small and thus can be negligible. When no external magnetic field is applied, magnetic moments in ferromagnetic or paramagnetic states arrange in disorder. When a magnetic field is applied adiabatically, the magnetic moments tend to be arranged in the direction of the external field, resulting in lower S_m and higher S_l and an increase in the system temperature. After the temperature recovers to the equilibrium state (release heat to environment), the adiabatic demagnetization process makes S_l decrease and S_m increase, thereby lowering the temperature of the system and achieving magnetic refrigeration.

Studies have shown that in addition to the above traditional MCE, another kind of anti-MCE exists, mainly in antiferromagnetic materials. The magnetic domains of general antiferromagnetic materials are divided into two subdomains. When applied in the direction of one domain magnetic moment, the magnetic moment of the other orientation disrupts the original equilibrium in the course of deflection along the external field direction, thus inducing the S_m to decrease and increase, and then the anti-MCE [119] appears. Regardless of the positive and negative magnetic heat, if the magnetic moment rotation process is accompanied by structural changes, it may produce a giant magnetocaloric effect (GMCE).

1.5.2.2 Characterization of Magnetocaloric Effect

MCE is expressed as a ΔS_m or adiabatic change (ΔT_{ad}) produced during addition/demagnetization. Moreover, the RC value of the material is also an important parameter to evaluate the MCE. The relationship between magnetic entropy S_m and temperature T and magnetic intensity M under magnetic field H can be deduced by using thermodynamic Maxwell relations:

$$\left(\frac{\partial S}{\partial H}\right)_T = \left(\frac{\partial M}{\partial T}\right)_H \tag{1.6}$$

Considering the solid-state properties of magnetic refrigeration materials, the influence of pressure on thermodynamic quantities is generally not considered. Therefore, the thermodynamic entropy of magnetic materials can be divided into three parts:

$$dS = \left(\frac{\partial S}{\partial T}\right)_H \mathrm{d}T + \left(\frac{\partial S}{\partial H}\right)_T \mathrm{d}H \tag{1.7}$$

Furthermore, the specific heat capacity C_H is defined as $C_H = T\left(\frac{\partial S}{\partial T}\right)_T$. So, the entropy change is:

$$\mathrm{d}S = \left(\frac{C_H}{T}\right)\mathrm{d}T + \left(\frac{\partial M}{\partial T}\right)_H \mathrm{d}H \tag{1.8}$$

Under the adiabatic condition, $\mathrm{d}S = 0$, so the expression can be obtained from ΔT_{ad} upper formula:

$$\Delta T_{ad} = -\int_{H_1}^{H_2} \left(\frac{T}{C_H}\right)\left(\frac{\partial M}{\partial T}\right)_H \mathrm{d}H \tag{1.9}$$

However, in addition to obtaining the ΔT_{ad} by measuring the specific heat capacity, it can also be obtained by direct adiabatic demagnetization measurement, which is also recognized as an accurate method for measuring the magnetocaloric properties of materials. Direct measurement requires high adiabatic performance in the test environment and the accuracy of temperature sensors.

Under isothermal conditions, $dT = 0$, the ΔS_m expression in the process of magnetic field change can be obtained by:

$$\Delta S_m = S_m(H_2, T) - S_m(H_1, T) = \int_{H_1}^{H_2} \left(\frac{\partial M}{\partial T}\right)_H dH \tag{1.10}$$

To facilitate measurement and calculation, the magnetic field and temperature are abstracted into discrete points:

$$\Delta S_m = \sum_j \frac{M_{i+1}(T, H_{j+1}) - M_i(T, H_j)}{T_{i+1} - T_i} \Delta H_i \tag{1.11}$$

where M_i and M_{i+1} are the magnetizations; when the magnetic field is H_j and H_{j+1}, the temperature is T_i and T_{i+1}, respectively. Although this method is simple, it has always been controversial on the applicability near the first-order magnetic phase change, because the use of isothermal magnetization curves to calculate the ΔS_m near the magnetic phase change will produce a virtual high value [120]. The results show, however, that for the first-order phase transition with weak magnetoelastic coupling ($\lambda_1/\lambda_0 < 2$), the above Maxwell relation is fully applicable to the calculation of ΔS_m [121]. While, for Ni–Mn–Ga alloys, with $\lambda_1 \sim 10^7$ and $\lambda_0 \sim 10^{10}$ erg cm^{-3} that does not meet the above condition [122], the magnetocaloric values can be over-estimated. In this case, the measurement of the specific heat capacity in a constant magnetic field can be adopted to determine the entropy change:

$$\Delta S_m = S_m(H_2, T) - S_m(H_1, T) = \int_{T_1}^{T_2} \frac{C(T, H)}{\partial T} dT \tag{1.12}$$

Moreover, magnetic refrigeration needs a wide operating temperature interval for practical operation. The working temperature interval is usually characterized by the full width at half maximum, ΔT_{FWHM} of the ΔS_m–T curve. The RC value of the material is used for evaluation MCE, expressed by integrating the ΔT_{FWHM} (T_1–T_2) of the ΔS_m–T curve:

$$RC = \int_{T_1}^{T_2} \Delta S_m(T)_H dT \tag{1.12}$$

Besides the aforementioned indirect measurement, direct measurement of the temperature change via an infrared camera is more convincing. However, the measured data are often underestimated when the non-adiabatic condition is not reached. The infrared test can also be coupled with digital image correlation (DIC) to get the temperature data and profile simultaneously, which helps in analyzing the MCE behavior and mechanisms.

1.5.2.3 Development Scenario and Perspectives of Magnetocaloric Refrigeration

Since Warburg [123] discovered MCE in 1881, solid-state refrigeration based on MCE has drawn more and more attention for the possible replacement of vapor compression technology. In this regard, the development of high efficient working materials for MCE plays a vital role. The studies showed that Gd, Gd–Si–Ge [124], La–Fe–Si [125] and Mn–Fe–P–As [126] have excellent magnetocaloric properties. However, rare earth-containing materials are expensive, and elements such as As are toxic, which renders limitations for their wide applications.

Ni–Mn–X (X = Ga, In, Sn, Sb) MSMAs may act as alternative high-performance MCE materials. Their phase transition temperature can be adjusted to room temperature, making ambient temperature refrigeration possible. Obviously, some problems need to be solved before the final application of these alloys:

(1) Coordination of ΔS_m and *RC* values. When the first- and second-order phase transitions are coupled, the change of the magnetic transition state accompanies a discontinuous change of the lattice structure, producing a greater magnetization difference ΔM and thus a large ΔS_m. There are two causes for the occurrence of such first-order magnetic phase change. One is that martensite and austenite have different magnetic states. The first-order structural transformation is coupled with the second-order magnetic transition to reach a so-called magnetostructural coupled state. This phenomenon mostly exists in alloys such as Ni–Mn–In, Ni–Mn–In–Co and Ni–Mn–Sn. Another case is that the magnetic transformation temperature of austenite is consistent with martensitic transformation temperature, mainly including Ni–Mn–Ga and Ni–Mn–Ga–Cu alloys.

However, although the obtained entropy change can be high, the working temperature interval is generally narrow under magnetostructural coupled state. For example, in Ni–Mn–Ga alloys, the ΔS_m reached ~ 86 J/kg K under magnetostructural coupled state, but the ΔT_{FWHM} is only 1–3 [127], resulting in a decrease in the *RC* value. ΔT_{FWHM} can be enhanced by mixing materials with successive transition temperatures. However, this method ultimately dilutes the ΔS_m value, thus limiting its application [128]. Furthermore, some scholars introduce the intermediate phase transition [129], adjust the distance between the two phase transitions in materials with high and low temperature ΔT_{FWHM} [130] or apply bias pressure to enhance the *RC*.

(2) Working temperature. From the practical point of view, the working temperature of magnetic refrigeration should be at or slightly higher than room temperature. Although by adjusting the phase transition temperature, large ΔS_{m} currently produced in Ni–Mn–Ga single crystals can appear near-room temperature [127], the decrease of the magnetic phase transition temperature also implies the loss of the magnetic properties of the alloy, thus reducing the magnetocaloric properties of the alloy. Nevertheless, for highly magnetic Mn-rich $Ni_{50}Mn_{25-x}Ga_x$ alloys, the T_{c} is mostly stable around 90–100 °C [55], so how to coordinate the relationship between ΔS_{m} and the working temperature is crucial.

(3) Reversibility and hysteresis loss. The first-order phase transition is also accompanied by phase transition hysteresis resulting in heat loss. In addition, the existence of hysteresis effect greatly reduces the reversibility and efficiency of magnetic refrigeration. The first-order MT has an intrinsic large hysteresis, while the second-order magnetic transition has a small hysteresis. The phase transition hysteresis of Ni–Mn–Ga alloy during first-order MT is 10 K [55]. The hysteresis can be reduced by producing small-sized materials such as powders, films, ribbons and wires, or by introducing pores into bulk materials [125]. The reduced hysteresis in small-sized materials is related to the increase in specific surface area and thus less resistance during phase transformation.

(4) Cycle life and heat dissipating capacity. The first-order phase transition is generally accompanied by volume change, so for brittle materials, cracking and failure are easy to occur after multiple cycles. In bulk polycrystalline materials, cracks arise from the restriction of grains during phase transition. So, this problem can be partially solved by increasing the free surface of grains. In addition, improving the plasticity of the material by alloying can improve the cycle life [58]. The small-sized materials have excellent heat dissipation properties due to their large surface area, which is important for a thermal cycling system.

To sum up, the preparation of the alloy into small size materials can reduce the phase change hysteresis and partially solve the problems of cycle life, heat dissipation. By effectively adjusting the first- and second-order phase transition temperatures and the distance between them, the absolute value of ΔS_{m} and working temperature interval (*RC*) can be optimized. Alloying is a feasible strategy to modify the properties of the alloy and adjusting the transition temperatures.

1.6 Conclusions

MSMA has become an important kind of smart material for active sensors and actuators. The driving force for MSMAs includes heat/stress and magnetic field. Unlike NiTi alloy that shows low SME frequency during cycling due to the limited heat dissipation rate, MSMAs show a much higher working frequency up to 5000 Hz under a high-frequency magnetic field.

MSMAs show first-order MT and second-order magnetic transition. The MT temperature is highly composition dependent; however, the magnetic transition point (T_c) is less sensitive to composition. Thus, magnetostructural coupling state can be achieved via adjusting the composition, which exhibits some unique properties such as giant MCE at coupled state. Various martensite structures, including 5M, 7M and NM, exist in MSMAs.

MSMA shows multifunctional properties such as SME, superelasticity, MCE and MFIS. The MFIS value has reached its theoretical value in single crystals. Very high MFIS has also been achieved in some specific polycrystals such as foams with dual pores. The multi-caloric effects in MSMAs are still under intensive investigation, which shows potential for possible replacement of the conventional vapor compression technique.

Small-sized MSMAs bring some novel properties such as low hysteresis, high work output and low inertia. The very high specific surface area also renders them high application potential in minor-sized sensors, actuators, dampers as well as solid-state cooling devices.

This book includes eight chapters. We place the emphasis on small-sized materials, including foam, film, wire and powder. This chapter aims to give a brief overview of the concepts, martensitic transformation and properties of MSMAs. Chapter 2 introduces the preparation, deformation and properties of bulk alloys. Chapters 3, 4, 5 and 6 discuss foams, film, wire preparation, microstructure and powder, respectively. Chapter 7 describes the application of MSMAs.

References

1. R.C. O'Handley, *Modern Magnetic Materials: Principles and Applications* (Wiley, Hoboken, 2000)
2. A. Sozinov, A.A. Likhachev, N. Lanska et al., 10% magnetic-field-induced strain in Ni-Mn-Ga seven-layered martensite. J. Phys. IV **112**, 955–958 (2003)
3. U. Gaitzsch, M. Potschke, S. Roth et al., A 1% magnetostrain in polycrystalline 5M Ni-Mn-Ga. Acta Mater. **57**(2), 365–370 (2009)
4. R. Tickle, R.D. James, T. Shield et al., Ferromagnetic shape memory in the NiMnGa system. IEEE Trans. Magn. **35**(5), 4301–4310 (1999)
5. H.E. Karaca, I. Karaman, B. Basaran et al., Magnetic field and stress induced martensite reorientation in NiMnGa ferromagnetic shape memory alloy single crystals. Acta Mater. **54**(1), 233–245 (2006)
6. S.J. Murray, M. Marioni, S.M. Allen et al., 6% magnetic-field-induced strain by twin-boundary motion in ferromagnetic Ni-Mn-Ga. Appl. Phys. Lett. **77**(6), 886–888 (2000)
7. Y. Ganor, D. Shilo, T.W. Shield et al., Breaching the work output limitation of ferromagnetic shape memory alloys. Appl. Phys. Lett. **93**(12), 122509 (2008)
8. A.A. Likhachev, A. Sozinov, K. Ullakko, Different modeling concepts of magnetic shape memory and their comparison with some experimental results obtained in Ni-Mn-Ga. Mater. Sci. Eng. A **378**(1–2), 513–518 (2004)

9. Z. Li, Y. Zhang, C.F. Sánchez-Valdés et al., Giant magnetocaloric effect in melt-spun Ni-Mn-Ga ribbons with magneto-multistructural transformation. Appl. Phys. Lett. **104**(4), 044101 (2014)
10. H.B. Wang, C. Liu, Y.C. Lei et al., Characterization of $Ni_{55.6}Mn_{11.4}Fe_{7.4}Ga_{25.6}$ high temperature shape memory alloy thin film. J. Alloy. Compd. **465**(1–2), 458–461 (2008)
11. S.M. Ueland, Y. Chen, C.A. Schuh, Oligocrystalline shape memory alloys. Adv. Funct. Mater. **22**(10), 2094–2099 (2012)
12. D.C. Dunand, P. Müllner, Size effects on magnetic actuation in Ni-Mn-Ga shape-memory alloys. Adv. Mater. **23**(2), 216–232 (2011)
13. C.V. Stager, C.C.M. Campbell, Antiferromagnetic order in the Heusler alloy, $Ni_2Mn(Mn_xSn_{1-x})$. Can. J. Phys. **56**(6), 674–677 (1978)
14. P.J. Webster, K.R.A. Ziebeck, S.L. Town et al., Magnetic order and phase transformation in Ni_2MnGa. Philos. Mag. B **49**(3), 295–310 (1984)
15. K. Ullakko, J.K. Huang, C. Kantner et al., Large magnetic-field-induced strains in Ni_2MnGa single crystals. Appl. Phys. Lett. **69**(13), 1966–1968 (1996)
16. Y. Boonyongmaneerat, M. Chmielus, D.C. Dunand et al., Increasing magnetoplasticity in polycrystalline Ni-Mn-Ga by reducing internal constraints through porosity. Phys. Rev. Lett. **99**(24), 247201 (2007)
17. M. Chmielus, X.X. Zhang, C. Witherspoon et al., Giant magnetic-field-induced strains in polycrystalline Ni-Mn-Ga foams. Nat. Mater. **8**(11), 863–866 (2009)
18. X.X. Zhang, C. Witherspoon, P. Müllner et al., Effect of pore architecture on magnetic-field-induced strain in polycrystalline Ni-Mn-Ga. Acta Mater. **59**(5), 2229–2239 (2011)
19. U. Gaitzsch, J. Romberg, M. Potschke et al., Stable magnetic-field-induced strain above 1% in polycrystalline Ni-Mn-Ga. Scripta Mater. **65**(8), 679–682 (2011)
20. Y. Sutou, Y. Imano, N. Koeda et al., Magnetic and martensitic transformations of NiMnX(X=In, Sn, Sb) ferromagnetic shape memory alloys. Appl. Phys. Lett. **85**(19), 4358–4360 (2004)
21. V.A. Chernenko, E. Cesari, V.V. Kokorin et al., The development of new ferromagnetic shape memory alloys in Ni-Mn-Ga system. Scr. Metall. Mater. **33**(8), 1239–1244 (1995)
22. D.Y. Cong, *Study on Crystal Structure and Microstructure of Novel Ni-Mn-Ga Magnetic Shape Memory Alloys* (Northeast University, Shenyang, 2008)
23. A. Ayuela, J. Enkovaara, K. Ullakko et al., Structural properties of magnetic Heusler alloys. J. Phys.-Condens. Matter **11**(8), 2017–2026 (1999)
24. M. Chmielus, *Composition, Structure and Magneto Mechanical Properties of Ni-Mn-Ga Magnetic Shape—Memory Alloys* (Boise State University, Boise, 2010)
25. L. Righi, F. Albertini, G. Calestani et al., Incommensurate modulated structure of the ferromagnetic shape-memory Ni_2MnGa martensite. J. Solid State Chem. **179**(11), 3525–3533 (2006)
26. J. Pons, R. Santamarta, V.A. Chernenko et al., Long-period martensitic structures of Ni-Mn-Ga alloys studied by high-resolution transmission electron microscopy. J. Appl. Phys. **97**(8), 083516 (2005)
27. C.B. Jiang, Y. Muhammad, L.F. Deng et al., Composition dependence on the martensitic structures of the Mn-rich NiMnGa alloys. Acta Mater. **52**(9), 2779–2785 (2004)
28. N. Lanska, O. Soderberg, A. Sozinov et al., Composition and temperature dependence of the crystal structure of Ni-Mn-Ga alloys. J. Appl. Phys. **95**(12), 8074–8078 (2004)
29. Y. Ge, H. Jiang, A. Sozinov et al., Crystal structure and macrotwin interface of five-layered martensite in Ni-Mn-Ga magnetic shape memory alloy. Mater. Sci. Eng. A **438**, 961–964 (2006)
30. L. Righi, F. Albertini, E. Villa et al., Crystal structure of 7 M modulated Ni-Mn-Ga martensitic phase. Acta Mater. **56**(16), 4529–4535 (2008)
31. M. Han, J.C. Bennett, M.A. Gharghouri et al., Microstructure characterization of the non-modulated martensite in Ni-Mn-Ga alloy. Mater. Charact. **59**(6), 764–768 (2008)
32. R. Kainuma, Y. Imano, W. Ito et al., Magnetic-field-induced shape recovery by reverse phase transformation. Nature **439**(7079), 957–960 (2006)

33. V.S. Bai, T. Rajasekharan, Evidence of a critical Mn-Mn distance for the onset of ferromagnetism in NiAs type compounds. J. Magn. Magn. Mater. **42**(2), 198–200 (1984)
34. V.D. Buchelnikov, P. Entel, S.V. Taskaev et al., Monte Carlo study of the influence of antiferromagnetic exchange interactions on the phase transitions of ferromagnetic Ni-Mn-X alloys (X=In, Sn, Sb). Phys. Rev. B **78**(18), 184427 (2008)
35. E. Wachtel, F. Henninger, B. Predel, Constitution and magnetic properties of Ni-Mn-Sn alloys-solid and liquid state. J. Magn. Magn. Mater. **38**(3), 305–315 (1983)
36. S. Yuan, P.L. Kuhns, A.P. Reyes et al., Magnetically nanostructured state in a Ni-Mn-Sn shape-memory alloy. Phys. Rev. B **91**(21), 214421 (2015)
37. H.H. Zhang, M.F. Qian, X.X. Zhang et al., Martensite transformation and magnetic properties of Fe-doped Ni-Mn-Sn alloys with dual phases. J. Alloy. Compd. **689**, 481–488 (2016)
38. T. Krenke, M. Acet, E.F. Wassermann et al., Martensitic transitions and the nature of ferromagnetism in the austenitic and martensitic states of Ni−Mn−Sn alloys. Phys. Rev. B **72**(1), 014412 (2005)
39. R.F. Hamilton, H. Sehitoglu, Y. Chumlyakov et al., Stress dependence of the hysteresis in single crystal NiTi alloys. Acta Mater. **52**(11), 3383–3402 (2004)
40. V. Sanchez-Alarcos, V. Recarte, J.I. Perez-Landazabal et al., Correlation between atomic order and the characteristics of the structural and magnetic transformations in Ni-Mn-Ga shape memory alloys. Acta Mater. **55**(11), 3883–3889 (2007)
41. V.A. Chernenko, J. Pons, C. Segui et al., Premartensitic phenomena and other phase transformations in Ni-Mn-Ga alloys studied by dynamical mechanical analysis and electron diffraction. Acta Mater. **50**(1), 53–60 (2002)
42. C. Segui, E. Cesari, J. Pons et al., Internal friction behaviour of Ni-Mn-Ga. Mater. Sci. Eng. A **370**(1–2), 481–484 (2004)
43. A. Zheludev, S.M. Shapiro, P. Wochner et al., Phonon anomaly, central peak, and microstructures in Ni_2MnGa. Phys. Rev. B **51**(17), 11310 (1995)
44. V.A. Chernenko, V. L'Vov, J. Pons et al., Superelasticity in high-temperature Ni-Mn-Ga alloys. J. Appl. Phys. **93**(5), 2394–2399 (2003)
45. C. Segui, V.A. Chernenko, J. Pons et al., Low temperature-induced intermartensitic phase transformations in Ni-Mn-Ga single crystal. Acta Mater. **53**(1), 111–120 (2005)
46. S.R. Yeduru, A. Backen, S. Fähler et al., Large superplastic strain in non-modulated epitaxial Ni–Mn–Ga films. Phys. Procedia **10**, 162–167 (2010)
47. W.H. Wang, Z.H. Liu, J. Zhang et al., Thermoelastic intermartensitic transformation and its internal stress dependency in $Ni_{52}Mn_{24}Ga_{24}$ single crystals. Phys. Rev. B **66**(5), 052411 (2002)
48. V.A. Chernenko, Compositional instability of β-phase in Ni-Mn-Ga alloys. Scr. Mater. **40**(5), 523–527 (1999)
49. S.K. Wu, S.T. Yang, Effect of composition on transformation temperatures of Ni-Mn-Ga shape memory alloys. Mater. Lett. **57**(26–27), 4291–4296 (2003)
50. X. Jin, M. Marioni, D. Bono et al., Empirical mapping of Ni-Mn-Ga properties with composition and valence electron concentration. J. Appl. Phys. **91**(10), 8222–8224 (2002)
51. M.F. Qian, X.X. Zhang, L.S. Wei et al., Effect of chemical ordering annealing on martensitic transformation and superelasticity in polycrystalline Ni-Mn-Ga microwires. J. Alloy. Compd. **645**, 335–343 (2015)
52. M. Kreissl, K.U. Neumann, T. Stephens et al., The influence of atomic order on the magnetic and structural properties of the ferromagnetic shape memory compound Ni_2MnGa. J. Phys. Condens. Matter **15**(22), 3831–3839 (2003)
53. X.X. Zhang, M.F. Qian, Z. Zhang et al., Magnetostructural coupling and magnetocaloric effect in Ni-Mn-Ga-Cu microwires. Appl. Phys. Lett. **108**(5), 052401 (2016)
54. C.B. Jiang, J.M. Wang, P.P. Li et al., Search for transformation from paramagnetic martensite to ferromagnetic austenite: NiMnGaCu alloys. Appl. Phys. Lett. **95**(1), 012501 (2009)
55. X.Z. Zhou, H. Kunkel, G. Williams et al., Phase transitions and the magnetocaloric effect in Mn rich Ni-Mn-Ga Heusler alloys. J. Magn. Magn. Mater. **305**(2), 372–376 (2006)

56. B. Jiang, W. Zhou, X. Qi et al., Recent progress of magnetically controlled shape memory materials. Mater. Sci. Forum **426**(pt. 3), 2285–2290 (2003)
57. C.H. Huang, Y. Wang, Z. Tang et al., Influence of atomic ordering on elastocaloric and magnetocaloric effects of a Ni-Cu-Mn-Ga ferromagnetic shape memory alloy. J. Alloy. Compd. **630**, 244–249 (2015)
58. J.M. Wang, H.Y. Bai, C.B. Jiang et al., A highly plastic $Ni_{50}Mn_{25}Cu_{18}Ga_7$ high-temperature shape memory alloy. Mater. Sci. Eng. A **527**(7–8), 1975–1978 (2010)
59. P.P. Li, J.M. Wang, C.B. Jiang, Martensitic transformation in Cu-doped NiMnGa magnetic shape memory alloys. Chin. Phys. B **20**(2), 028104 (2011)
60. I. Glavatskyy, N. Glavatska, A. Dobrinsky et al., Crystal structure and high-temperature magnetoplasticity in the new Ni-Mn-Ga-Cu magnetic shape memory alloys. Scripta Mater. **56**(7), 565–568 (2007)
61. V. Sanchez-Alarcos, J.I. Perez-Landazabal, V. Recarte et al., Correlation between composition and phase transformation temperatures in Ni-Mn-Ga-Co ferromagnetic shape memory alloys. Acta Mater. **56**(19), 5370–5376 (2008)
62. D. Kikuchi, T. Kanomata, Y. Yamaguchi et al., Magnetic properties of ferromagnetic shape memory alloys $Ni_2Mn_{1-x}Fe_xGa$. J. Alloy. Compd. **383**(1–2), 184–188 (2004)
63. K. Tsuchiya, A. Tsutsumi, H. Ohtsuka et al., Modification of Ni-Mn-Ga ferromagnetic shape memory alloy by addition of rare earth elements. Mater. Sci. Eng. A **378**(1–2), 370–376 (2004)
64. L. Gao, W. Cai, A.L. Liu et al., Martensitic transformation and mechanical properties of polycrystalline $Ni_{50}Mn_{29}Ga_{21-x}Gd_x$ ferromagnetic shape memory alloys. J. Alloy. Compd. **425**(1–2), 314–317 (2006)
65. O. Soderberg, K. Koho, T. Sammi et al., Effect of the selected alloying on Ni-Mn-Ga alloys. Mater. Sci. Eng. A **378**(1–2), 389–393 (2004)
66. X. Lu, X. Chen, L. Qiu et al., Martensitic transformation of Ni-Mn-Ga(C, Si, Ge) Heusler alloys. J. Phys. IV **112**, 917–920 (2003)
67. I. Glavatskyy, N. Glavatska, O. Soderberg et al., Transformation temperatures and magnetoplasticity of Ni-Mn-Ga alloyed with Si, In, Co or Fe. Scripta Mater. **54**(11), 1891–1895 (2006)
68. J.D. Callaway, R.F. Hamilton, H. Sehitoglu et al., Shape memory and martensite deformation response of Ni_2MnGa. Smart Mater. Struct. **16**(1), S108–S114 (2007)
69. Y. Zhang, M. Li, Y.D. Wang et al., Superelasticity and serration behavior in small-sized NiMnGa alloys. Adv. Eng. Mater. **16**(8), 955–960 (2014)
70. V.A. Chernenko, E. Villa, S. Besseghini et al., Giant two-way shape memory effect in high-temperature Ni–Mn–Ga single crystal. Phys. Procedia **10**, 94–98 (2010)
71. Y.Q. Ma, S.Y. Yang, W.J. Jin et al., $Ni_{56}Mn_{25-x}Cu_xGa_{19}$ ($x = 0, 1, 2, 4, 8$) high-temperature shape-memory alloys. J. Alloy. Compd. **471**(1–2), 570–574 (2009)
72. W.H. Wang, G.H. Wu, J.L. Chen et al., Magnetic field-controlled shape memory in $Ni_{52.5}Mn_{23.5}Ga_{24}$ single crystals. Adv. Eng. Mater. **3**(5), 330–333 (2001)
73. Y.X. Li, H.Y. Liu, F.B. Meng et al., Magnetic field-controlled two-way shape memory in CoNiGa single crystals. Appl. Phys. Lett. **84**(18), 3594–3596 (2004)
74. A.A. Likhachev, A. Sozinov, K. Ullakko, Influence of external stress on the reversibility of magnetic-field-controlled shape memory effect in Ni-Mn-Ga, in *Smart Structures and Materials 2001: Active Materials: Behavior and Mechanics*, vol. 4333 (International Society for Optics and Photonics, 2001), pp. 197–206
75. O. Heczko, L. Straka, Magnetic properties of stress-induced martensite and martensitic transformation in Ni-Mn-Ga magnetic shape memory alloy. Mater. Sci. Eng. A **378**(1–2), 394–398 (2004)
76. I. Aaltio, O. Soderberg, Y.L. Ge et al., Twin boundary nucleation and motion in Ni-Mn-Ga magnetic shape memory material with a low twinning stress. Scripta Mater. **62**(1), 9–12 (2010)
77. A. Sozinov, N. Lanska, A. Soroka et al., Highly mobile type II twin boundary in Ni-Mn-Ga five-layered martensite. Appl. Phys. Lett. **99**(12), 124103 (2011)

78. O. Heczko, L. Straka, K. Ullakko, Relation between structure, magnetization process and magnetic shape memory effect of various martensites occurring in Ni-Mn-Ga alloys. J. Phys. IV **112**, 959–962 (2003)
79. A. Sozinov, A.A. Likhachev, N. Lanska et al., Giant magnetic-field-induced strain in NiMnGa seven-layered martensitic phase. Appl. Phys. Lett. **80**(10), 1746–1748 (2002)
80. A. Sozinov, A.A. Likhachev, N. Lanska et al., Stress- and magnetic-field-induced variant rearrangement in Ni-Mn-Ga single crystals with seven-layered martensitic structure. Mater. Sci. Eng. A **378**(1–2), 399–402 (2004)
81. A. Sozinov, A.A. Likhachev, K. Ullakko, Magnetic and magnetomechanical properties of Ni-Mn-Ga alloys with easy axis and easy plane of magnetization, in *Smart Structures and Materials 2001: Active Materials: Behavior and Mechanics*, vol. 4333 (International Society for Optics and Photonics, 2001), pp. 189–196
82. V.A. Chernenko, M. Chmielus, P. Müllner, Large magnetic-field-induced strains in Ni-Mn-Ga nonmodulated martensite. Appl. Phys. Lett. **95**(16), 104103 (2009)
83. O. Heczko, L. Straka, V. Novak et al., Magnetic anisotropy of nonmodulated Ni-Mn-Ga martensite revisited. J. Appl. Phys. **107**(9), 09A914 (2010)
84. L. Straka, O. Heczko, H. Seiner et al., Highly mobile twinned interface in 10 M modulated Ni-Mn-Ga martensite: analysis beyond the tetragonal approximation of lattice. Acta Mater. **59**(20), 7450–7463 (2011)
85. K. Ullakko, Magnetically controlled shape memory alloys: a new class of actuator materials. J. Mater. Eng. Perform. **5**(3), 405–409 (1996)
86. R.C. O'Handley, Model for strain and magnetization in magnetic shape-memory alloys. J. Appl. Phys. **83**(6), 3263–3270 (1998)
87. P. Müllner, D. Mukherji, M. Aguirre et al., Micromechanics of magnetic-field-induced twin-boundary motion in Ni-Mn-Ga magnetic shape-memory alloys. Solid-Solid Phase Transformations Inorg. Mater. **2**, 171–185 (2005)
88. A. Pramanick, X.L. Wang, Characterization of magnetoelastic coupling in ferromagnetic shape memory alloys using neutron diffraction. JOM **65**(1), 54–64 (2013)
89. P. Müllner, Z. Clark, L. Kenoyer et al., Nanomechanics and magnetic structure of orthorhombic Ni-Mn-Ga martensite. Mater. Sci. Eng. A **481**, 66–72 (2008)
90. J. Pons, E. Cesari, C. Segui et al., Ferromagnetic shape memory alloys: alternatives to Ni-Mn-Ga. Mater. Sci. Eng. A **481**, 57–65 (2008)
91. Z.H. Liu, M. Zhang, W.Q. Wang et al., Magnetic properties and martensitic transformation in quaternary Heusler alloy of NiMnFeGa. J. Appl. Phys. **92**(9), 5006–5010 (2002)
92. A. Fujita, K. Fukamichi, F. Gejima et al., Magnetic properties and large magnetic-field-induced strains in off-stoichiometric Ni-Mn-Al Heusler alloys. Appl. Phys. Lett. **77**(19), 3054–3056 (2000)
93. H. Morito, A. Fujita, K. Fukamichi et al., Magnetic-field-induced strain of Fe-Ni-Ga in single-variant state. Appl. Phys. Lett. **83**(24), 4993–4995 (2003)
94. H. Morito, K. Oikawa, A. Fujita et al., Enhancement of magnetic-field-induced strain in Ni-Fe-Ga-Co Heusler alloy. Scripta Mater. **53**(11), 1237–1240 (2005)
95. H. Morito, A. Fujita, K. Fukamichi et al., Magnetocrystalline anisotropy in single-crystal Co-Ni-Al ferromagnetic shape-memory alloy. Appl. Phys. Lett. **81**(9), 1657–1659 (2002)
96. J. Liu, H.X. Zheng, Y.L. Huang et al., Microstructure and magnetic field induced strain of directionally solidified ferromagnetic shape memory CoNiAl alloys. Scripta Mater. **53**(1), 29–33 (2005)
97. M. Sato, T. Okazaki, Y. Furuya et al., Magnetostrictive and shape memory properties of Heusler type CO_2NiGa alloys. Mater. Trans. **44**(3), 372–376 (2003)
98. T. Kakeshita, T. Takeuchi, T. Fukuda et al., Giant magnetostriction in an ordered Fe_3Pt single crystal exhibiting a martensitic transformation. Appl. Phys. Lett. **77**(10), 1502–1504 (2000)
99. T. Fukuda, T. Sakamoto, T. Kakeshita et al., Rearrangement of martensite variants in iron-based ferromagnetic shape memory alloys under magnetic field. Mater. Trans. **45**(2), 188–192 (2004)

100. H.Y. Yasuda, N. Komoto, M. Ueda et al., Microstructure control for developing Fe-Pd ferromagnetic shape memory alloys. Sci. Technol. Adv. Mater. **3**(2), 165–169 (2002)
101. Y. Furuya, N.W. Hagood, H. Kimura et al., Shape memory effect and magnetostriction in rapidly solidified Fe-29.6 at% Pd alloy. Mater. Trans. JIM **39**(12), 1248–1254 (1998)
102. M. Pőtschke, S. Weiss, U. Gaitzsch et al., Magnetically resettable 0.16% free strain in polycrystalline Ni-Mn-Ga plates. Scripta Mater. **63**(4), 383–386 (2010)
103. S. Besseghini, E. Villa, F. Passaretti et al., Plastic deformation of NiMnGa polycrystals. Mater. Sci. Eng. A **378**(1–2), 415–418 (2004)
104. H. Morawiec, J. Lelątko, T. Goryczka et al., Extruded rods with <001> axial texture of polycrystalline Ni-Mn-Ga alloys, in *Materials Science Forum*, vol. 635 (Trans Tech Publications Ltd., 2010), pp. 189–194
105. R. Chulist, W. Skrotzki, C.G. Oertel et al., Microstructure and texture in $Ni_{50}Mn_{29}Ga_{21}$ deformed by high-pressure torsion. Scripta Mater. **62**(9), 650–653 (2010)
106. N. Scheerbaum, D. Hinz, O. Gutfleisch et al., Textured polymer bonded composites with Ni-Mn-Ga magnetic shape memory particles. Acta Mater. **55**(8), 2707–2713 (2007)
107. S. Kauffmann-Weiss, N. Scheerbaum, J. Liu et al., Reversible magnetic field induced strain in Ni_2MnGa-polymer-composites. Adv. Eng. Mater. **14**(1–2), 20–27 (2012)
108. X. Wang, M.J. Dapino, Behavior of NiMnGa under dynamic magnetic fields considering magnetic diffusion and eddy current power loss, in *SME 2007 International Mechanical Engineering Congress and Exposition* (American Society of Mechanical Engineers Digital Collection, 2007), pp. 381–387
109. J. Gutierrez, J.M. Barandiaran, P. Lazpita et al., Magnetic properties of a rapidly quenched Ni-Mn-Ga shape memory alloy. Sens. Actuators A Phys. **129**(1–2), 163–166 (2006)
110. P. Lazpita, G. Rojo, J. Gutierrez et al., Correlation between magnetization and deformation in a NiMnGa shape memory alloy polycrystalline ribbon. Sens. Lett. **5**(1), 65–68 (2007)
111. S.H. Guo, Y.H. Zhang, B.Y. Quan et al., Martensitic transformation and magnetic-field-induced strain in magnetic shape memory alloy NiMnGa melt-spun ribbon, in *Materials Science Forum*, vol. 475 (Trans Tech Publications Ltd., 2005), pp. 2009–2012
112. Y. Chen, X.X. Zhang, D.C. Dunand et al., Shape memory and superelasticity in polycrystalline Cu-Al-Ni microwires. Appl. Phys. Lett. **95**(17), 171906 (2009)
113. N. Scheerbaum, O. Heczko, J. Liu et al., Magnetic field-induced twin boundary motion in polycrystalline Ni–Mn–Ga fibres. New J. Phys. **10**(7), 073002 (2008)
114. F. Bernard, P. Delobelle, C. Rousselot et al., Microstructural, mechanical and magnetic properties of shape memory alloy $Ni_{55}Mn_{23}Ga_{22}$ thin films deposited by radio-frequency magnetron sputtering. Thin Solid Films **518**(1), 399–412 (2009)
115. R. Kainuma, Y. Imano, W. Ito et al., Metamagnetic shape memory effect in a Heusler-type $Ni_{43}Co_7Mn_{39}Sn_{11}$ polycrystalline alloy. Appl. Phys. Lett. **88**(19), 192513 (2006)
116. T. Krenke, E. Duman, M. Acet et al., Magnetic superelasticity and inverse magnetocaloric effect in Ni-Mn-In. Phys. Rev. B **75**(10), 104414 (2007)
117. J. LIU, S. AKSOY, N. SCHEERBAUM, et al. Large magnetostrain in polycrystalline Ni-Mn-In-Co. Applied Physics Letters, **95**(23), 232515 (2009)
118. K.A. Gschneidner, V.K. Pecharsky, A.O. Tsokol, Recent developments in magnetocaloric materials. Rep. Progr. Phys. **68**(6), 1479–1539 (2005)
119. A. Biswas, T. Samanta, S. Banerjee et al., Observation of large low field magnetoresistance and large magnetocaloric effects in polycrystalline $Pr_{0.65}(Ca_{0.7}Sr_{0.3})_{0.35}MnO_3$. Appl. Phys. Lett. **92**(1), 012502 (2008)
120. G.J. Liu, J.R. Sun, J. Shen et al., Determination of the entropy changes in the compounds with a first-order magnetic transition. Appl. Phys. Lett. **90**(3), 032507 (2007)
121. L. Tocado, E. Palacios, R. Burriel, Entropy determinations and magnetocaloric parameters in systems with first-order transitions: study of MnAs. J. Appl. Phys. **105**(9), 093918 (2009)
122. V.A. Chernenko, R.L. Anton, M. Kohl et al., Magnetic domains in Ni-Mn-Ga martensitic thin films. J. Phys. Condens. Matter **17**(34), 5215–5224 (2005)
123. E. Warburg, Magnetiche Untersuchungen Über Einige Wirkungen Der Koerzitivkraft. Annu. Phys. **13**, 141–164 (1881)

124. V.K. Pecharsky, K.A. Gschneidner, Giant magnetocaloric effect in $Gd_5(Si_2Ge_2)$. Phys. Rev. Lett. **78**(23), 4494–4497 (1997)
125. J. Lyubina, R. Schafer, N. Martin et al., Novel design of La(Fe, $Si)_{13}$ alloys towards high magnetic refrigeration performance. Adv. Mater. **22**(33), 3735–3739 (2010)
126. O. Tegus, E. Bruck, L. Zhang et al., Magnetic-phase transitions and magnetocaloric effects. Phys. B Condens. Matter **319**(1–4), 174–192 (2002)
127. M. Pasquale, C.P. Sasso, L.H. Lewis et al., Magnetostructural transition and magnetocaloric effect in $Ni_{55}Mn_{20}Ga_{25}$ single crystals. Phys. Rev. B **72**(9), 094435 (2005)
128. A. Smaili, R. Chahine, Composite materials for Ericsson-like magnetic refrigeration cycle. J. Appl. Phys. **81**(2), 824–829 (1997)
129. Q. Zhang, J.H. Cho, B. Li et al., Magnetocaloric effect in Ho_2In over a wide temperature range. Appl. Phys. Lett. **94**(18), 182501 (2009)
130. B.J. Korte, V.K. Pecharsky, K.A. Gschneidner, The correlation of the magnetic properties and the magnetocaloric effect in $(Gd_{1-x}Er_x)NiAl$ alloys. J. Appl. Phys. **84**(10), 5677–5685 (1998)

Chapter 2
Preparation and Properties of Bulk Magnetic Shape Memory Alloys

2.1 Introduction

Bulk alloys consist of single crystals and polycrystals. Single crystalline Ni–Mn–Ga alloy is the most widely studied MSMA. In 1996, Ullakko et al. [1] found 0.19% MFIS in a ferromagnetic Ni–Mn–Ga single crystal. The obtained MFIS reached the theoretical value (6% for 5M and 9.4% for 7M martensite) in single crystals in about six years. However, it was not until 2009 that giant MFIS was reached (8.7%) in polycrystalline alloys with carefully created dual pores with a size similar to grains. In 2006, Kainuma et al. [2] reported the large stress output in a Ni–Mn–Sn–Co alloy via meta-magnetic transition (magnetic-field-induced backward martensitic transformation). Then other various Ni–Mn–X (X = In, Sn, Sb) alloys received attention because of their high magnetic entropy change under an affordable magnetic field by a permanent magnet.

Bulk MSMAs can be prepared by element melting and casting. Mold casting using copper is often adopted to avoid the element segregation during solidification. Fabrication of single crystals is, however, much challenging since composition segregation seriously changes the martensitic transformation behavior and properties. The application of polycrystalline alloys has to increase its small MFIS, ductility and deformation capacity to shape complex components.

Plastic deformation is effective in shaping, microstructure modification and the creation of favorable crystal orientation. MSMAs are intermetallic compounds with intrinsic brittleness due to the ordered crystal structure. So, it is only possible to process at high temperatures. Interestingly, Ni–Mn–Ga exhibit a transition from $L2_1$ ordered structure to B2 semiordered structure during at 700–900 °C heating. The plastic deformation capacity of B2 phase is much higher than that of the $L2_1$ phase. This makes it possible for these MSMAs to be processed by extrusion, rolling or forging in B2 phase temperatures [2–7].

X. Zhang and M. Qian, *Magnetic Shape Memory Alloys*,
https://doi.org/10.1007/978-981-16-6336-9_2

2.2 Preparation of Polycrystalline Alloy

2.2.1 Melting and Casting

MSMAs are usually prepared by arc or induction melting and then cast into a metal mold. To prevent oxidation during melting at high temperatures, vacuum or inert gas is often used. For Mn-containing alloys, an extra 1–3% Mn element is needed to compensate the Mn loss because of the high saturation vapor pressure of Mn. To ensure a homogeneous composition distribution, the billet is often remelted several times (3–4 times) before casting into a copper mold. Nevertheless, it is necessary to anneal the ingot for a long time at high temperatures to enhance the composition homogenization [8].

Composition homogenization temperature is as high as 800–1000 °C for most MSMAs, because a B2 single-phase state can be reached at such temperature. This high-temperature heat treatment can cause a huge change in microstructure. For example, as-cast $Ni_{50}Mn_{37}Sn_{13}$ alloy contains a multiphase, non-uniform structure. After heat treatment at 950 °C for 72 h, the alloy shows a single-phase state without composition segregation. When the annealing time was extended to 4 weeks, the martensitic transformation peak becomes sharp. This indicates that the composition homogenization heat treatment has a great influence on the microstructure as well as phase transitions [8].

2.2.2 Solid-State Sintering

Solid-state sintering makes use of element or pre-alloy powders for densification at high temperatures. Under pressureless sintering conditions, $Ni_{43}Co_7Mn_{39}Sn_{11}$ pre-alloyed powder shows porosity 65% after sintering at 900 °C for 12 h; the porosity was reduced to 5% after 144 h [9]. So, the sintering is often carried out under stress to enhance densification. Some fast sintering techniques, such as spark plasma sintering (SPS), can also be used. For a Ni–Co–Mn–Sn alloy prepared by SPS, C pollution caused by graphite mold existed, which diminish the MT [10]. Obviously, post-sintering annealing is required for expected properties.

The advantage of the solid-state sintering is that it is feasible to adjust the composition and can use element powder instead of pre-alloying powder as raw material, which is cost-effective. In addition, element powder has a lower hardness than pre-alloyed powder and thus helpful for mixing [11].

2.3 Preparation of Single Crystalline Alloy

Czochralski (pulling method) and Bridgman methods are widely used for the preparation of single crystals. The pulling method is to heat and melt the metal in a crucible, and then the seed crystal attacks the melt and pull from above. In Bridgeman process, the alloy is melted in a crucible, and then the crucible slowly moves from the high-temperature region to the low-temperature region, in which unidirectional solidification occurs to get the crystal. The growth rate of single crystals by Bridgeman method is slower than that by the pulling.

A $Ni_{40.6}Co_{8.5}Mn_{40.9}Sn_{10}$ alloy prepared by Bridgman method is shown in Fig. 2.1 [12]. No secondary γ phase appears (Fig. 2.1a). The four elements distribute fairly homogeneous within 30 mm from seed crystal position; however, obvious macro-composition segregation exists along the single-crystal growth direction (Fig. 2.1b): Mn and Co elements increase with the increase of single-crystal length, while the content of Ni and Sn elements decreases. As a result, step-like MT and broadened MT temperature range appears as a result of composition segregation (Fig. 2.1c).

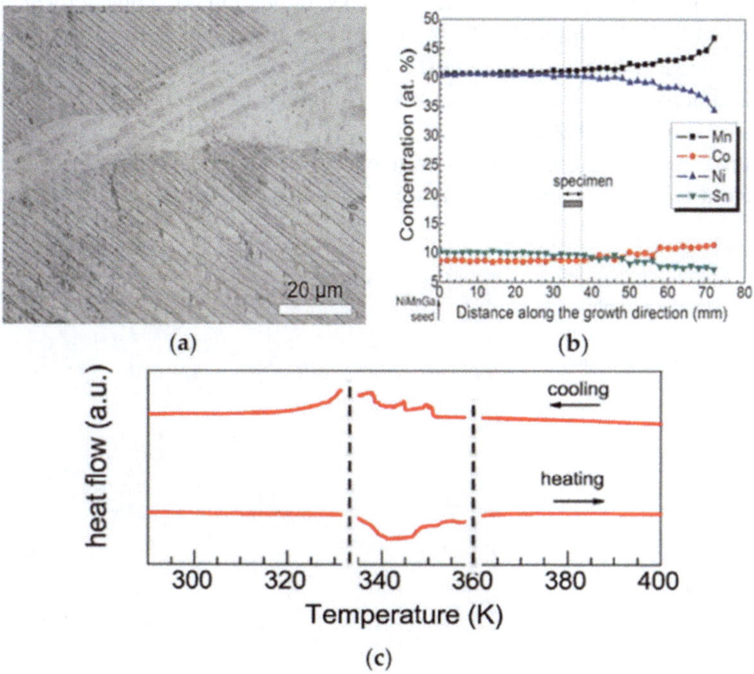

Fig. 2.1 $Ni_{40.6}Co_{8.5}Mn_{40.9}Sn_{10}$ alloy prepared by Bridgman method [12]. **a** Microstructure; **b** components along crystal growth direction; **c** DSC curves

2.4 Plastic Deformation Behavior of Polycrystals

MSMAs are usually brittle and thus difficult to sustain a large plastic deformation. For the application, complex shapes by plastic shaping such as plate, tube and rod are necessary. The plastic deformation is also a feasible strategy to modify the microstructure and thus properties. For example, by rolling, extrusion and texture can be formed. Rearrangement of martensite variant structure can also be formed, thus creating in-plane plastic deformation [13–15]. For plastic deformation controlled by dislocation, the dynamic material model is often adopted to evaluate the deformation ability [2–7].

2.4.1 High-Temperature Compressive Behavior

Figure 2.2 is a photograph of the polycrystalline Ni–Mn–Ga alloy after compression at temperatures > 600 °C. After compression to a ratio of 60%, the samples compressed at 900 and 1000 °C have no cracks, indicating good plasticity at temperatures > 900 °C. At strain rate 1 s^{-1} and temperature 800 °C, macroscopic cracks begin to appear; while at strain rates 0.1 s^{-1} and 1 s^{-1} and temperature 700 °C, more obvious macroscopic cracks are formed. The crack occurs in all 600 °C samples except for strain rate 0.001 s^{-1}.

Figure 2.3 shows the true stress–strain curves of polycrystalline Ni–Mn–Ga alloy at different strain rates and temperatures. The alloy subjects to elastic deformation

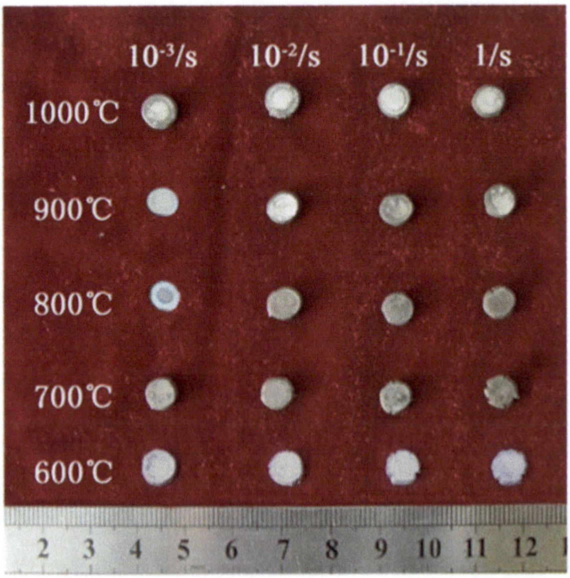

Fig. 2.2 Macro-photograph of the Ni–Mn–Ga sample after compression at various temperatures and strain rates

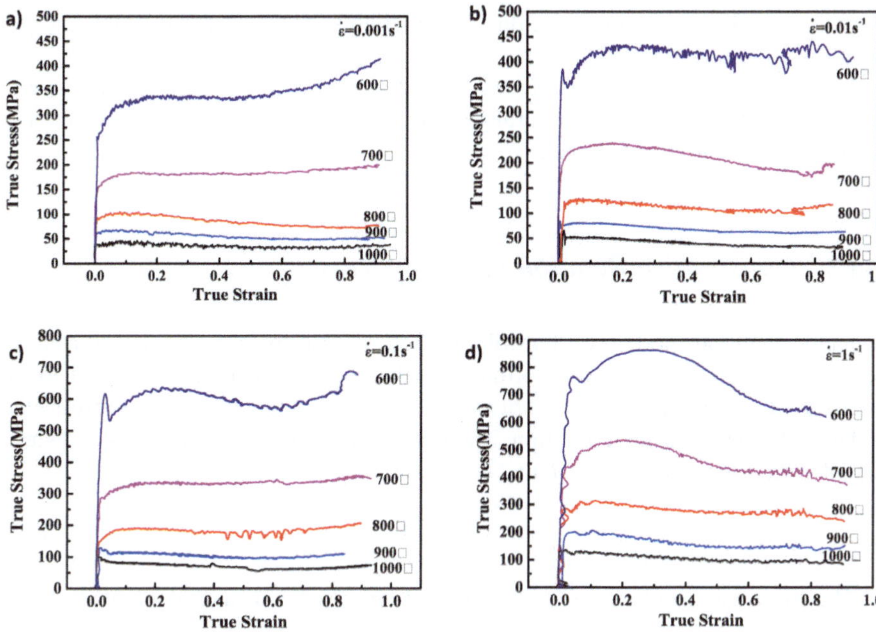

Fig. 2.3 High-temperature compression true stress–strain curves at different strain rates **a** 0.001 s^{-1}; **b** 0.01 s^{-1}; **c** 0.1 s^{-1}; **d** 1 s^{-1}

at first and then plastic deformation by dynamic recovery and recrystallization. As strain increases, softening and hardening reach equilibrium to meet a stress plateau. At a true strain 0.4, the alloy enters a steady deformation state, with a slight change of metal flow stress. The peak stress under different deformation conditions is listed in Table 2.1.

At the same strain rate, the peak value of full stress decreases with increasing temperature. At a strain rate 0.001 s^{-1}, the peak stress at the adjacent temperature above 800 °C is about 40 MPa. While at temperatures < 800 °C, the adjacent stress difference is about 100 MPa, indicating that the softening effect of the alloy is not obvious at low temperatures. At the strain rate 1 s^{-1}, the difference of flow stress

Table 2.1 Rheological stress peaks of polycrystalline Ni–Mn–Ga alloys under different deformation parameters (MPa)

$\dot{\varepsilon}/s^{-1}$	T/°C				
	600	700	800	900	1000
0.001	335.7	184.4	105.1	67.8	46.1
0.01	434.1	239.8	130.3	85.6	71.9
0.1	638.1	338.6	192.1	129.8	99.1
1.0	864.4	536.9	315.8	209	136.9

increases at adjacent temperature due to the strain hardening effect, which exceeds the alloy softening effect.

Moreover, at a high strain rate 1 s^{-1}, due to dynamic instability, the peak flow stress at 600 and 700 °C appears in the range of true strain 0.2–0.3. The stress decreases since cracks occur after the peak.

2.4.2 Constitutive Relation of Flow Stress

Dynamic material model is a feasible method to describe the plastic deformation behavior. In this model, the deformation behavior of a material is related to the microstructure and the history of hot working, as well as the deformation stress state and external deformation condition. The relationship between flow stress and strain rate, deformation temperature T and strain ε is:

$$\sigma = f(\dot{\varepsilon}, T, \varepsilon) \tag{2.1}$$

The strain rate can be expressed as a function of stress σ and temperature T:

$$\dot{\varepsilon} = AF(\sigma)\exp(-Q/RT) \tag{2.2}$$

At different stress levels, stress functions $F(\sigma)$ take different forms:

$$\text{At low stress, } F(\sigma) = \sigma^n \tag{2.3}$$

$$\text{At high stress, } F(\sigma) = \exp(\beta\sigma) \tag{2.4}$$

$$\text{At all stresses, } F(\sigma) = [\sinh(\alpha\sigma)]^n \tag{2.5}$$

where $\alpha = \beta/n$, α, β, n and A are constants, Q is deformation activation energy (kJ/mol), R is gas constant (8.314 J mol^{-1} K^{-1}), $\dot{\varepsilon}$ is a rate of strain (s^{-1}) and T is the absolute temperature (K).

The strain rate and temperature can be expressed as Zener–Hollomon parameter (Z):

$$Z = \dot{\varepsilon}\exp(Q/RT) \tag{2.6}$$

By substitution Eqs. (2.3) and (2.4) into (2.2), the strain rate can be expressed:

$$\ln(\dot{\varepsilon}) = \ln(B_1) + n_1\ln(\sigma) \tag{2.7}$$

$$\ln(\dot{\varepsilon}) = \ln(B_2) + \beta\sigma \tag{2.8}$$

Table 2.2 All parameter values of rheological stress equation of polycrystalline Ni–Mn–Ga alloy under peak rheological stress

Parameters	T/°C				
	600	700	800	900	1000
n_1	7.1449	6.4842	6.2422	6.0694	6.4205
β	0.01286	0.01991	0.03318	0.04922	0.07686
α	0.00593				

It shows that $\ln(\dot{\varepsilon})$ is linear with $\ln\sigma$ at low stress and linear at high stress level. The parameters n_1 and β can be determined by substituting the peak flow stresses of Table 2.1 into the Eqs. (2.7) and (2.8), respectively. By linear regression of the plot, the result is listed in Table 2.2. The α value can be obtained using n_1 and β values. Here, n_1 acts as a stress index at high temperature, which is used to analyze the deformation mechanism in creep deformation. When n_1 is 1, it is diffusion creep; when n_1 is 2, it is grain boundary slip creep; when n_1 is 3, it is dislocation slip creep and when n_1 is 4–5 or higher, it belongs to dislocation climbing creep [16]. As n_1 is 6.47 in polycrystalline Ni–Mn–Ga alloy, the deformation is controlled by dislocation climbing.

By substituting the $\alpha = 0.00593$, the equation at all stress levels is:

$$\dot{\varepsilon} = A[\sinh(\alpha\sigma)]^{n_2} \exp\left(-\frac{Q}{RT}\right) \tag{2.9}$$

By means of temperature compensation factor Z, the flow stress is:

$$\sigma = \frac{1}{\alpha} \ln\left\{\left[\frac{Z}{A}\right]^{\frac{1}{n_2}} + \left\{\left[\frac{Z}{A}\right]^{\frac{2}{n_2}} + 1\right\}^{\frac{1}{2}}\right\} \tag{2.10}$$

So the constitutive materials model is:

$$\ln(\dot{\varepsilon}) = \ln A - \frac{Q}{RT} + n_2 \ln[\sinh(\alpha\sigma)] \tag{2.11}$$

Using the flow stress in Table 2.1, the $\ln[\sinh(\alpha\sigma)] \sim \ln(\dot{\varepsilon})$ and $\ln[\sinh(\alpha\sigma)]\cdot\ln[\sinh(\alpha\sigma)] \sim \ln 10{,}000/T$ curves can be obtained. After linear regression, the deformation activation energy Q as well as the n_2 value and the α value can be obtained, as shown in Table 2.3. The obtained Q is 321.39 kJ/mol, which is higher than Ni, Mn and Ga self-diffusion activation energy in Ni–Mn–Ga alloy [17, 18], indicating that the deformation process is not controlled by diffusion, but by dynamic recrystallization.

The flow stress and Z parameters can be expressed by:

$$Z = \dot{\varepsilon} \exp(321.39/RT) \tag{2.12}$$

Table 2.3 Parameters of polycrystalline Ni–Mn–Ga alloy under peak flow stress

Parameters	$\dot{\varepsilon}$/s^{-1}	T/°C				
		600	700	800	900	1000
Q kJ/mol	0.001	245.16				
	0.01	264.20				
	0.1	340.34				
	1	435.86				
Q average		321.39				
n_2		2.1551	3.1739	4.4334	5.1161	5.9060
n_2 average		4.29203				
A		1.243×10^{13}				

$$\sigma = 186.634 \times \ln\{[Z/(1.243 \times 10^{13})]^{1/4.292} + \{[Z/(1.243 \times 10^{13})]^{2/4.292} + 1\}^{1/2}\} \tag{2.13}$$

These equations can estimate the flow stress at 600–000 °C with strain rate 0.001–1 s^{-1}, which can provide the constitutive equation for the numerical simulation Ni–Mn–Ga alloys.

2.4.3 *Hot Working Diagram*

2.4.3.1 Deformation Temperature and Strain Rate

The ability of materials to change shape without fracture through plastic deformation is called plasticity [19, 20]. The plasticity is related to microstructure, temperature and strain rate as well as stress state. So plasticity consists of stress state plasticity and inherent plasticity. For intrinsic plasticity, a hot working diagram is usually established based on the dynamic material model (DMM), which is actually overlapped by the superposition of power dissipation and instability diagrams, where the instability criterion is the continuous instability criterion here. Using true stress–strain curves (Fig. 2.3), a hot working diagram at total strains 0.4 and 0.5 is established for a Ni–Mn–Ga alloy, as shown in Fig. 2.4.

The solid lines in the diagram represent the percentage of energy dissipation efficiency, and the shadow region represents the flow instability areas. There are two power dissipation peaks D_1 and D_2: D_1 corresponds to temperature 750–780 °C and strain rate 0.03–0.3; while D_2 to temperature 950–1000 °C and strain rate 0.03–0.3. The deformation mechanisms at these two regions may be dynamic recovery or recrystallization.

It can also be seen from Fig. 2.4 that a valley region between the peak dissipation efficiency exists corresponding to temperature about 900 °C, which may be related to

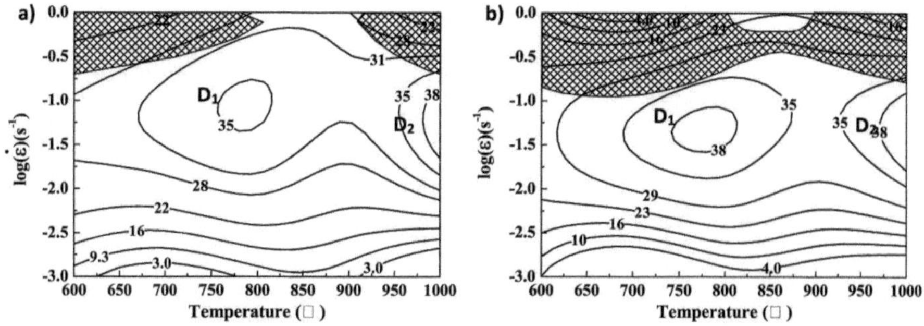

Fig. 2.4 Hot working diagrams of Ni–Mn–Ga alloy under different strains. **a** $\varepsilon = 0.4$; **b** $\varepsilon = 0.5$

the order–disorder transition of polycrystalline Ni–Mn–Ga alloys. Compared with the instability regions at different strains, it can be seen that with the increase of deformation strain, different instability phenomena occur. The instability region is larger at higher strains. At a lower strain 0.4, two instability regions appear at low and high temperatures. However, the instability regions overlapped at a high strain 0.5. The ideal thermal deformation parameters of polycrystalline Ni–Mn–Ga alloy are 750–800 and 950–1000 °C, with strain rate 0.03–0.3 s^{-1}.

2.4.3.2 Microstructure After High-Temperature Deformation

After the sample was subjected to compressive deformation, the sample is quenched by water to retain the deformed microstructure. Considering that the compressed sample changes into a drum shape, the maximum sized cross-section (center at the middle of height) was selected for microstructure observation.

Figure 2.5 is the microstructure after compression at temperatures of 700, 800, 900 and 1000 °C at a strain rate 0.001 s^{-1}. The original alloy exhibits columnar grains formed during casting. At temperature 700 °C (Fig. 2.5a), the grain morphology keeps unchanged; but a few fine grains (diameter 1–8 μm) start to be formed at the grain boundaries and the grain boundaries show a slight zigzag shape, implying the occurrence of recrystallization. From Fig. 2.5b, it can be seen that at 800 °C, more fine grains appear in the elongated deformed grain boundary region, showing the obvious zigzag shape and the grain diameter is about 10 μm. When the deformation temperature reaches 900 °C, the original columnar grains disappear and are replaced by equiaxed grain structure (Fig. 2.5c). The grains grow to 30–100 μm, indicating that the Ni–Mn–Ga alloy has undergone obvious dynamic recrystallization. The recrystallization process of polycrystalline Ni–Mn–Ga alloys is more sufficient at 1000 °C, and the recrystallized grains become larger and the diameter is about 200 μm. At the same time, it can be seen from Fig. 2.5d that the twins in most grains are along the vertical direction, which indicates that the material produces texture after deformation, and the grains have similar orientation, which makes the direction of martensite twin variants basically consistent.

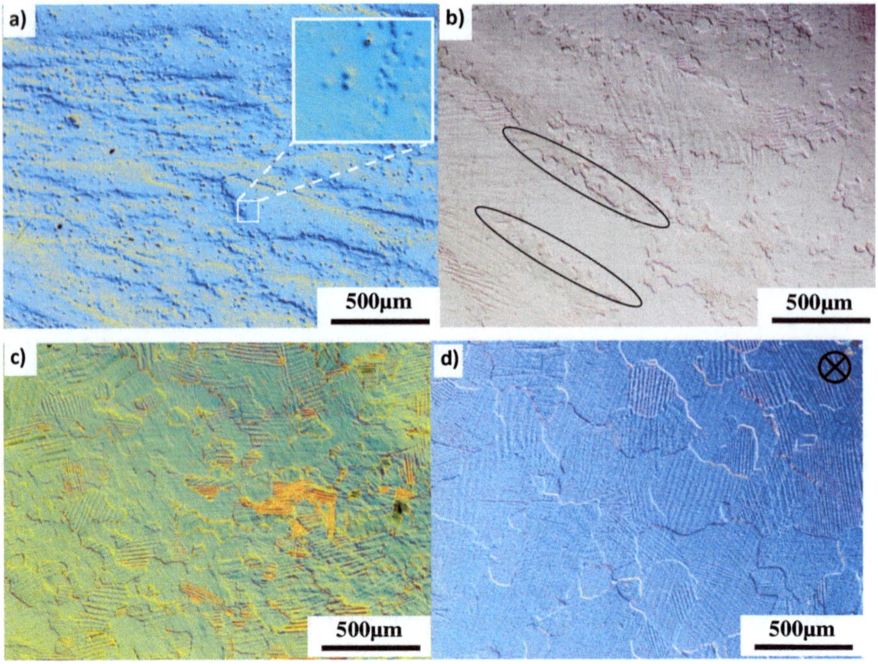

Fig. 2.5 Microstructure of Ni–Mn–Ga alloy after compressive tests at different temperatures ($\varepsilon = 0.001\ s^{-1}$) (⊕ indicates compression direction). **a** 700 °C; **b** 800 °C; **c** 900 °C; **d** 1000 °C

At a strain rate 0.1 s^{-1}, the alloy shows similar microstructure change with temperature. But the deformation structure at temperature 800 °C and strain rate 0.001/s is similar to that at temperature 900 °C and strain rate 0.1 s^{-1}, which indicates that the deformation temperature and strain rate are complementary at deformation. This can be explained from the stress–strain curve that the strain rate compensation factor has the same rheological stress when the Zener–Hollomon parameters are the same at high-temperature deformation.

The microstructure of the alloy was affected by the strain rate. Here, the microstructure of the alloy deformed at 1000 °C (Fig. 2.6) was analyzed. The grain size is 200, 100 and 50 μm at strain rates 0.001, 0.01 and 0.1 s^{-1}, respectively. At a high strain rate 1 s^{-1} (Fig. 2.6d), the zigzag grain boundaries formed in Fig. 2.6b disappeared; instead, fine grains of size < 10 μm are formed by dynamic recrystallization. At a high strain rate, the alloy softening is combined by dynamic recovery and dynamic recrystallization: Dynamic recovery occurs within the grain while dynamic recrystallization occurs at grain boundaries. Both processes reduce the dislocation density and thus work hardening effect.

From the microstructure analysis, it is proved that the softening effect in polycrystalline Ni–Mn–Ga alloys is mainly dynamic recovery and dynamic recrystallization. At low temperatures, the dynamic recovery was created by edge dislocation climb, screw dislocation intersecting slip, positive and negative dislocation cancel,

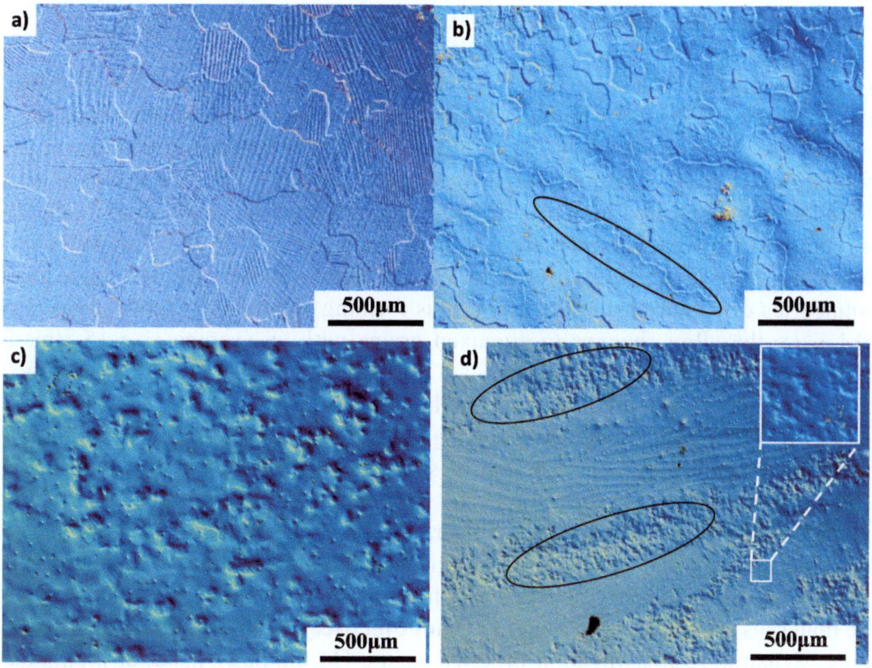

Fig. 2.6 Microstructure of Ni–Mn–Ga alloys after compressive tests at different strain rates (T = 1000 °C). **a** 0.001 s^{-1}, **b** 0.01 s^{-1}, **c** 0.1 s^{-1}, **d** 1 s^{-1}

and thus, dislocation density decreases in the deformed grains. When the temperature increases, it changes from dynamic recovery to dynamic recrystallization. This is realized when the dislocation accumulates to a high enough density. During dynamic recrystallization, new grains are formed, and the grain nucleation rate does not change much with the increase of temperature, while the grain growth rate increases with deformation temperature. So, the recrystallization grain size becomes larger with the increase of temperature.

Dynamic recrystallization is generally divided into two categories: (1) Discontinuous recrystallization. It is related to nucleation and large angle migration occur at deformed grain boundaries to eliminate deformation defects such as dislocations and subgrain boundaries in deformed tissues; (2) continuous dynamic recrystallization. Subgrain boundaries absorb continuously dislocations, resulting in continuously increase of the grain boundary angle, and finally changes from small angle grain boundary to large angle grain boundary (subgrain to real grain). In this process, dislocations are consumed. In polycrystalline Ni–Mn–Ga alloys, the recrystallized grains nucleate at grain boundaries, making the original columnar crystal boundaries serrate and then grow up. So, discontinuous recrystallization is the main formation mechanism in the formation of recrystallized grains in polycrystalline Ni–Mn–Ga alloy.

2.4.3.3 High-Temperature Deformation Mechanism

The alloys processed at strain rates 0.1 and 1 s^{-1} at 1000 °C were selected for EBSD analysis to reveal the deformation mechanism of polycrystalline Ni–Mn–Ga alloy at high temperature, as shown in Fig. 2.7. The orientation difference of adjacent two points greater than 2° is considered to be grain boundary. The small angle grain boundary is represented by white line (2–15°), and the large angle grain boundary is represented by black line (0–80°). At high strain rate (1 s^{-1}), the alloy still keeps columnar grain structure (aspect ratio > 2). A large number of small angle grain boundaries distribute inside the grain, thus forming subgrains. At low strain rates (0.1 s^{-1}), the alloy changes to equiaxed grains with an average size 15 μm. In addition, large angle grain boundaries are formed between fine grains, while almost no small angle grain boundaries exist in the fine grains.

Figure 2.8 is a statistical chart showing the fractions grain boundary orientations at different strain rates. The fraction of small angle grain boundaries is less than that at high strain rate at low strain rate. However, at high strain rate, the fraction of large angle grain boundaries is higher. This may be explained by the effect of dynamic recovery and dynamic recrystallization on the softening of polycrystalline Ni–Mn–Ga alloys during compressive deformation. At low temperature (< 700 °C), the materials are mainly subjected to dynamic recovery. With the increase of temperature, dynamic recrystallization process occurs. On the other hand, at high strain rate, the alloy exhibits mainly dynamic recovery because the dynamic recrystallization process is inhibited. Therefore, the alloy can be fully recrystallized at a high temperature 1000 °C and strain rate 0.1 s^{-1}.

Figure 2.9 is the grain orientation map at a strain rate of 1 s^{-1} at 1000 °C. It can be seen that the grain shows a preferential orientation. The color matching of the reverse pole diagram (yellow) shows the normal direction of the sample (i.e., compression direction) is parallel to the <111>$_c$ direction of martensite. A blue line is used to

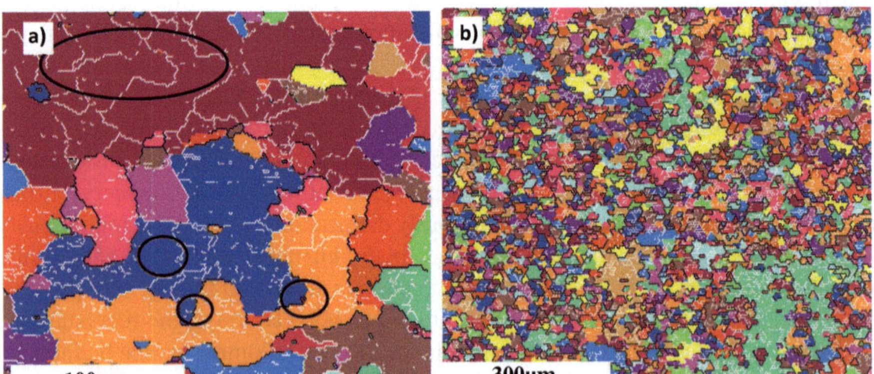

Fig. 2.7 Grain orientation of Ni–Mn–Ga alloy after a compressive test at 1000 °C for different strain rates. **a** 1 s^{-1}, **b** 0.1 s^{-1}

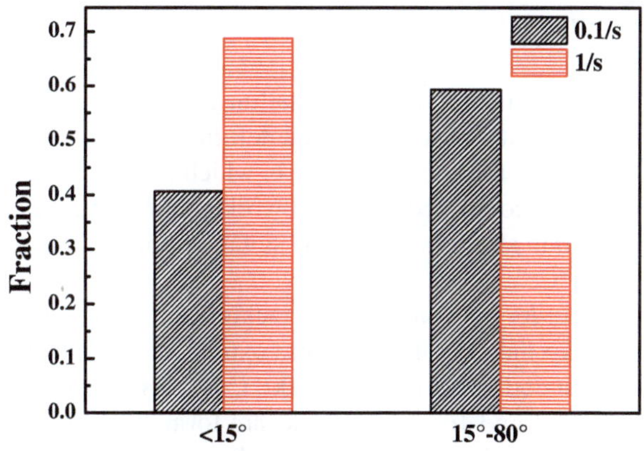

Fig. 2.8 Fraction of grain boundary at different strain rates after the alloy was compressed at 1000 °C

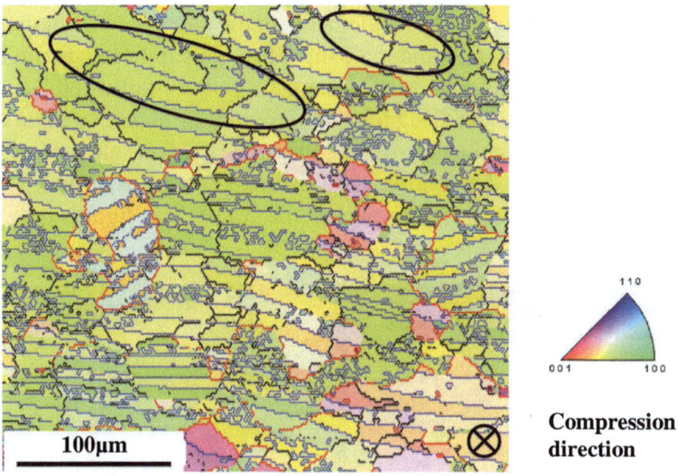

Fig. 2.9 Grain orientation diagram of Ni–Mn–Ga after compressive test at 1 s^{-1} at 1000 °C

represent the martensite grain boundary, which is different from the undeformed alloy martensite. At the same time, the small angle grain boundary is represented by black line and the large angle grain boundary by red line. The martensite twin energy passes through the small angle grain boundary and ends at the large angle grain boundary. This means that the small angle grain boundary is weaker than the large angle grain boundary.

2.4.4 Hot Extrusion

Hot deformation of polycrystalline Ni–Mn–Ga alloy is difficult at low temperature. However, it showed enhanced deformation capacity at temperatures higher than the chemical ordering temperature, which is 700–1100 °C for Ni–Mn–Ga alloys. As it has been confirmed in the hot working diagram, the temperatures around 1000 °C at moderate strain rates may be suitable. So, the hot extrusion is selected at 1000 and 1050 °C.

In order to avoid surface cracking and reduce oxidation, low carbon steel is used to wrap the alloy ingot before extrusion. The macroscopic photos of Fig. 2.10a, b correspond to extrusion at 1000 °C (extrusion ratio 9:1) and 1050 °C (extrusion ratio 9:1, 12:1 and 16:1). No cracks are formed in the extruded rods. The alloy interface between clad low carbon steel and polycrystalline Ni–Mn–Ga is shown in Fig. 2.10c. There is an oxidation layer between low carbon steel and steel coating. However, EDS test shows that the content of Fe elements in the alloy can be ignored, showing that the Fe element does not contaminate the alloy. This indicates that it is feasible to use a low carbon ladle steel for the extrusion of polycrystalline Ni–Mn–Ga alloy.

Statistical grain size using EBSD for samples with different extrusion conditions is shown in Fig. 2.11. From the inverse pole diagram (IPF), the alloy shows equiaxed grains, implying sufficient dynamic recrystallization after hot extrusion. The width of the columnar grains is 200–500 μm; however, the grain size is significantly decreased

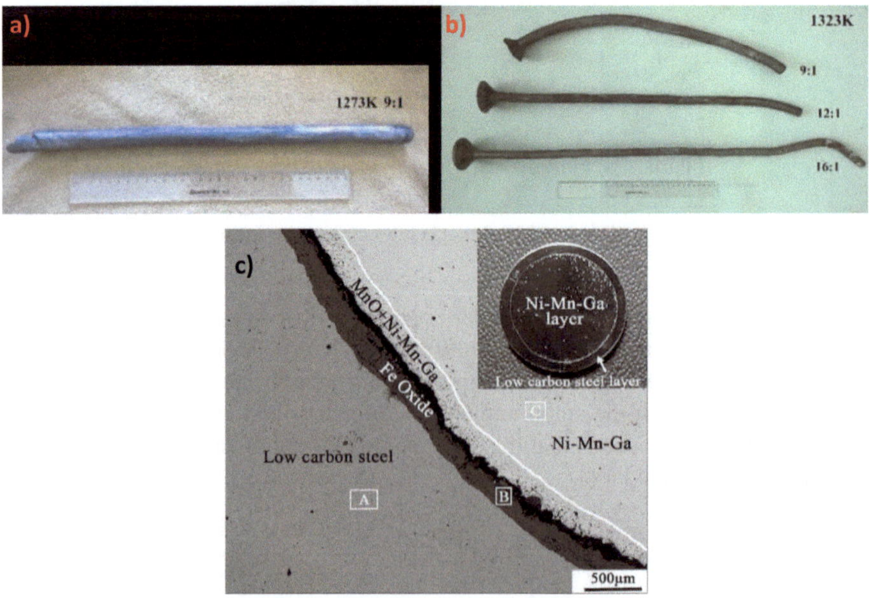

Fig. 2.10 Macroscopic and BSE photographs of polycrystalline Ni–Mn–Ga alloys under different extrusion ratios and temperatures **a** 1000 °C, **b** 1050 °C. **c** is BSE image showing cross-section of extruded sample extruded at 1050 °C with extrusion ratio 12:1

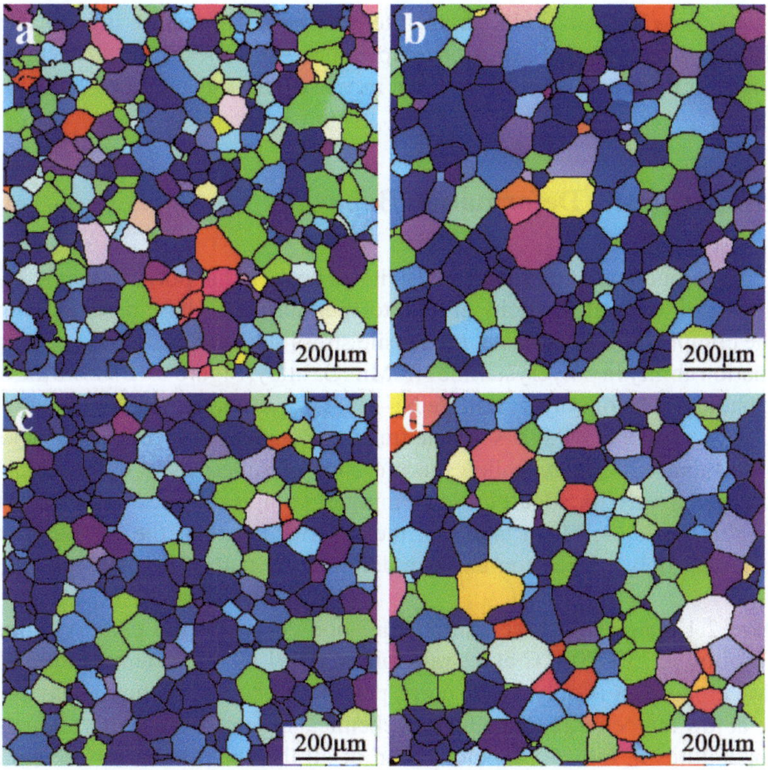

Fig. 2.11 Inverse pole diagram (IPF) diagram of polycrystalline Ni–Mn–Ga alloy under different extrusion conditions. **a** 1000 °C 9:1; **b** 1050 °C 9:1; **c** 1050 °C 12:1; and **d** 1050 °C 16:1

after extrusion, i.e., 60–70 μm. The statistical data are shown in Table 2.4. The grain size increases with the increase of extrusion temperature, while the grain size decreases slightly with the increase of extrusion ratio. In addition, the polycrystalline

Table 2.4 Grain sizes in polycrystalline Ni–Mn–Ga alloy with different extrusion temperature (T) and extrusion ratios (R)

Status		Average grain size (μm)
As-cast		200–500 (width)
Extrusion state	T = 1000 °C, R = 9:1	61.8 ± 5.6 (diameter)
	T = 1050 °C, R = 9:1	75.7 ± 1.0 (diameter)
	T = 1050 °C, R = 12:1	66.3 ± 2.3 (diameter)
	T = 1050 °C, R = 16:1	69.5 ± 1.0 (diameter)

Ni–Mn–Ga alloy formed <1 1 1> texture after extrusion. The texture strength increases slightly with the increase of extrusion temperature, while the texture strength remains unchanged or decreased slightly with the increase of extrusion ratio.

2.5 Superplastic Deformation Behavior

From high-temperature compressive test, it can be seen that polycrystalline Ni–Mn–Ga alloy exhibits good plasticity at high temperature (B2 phase region). Compared with the compressive stress state, the tensile stress state can better reveal the plastic deformation ability in polycrystalline Ni–Mn–Ga alloy. Figure 2.12 shows the macroscopic photographs of polycrystalline Ni–Mn–Ga alloys after tensile test at 800 and 900 °C at a strain rate of 0.001 s^{-1}. The fracture elongation exceeds 100% at temperatures 800 and 900 °C, revealing the presence of superplastic deformation. This provides a new way for superplastic forming of complex shapes by polycrystalline Ni–Mn–Ga alloys [4, 7]. In addition, the deformed part of the sample is mostly uniform, with only a little non-uniform deformation part of the necking region [4, 7].

Figure 2.13 shows the true stress-elongation curves of the polycrystalline Ni–Mn–Ga alloy at 750–900 °C at a strain rate of 0.001 s^{-1}. At a temperature 750 °C, the alloy elongation is only 5.4%, corresponding to brittle fracture behavior of the sample at low temperature (the alloy is a chemical ordered state at this temperature). At temperatures > 750 °C, the elongation of all the samples is > 40% and shows good plasticity of the polycrystalline Ni–Mn–Ga alloy. The enhanced plasticity is attributed to the partially ordered state in the B2 phase.

From Fig. 2.13, it can also be seen that the flow stress under tensile condition is similar to that in compression state in polycrystalline Ni–Mn–Ga alloy. At the initial stage of deformation, the flow stress increases rapidly due to elastic deformation. When the elongation reaches 10%, the flow stress begins to enter the steady state, in

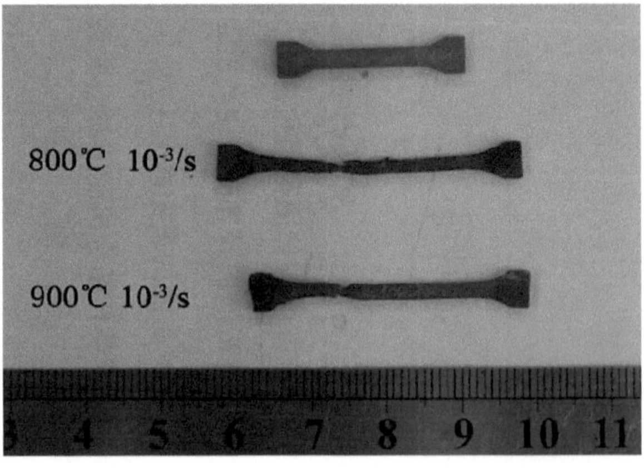

Fig. 2.12 Macro-morphology of Ni–Mn–Ga samples after high-temperature tensile test at various temperatures and strain rates

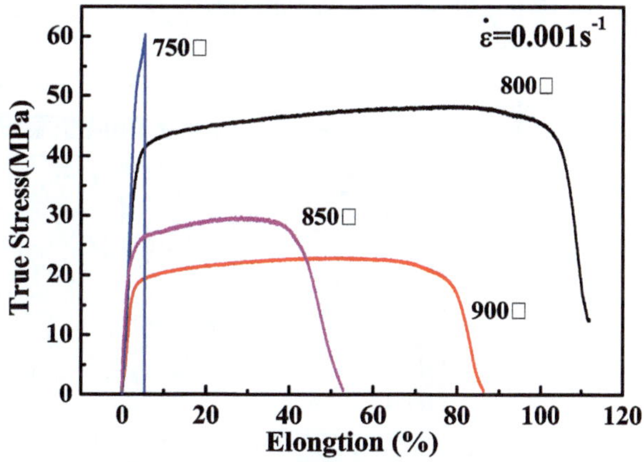

Fig. 2.13 True stress-elongation curve of Ni–Mn–Ga alloy at different temperatures (strain rate 0.001 s^{-1})

which the flow stress increases slightly in the steady state, showing a work hardening effect. The stress decreases rapidly at the final stage due to necking. In addition, as the deformation temperature increases, the steady-state flow stress and peak stress in polycrystalline Ni–Mn–Ga alloy decrease continuously. The flow stress in tensile state is smaller than that in compressive state (about 1/2 in compression state).

Figure 2.14 shows the fracture morphology of polycrystalline Ni–Mn–Ga alloy after tensile test at different temperatures. At 750 °C, the alloy shows obvious brittle fracture characteristics with fracture elongation of only 5.4%. The fracture mode is mainly intergranular fracture with transgranular fracture characteristic in a few regions. When the deformation temperature rises to 900 °C, the alloy exhibits ductile fracture with an obvious neck effect.

Figure 2.15 shows the cross-section and longitudinal section microstructure of Ni–Mn–Ga alloy after deformation at temperatures 800 and 900 °C. From Fig. 2.15a, the grains in the deformed region are all elongated at 800 °C with only 3–4 grains in the width direction (marked *A*, *B*, *C*, *D*, in which grain B has a width ~ 1 mm). Figure 2.15b shows the cross-section on the other side of the tensile sample. The

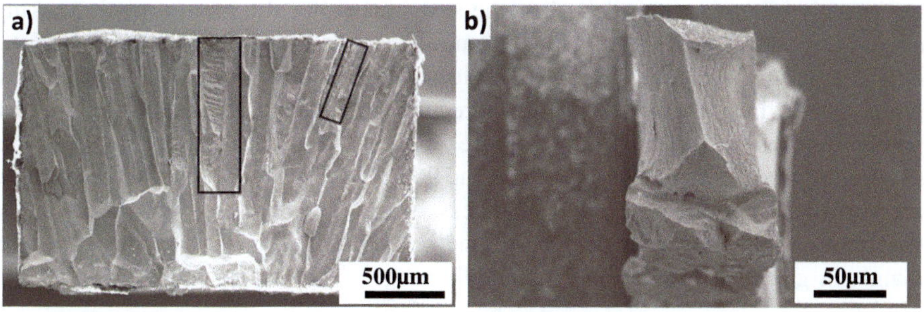

Fig. 2.14 SEM images of the tensile fracture surface in Ni–Nn–Ga alloy at different temperatures (strain rate 0.001 s^{-1}). **a** 750 °C, **b** 900 °C

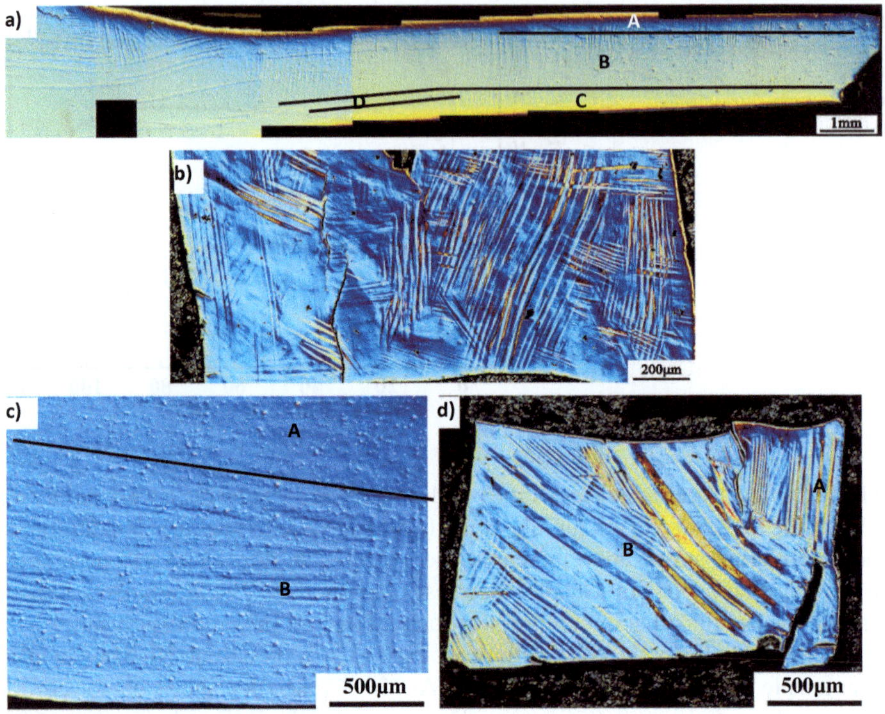

Fig. 2.15 Cross-section microstructure of Ni–Mn–Ga alloy after tensile deformation at different temperatures under a strain rate of 0.001 s^{-1}. **a** 800 °C longitudinal section, **b** 800 °C cross-section, **c** 900 °C longitudinal section, and **d** 900 °C cross-section

twins span the whole thickness direction, indicating that there is only one grain in the thickness direction. On the whole, the sample keeps a polycrystalline character after high-temperature deformation.

Studies show that the grain growth temperature in polycrystalline Ni–Mn–Ga alloy is > 900 °C under no stress. This implies, at 800 °C without stress, the grain does not grow, while under tensile stress, the stress promotes the grain growth, making a decrease of the alloy grain growth temperature. In Fig. 2.15c tested at 900 °C, only two grains (*A*, *B*) exist in the sample width direction. In Fig. 2.15d, only one grain appears in the sample thickness direction. In addition, the resulting twins are relatively coarse. So, grain growth is more obvious at 900 °C. The fact that the grains show elongated deformed grains at 800 and 900 °C indicates the absence of dynamic recrystallization in the deformation process, but the dynamic recovery process. The deformation mechanism of polycrystalline Ni–Mn–Ga alloys at 800 and 900 °C is mainly in-granular dislocation slip and climbing.

2.6 Magnetic-Field-Induced Strain

The magnetic-field-induced strain (MFIS) obtained in Ni–Mn–Ga alloy is derived from the magnetic field driving twin boundary motion. That is, the driving force of the magnetic field on the twin boundary is larger than the twin boundary motion resistance. The driving force is related to the magnetic anisotropy energy, while the twin movement resistance depends on the martensite structure.

For the driving force of the magnetic field acting on the twin boundary, it begins to increase with the increase of the magnetic field and then reaches the maximum value at a certain magnetic field. This maximum driving force is determined by the magnetocrystalline anisotropy energy (K_u) and the strain of the twin during shear deformation. The maximum driving stress of the magnetic field acting on the twin is not sensitive to the microstructure but is determined by the crystal structure. However, the critical stress of twinning is sensitive to microstructure. Hence, in order to obtain the magnetically induced strain of Ni–Mn–Ga alloy, one will focus on how to reduce the resistance of twin motion, which can be effectively reduced by training. For example, in single-crystal materials, the critical stress of twin boundary motion of 5M and 7M martensites can be reduced to 1–2 MPa; especially for 5M martensite, the critical stress can be reduced to 0.1–0.3 MPa [21, 22] in single crystal of type II twin. The maximum stress of magnetic field acting on twins is generally around 2 MPa under a magnetic field of 2 T. The theoretical maximum strain that is close to 6% and 10% is obtained in 5M and 7M martensite single crystals. While the magnetic force that NM martensite can achieve under the action of magnetic field is much smaller than the critical stress of its twin boundary motion, so the MFIS obtained under the commonly used magnetic field intensity is still very small.

Training methods in single crystalline Ni–Mn–Ga alloy include thermal–mechanical training, magnetic–mechanical training, mechanical training, thermal–magnetic training and thermal–magnetic–mechanical training. (1) Thermal–mechanical training is the treatment of martensitic transformation in which the material cools from the austenite state under constant stress. This training treatment can prompt twin variants to grow in one direction so that the number of twin variants is reduced, and the critical stress of the twin boundary motion decreases. (2) Magnetic–mechanical treatment is usually realized by repeated cyclic deformation of materials in a rotating magnetic field or by applying a constant bias-stress and a constant magnetic field perpendicular to the stress. During these processes, the twin boundaries are driven to undergo reciprocating motion to form twin variants along the a and c axes. (3) Mechanical training is to apply the load in the c axis direction; after unloading, the stress is applied in the direction perpendicular to the c axis. By this way, the critical stress of the twin boundary motion is reduced by the cyclic deformation. (4) Thermal–magnetic training is to heat the material above MT temperature, apply a magnetic field and then cool the material to induce MT. In this process, the orientation of martensite twin variants is reorientated along the axis c easy magnetization because the magnetic moment of ferromagnetic austenite is easy to turn along under the external field. (5) Thermal–magnetic–mechanical training is to heat the alloy to austenite state and cool it in a rotating magnetic field. During the training, the twin boundary moves back and forth, and preferential orientated twin variants are formed.

Through the above training methods, the twinning reciprocating motion of martensite can reduce the number of variants, form preferred orientation variants and reduce the pinning effect of point defects on twins. These effects can improve the mobility of twins in single and polycrystals.

The polycrystalline Ni–Mn–Ga alloy with texture can be obtained by directional solidification technology [23, 24]. Preferential texture formed by the directional solidification can reduce the orientation difference between adjacent grains, reduce the resistance of the twin boundary movement and thus facilitate the production of MFIS. Potschke et al. [23] prepared polycrystalline Ni–Mn–Ga alloys with $<100>_A$ wire texture by directional solidification. After 1000 °C heat treatment, the alloy exhibits uniform composition and grain in mm grade. The alloy is further subjected to three-way alternating compression training to obtain an approximate single variant structure to achieve a maximum strain of 8% in a 7M martensite. Gaitzsch et al. [25] also make use of directional solidification to induce 5M martensite structure in polycrystalline Ni–Mn–Ga alloy; the structure remains stable after 10^6 cycles.

2.7 Magnetic and Magnetocaloric Properties

2.7.1 Ni–Mn–Ga Alloy

Ni–Mn–Ga alloy exhibits first-order structural MT and second-order magnetic transition, resulting in significant magnetic and caloric change. By adjusting the composition to induce the alloy magnetic–structural coupling, a larger magnetocaloric effect can be obtained. However, the hysteresis effect of the first-order MT adversely affects the magnetocaloric performance, which is an important problem to be considered in research and application.

2.7.1.1 Thermal and Magnetic Hysteresis

Thermal hysteresis is one of the main characteristics of the first-order MT, which is mainly related to the resistance of phase interface motion during the phase transition. The resistance mainly comes from the irreversible elastic strain energy as well as the friction energy losses by defects and interface movement. *M–T* curves of $Ni_{54}Mn_{21.2}Ga_{24.8}$ and $Ni_{54.8}Mn_{20.7}Ga_{24.5}$ alloys are shown in Fig. 2.16 [26]. Both alloys show a magnetostructural coupling state at high magnetic field. The saturation magnetization of both alloys is comparable, ~ 70 emu/g. The first derivative maximum of the heating/cooling curves is taken as the transition peak temperature A_p and M_p; The *MT* temperature is obtained by the tangent method. Using the temperature difference between the A_s and the M_f to indicate the thermal hysteresis,

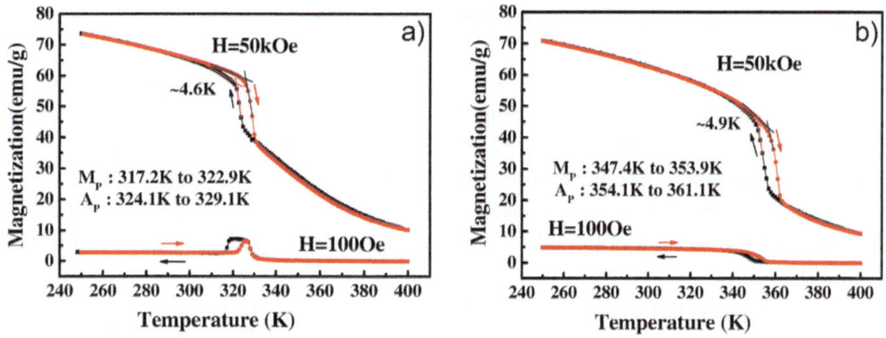

Fig. 2.16 Magnetization curves of Ni–Mn–Ga alloys exhibiting magnetocaloric coupled state. **a** $Ni_{54}Mn_{21.2}Ga_{24.8}$, **b** $Ni_{54.8}Mn_{20.7}Ga_{24.5}$

the thermal hysteresis of the $Ni_{54}Mn_{21.2}Ga_{24.8}$ alloy at 50 kOe is ~ 4.6 K, while the thermal hysteresis of the $Ni_{54}Mn_{21.2}Ga_{24.8}$ alloy at 50 kOe is 4.9 K, which is smaller than the hysteresis value of the Ni–Mn–Ga alloy reported in most literature (~ 10 K [27]) due to the magnetostructural coupling.

In addition to thermal hysteresis, magnetic hysteresis is also an important factor affecting magnetocaloric performance. Figure 2.17 shows the isothermal magnetization curves near the *MT* temperatures [26]. The area surrounded by the curves denotes the magnetic hysteresis. The average magnetic hysteresis of $Ni_{54}Mn_{21.2}Ga_{24.8}$ and $Ni_{54.8}Mn_{20.7}Ga_{24.5}$ alloys at 50 kOe is ~ 9.69 J/kg and 13.64 J/kg, respectively, similar to the hysteresis of $Ni_{52}Mn_{26}Ga_{22}$ alloy [28]. It can be seen that the slope of the magnetization and demagnetization curves changes obviously, implying the presence of magnetic-field-induced phase transition between ferromagnetic martensite and paramagnetic austenite. The magnetotransition is the main reason for the presence of hysteresis during phase transformation [28].

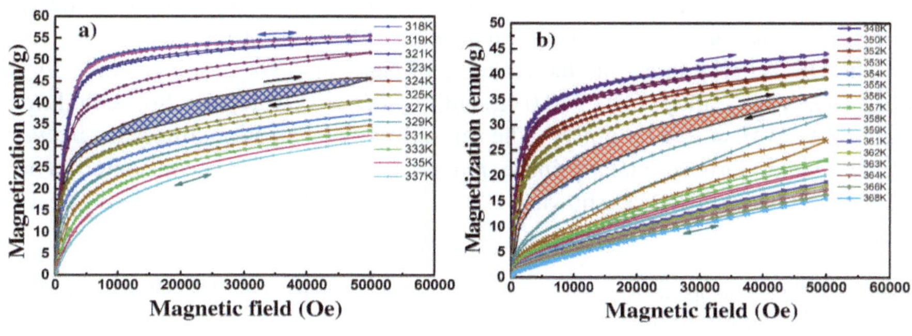

Fig. 2.17 Isothermal magnetization curves at different temperatures. **a** $Ni_{54}Mn_{21.2}Ga_{24.8}$, **b** $Ni_{54.8}Mn_{20.7}Ga_{24.5}$

2.7.1.2 Magnetic-Field-Induced Phase Transition

In Ni–Mn–Ga alloy, the martensite is ferromagnetic and austenite is paramagnetic. As a result, the phase transition temperature moves to the high-temperature range under bias-magnetic field. Here, the MT peak temperature A_p and M_p is used to evaluate the transition temperature change under magnetic field. The peak phase transition temperatures of $Ni_{54}Mn_{21.2}Ga_{24.8}$ and $Ni_{54.8}Mn_{20.7}Ga_{24.5}$ alloys under fields of 100 Oe and 50 kOe are obtained from the derivation of Fig. 2.16 as shown in Table 2.5 [26].

The change ΔT_{0h} and ΔT_{0c} of $Ni_{54}Mn_{21.2}Ga_{24.8}$ alloy is 5.0 and 5.7 K, while the change of ΔT_{0h} and ΔT_{0c} of $Ni_{54.8}Mn_{20.7}Ga_{24.5}$ alloy is 7.0 K and 6.5 K. The magnetocaloric effect here only considers the heating process. So, the $Ni_{54}Mn_{21.2}Ga_{24.8}$ alloy temperature $dA_p/dH = \sim 0.10$ K/kOe; while the change of $Ni_{54.8}Mn_{20.7}Ga_{24.5}$ alloy is slightly larger than that of $Ni_{54}Mn_{21.2}Ga_{24.8}$ alloy. Marcos et al. [29] reported that the magnetocaloric effect near the first-order *MT* temperature mainly depends on dA_p/dH, which reflects the intensity of phonon spin coupling that increases with the increase of *e*/*a*.

The maximum saturation magnetization difference near the phase transition point can be calculated according to the Clausius–Clapeyron equation:

$$\frac{\Delta H}{\Delta T} = \frac{Q}{T_0 \Delta M} \tag{2.14}$$

An item on the left side can be obtained by:

$$\frac{dA_p}{dH} = \frac{T_0 \Delta M}{Q} \tag{2.15}$$

where Q and T_0 are the enthalpy change of the transformation process and the theoretical transformation temperature between austenite and martensite, respectively, T_0 is $(A_p + M_p)/2$ and ΔM is the theoretical maximum saturation magnetization difference between austenite and martensite near the transformation point.

The DSC curves show that the enthalpy of $Ni_{54}Mn_{21.2}Ga_{24.8}$ alloy is 8.92 J/g. In $Ni_{54}Mn_{21.2}Ga_{24.8}$ alloy, T_0 is 320.0 K, and $dA_p/dH = 0.10$ K/kOe. So, according to Eq. (2.14), $\Delta M = \sim 27.9$ emu/g. The saturation magnetization difference between austenite and martensite near primary magnetic phase change is considered to be ~ 27.0 emu/g under 50 kOe as shown in Fig. 2.16a, while 32.3 emu/g during the heating

Table 2.5 Peak temperature of martensitic transformation in $Ni_{54}Mn_{21.2}Ga_{24.8}$ and $Ni_{54.8}Mn_{20.7}Ga_{24.5}$ alloys at 100 Oe and 50 kOe (K)

Alloy	A_p (100 Oe)	A_p (50 kOe)	M_p (100 Oe)	M_p (50 kOe)	ΔT_{0h}	ΔT_{0c}
$Ni_{54}Mn_{21.2}Ga_{24.8}$	324.1	329.1	317.2	322.9	5.0	5.7
$Ni_{54.8}Mn_{20.7}Ga_{24.5}$	354.1	361.1	347.4	353.9	7.0	6.5

process in $Ni_{54.8}Mn_{20.7}Ga_{24.5}$ alloy. The saturation magnetization difference between austenite and martensite near the first-order magnetic phase change is obtained to be ~ 31.0 emu/g from Fig. 2.16b. The two data are basically consistent.

2.7.1.3 Magnetocaloric Effect

The driving force of magnetic-induced martensitic transformation is Zeeman energy ($E_{zeeman} = \mu_0 \Delta MH$). Due to the limitation of the size of the external magnetic field (currently about 2 T of the permanent magnet ferromagnetic field), improving the magnetization difference (ΔM) between the parent and martensitic phases is an effective method to induce the magnetically induced martensitic transformation.

The magnetocaloric effect is often characterized by entropy change ΔS_m. ΔS_m can be obtained by testing isothermal magnetization curves at different temperatures and calculating by Maxwell relation. The applicability of this method in the vicinity of first-order magnetic phase change has been controversial in recent years because the use of isothermal magnetization curves to calculate the ΔS_m near magnetic phase change produces a virtual high value [29]. Nevertheless, studies show that the Maxwell relation is fully applicable for the first-order phase transition with weak magnetoelastic coupling ($\lambda_1/\lambda_0 < 2$) [30]. For Ni–Mn–Ga alloy, $\lambda_1 \sim 10^7$ erg cm^{-3} and $\lambda_0 \sim 10^{10}$ erg cm^{-3}, so it fully meets the above criterion [31].

In Ni–Mn–Ga alloy, the MT temperature is sensitive to composition, while the composition has little effect on Curie point. Therefore, by adjusting the composition of the alloy, the coupling of the first-order MT and the second-order magnetic phase change can be realized, that is, the magnetic structure coupling occurs. The magnetic structure coupling state is accompanied by the discontinuous change of the lattice structure to produce a large magnetization difference, resulting in a huge magnetic entropy change [26].

(1) Partially coupled magnetostructural state

The relationship between ΔS_m and temperature under different magnetic fields is calculated as shown in Fig. 2.18. It can be seen that the peak value ΔS_m near the phase change point A_p ~ 324 K, and ΔT_{FWHM} is only ~ 2.0 K under 50 kOe. The values of alloy ΔS_m under 20 and 50 kOe are −16.63 and −40.2 J/(kg K), respectively. This value is comparable to that of the $Ni_{55.5}Mn_{20}Ga_{24.5}$ bulk alloy at 20 kOe (−15.10 J/(kg K)) [32] and is higher than that of $Ni_{55}Mn_{20.6}Ga_{24.4}$ and $Ni_{55}Mn_{19.6}Ga_{25.4}$ single crystals at 20 kOe (−9.5 and −10.4 (kg K)) [33].

In order to evaluate the magnetocaloric properties of magnetic refrigeration materials, it is necessary to consider not only the value of ΔS_m, but also its temperature dependence. The magnetocaloric effect is comprehensively evaluated by the *RC* value of the material, expressed by subtracting the hysteresis value from the integral in the ΔT_{FWHM} (T_1–T_2) under the ΔS_m–T curve. The magnetic hysteresis is large for $Ni_{54.8}Mn_{20.7}Ga_{24.5}$ alloy, despite having a high ΔS_m value of ~ 9.69 J/kg, and the RC_{net} value at 50 kOe is only 49.9 J/kg because the ΔT_{FWHM} value is only 2.0 K.

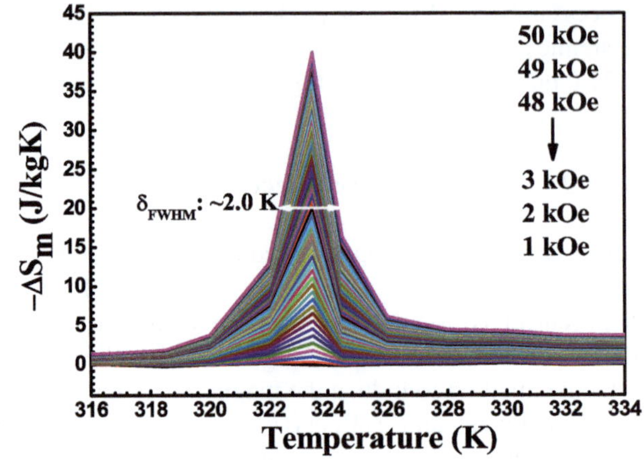

Fig. 2.18 Temperature dependence of ΔS_m under different magnetic field in $Ni_{54}Mn_{21.2}Ga_{24.8}$ alloy

(2) Coupled magnetostructural state

The relationship between ΔS_m and temperature under different magnetic fields is obtained by Maxwell equation calculation as shown in Fig. 2.19. It can be seen that the ΔS_m peaks near the phase transition point A_p ~ 354 K, and the ΔT_{FWHM} is only ~ 3.0 K under 50 kOe. The values of alloy ΔS_m at 20 kOe and 50 kOe are −13.58 and −33.38 J/(kg K), respectively. The value is comparable to that of the $Ni_{54.9}Mn_{20.5}Ga_{24.6}$ bulk alloy [32] at 20 kOe (−13.10 J/(kg K)). The *RC* is represented by the integral minus the hysteresis value in the ΔT_{FWHM} (T_1–T_2) under the ΔS_m–T curve. The RC_{net} is 60.43 J/kg under 50 kOe.

The isothermal magnetic entropy change ΔS_m values of both $Ni_{54}Mn_{21.2}Ga_{24.8}$ and $Ni_{54.8}Mn_{20.7}Ga_{24.5}$ alloys are very large, but due to the narrow ΔT_{FWHM} and large magnetic hysteresis, the value of *RC* is small. The results show that ΔS_m of $Ni_{54.8}Mn_{20.7}Ga_{24.5}$ alloy in the magnetic structure coupling state is smaller, the

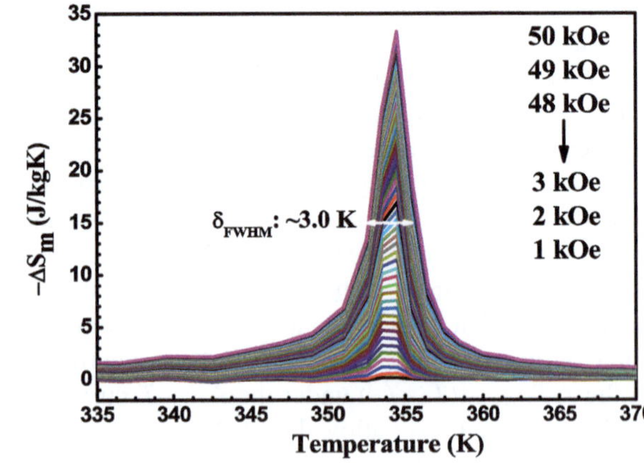

Fig. 2.19 Temperature dependent ΔS_m in $Ni_{54.8}Mn_{20.7}Ga_{24.5}$ alloy under different magnetic fields

magnetocaloric peak is half-height δT_{FWHM} wide and the magnetic refrigeration capacity is larger.

2.7.2 Ni–Mn–In–(Co) Alloy

2.7.2.1 Ni–Mn–In Alloy

Ni-Mn-In alloys are valued for the presence of positive and negative magnetocaloric effects near room temperature [34]. Figure 2.20 is the magnetization curves (M–T) of $Ni_{48.87}Mn_{37.07}In_{14.05}$ (In14) and $Ni_{47.75}Mn_{37.05}In_{15.20}$ (In15) alloys at low magnetic field (200 Oe) and high magnetic field (50 kOe). The phase transition temperature is shown in Table 2.6. At low magnetic field, the transition from antiferromagnetic martensite to ferromagnetic austenite occurs in In14 alloy with the increase of temperature. Then the second-order magnetic transition from ferromagnetic austenite to paramagnetic austenite occurs accompanied by the magnetization decreases. Due to the coupling of the first-order structural transition and the second-order magnetic

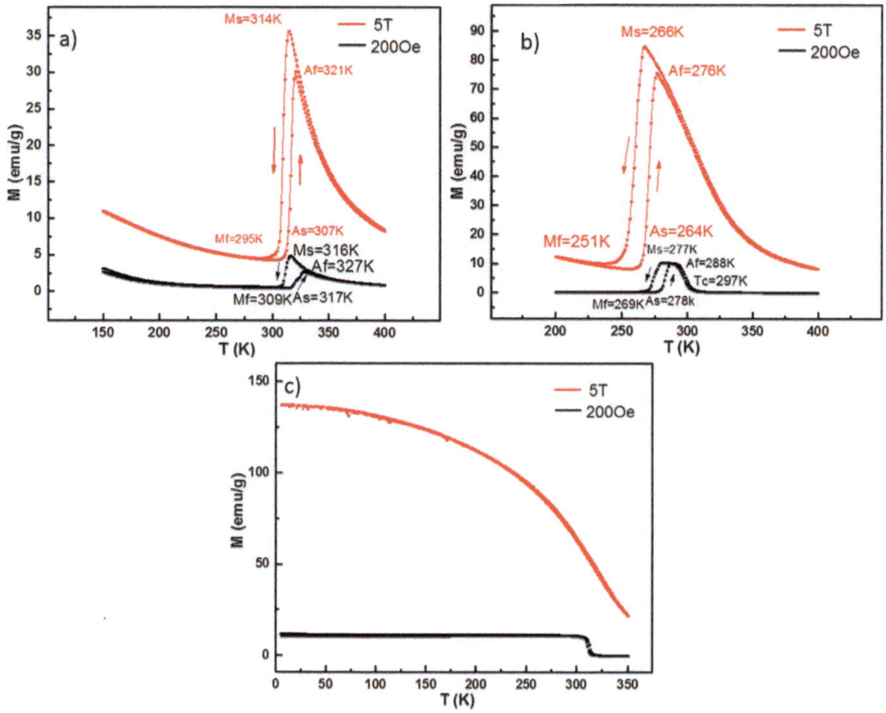

Fig. 2.20 Magnetization curves of Ni–Mn–In alloys under magnetic fields 200 Oe and 50 kOe. **a** In14, **b** In15, **c** In16

Table 2.6 Phase transition temperatures and Curie point of Ni–Mn–In alloys

	Magnetic field (Oe)	M_s (K)	M_f (K)	A_s (K)	A_f (K)	T_c (K)
In14	200	316	309	317	327	–
	50,000	314	295	307	321	–
In15	200	277	269	278	288	297
	50,000	266	251	264	276	–

transition, the magnetization decreases without reaching the maximum value. The phase change point of alloy decreases and the A_s point decreases for about 10 K under high magnetic field. It shows that the applied magnetic field can promote the inverse transformation of martensite.

As temperature increases, a first-order phase transition occurs in In15 alloy from the antiferromagnetic martensite to the ferromagnetic austenite, and then the second-order magnetic phase change occurs from ferromagnetic austenite to paramagnetic (antiferromagnetic) austenite. The phase transition temperature A_s of In15 alloy decreases from 317 to 278 K. So, with the content of In elements increases, the phase transition temperature decreases. Meanwhile, the transition temperature of In15 alloy at high magnetic field to low-temperature shift is more obvious than that of In14 alloy, reaching about 20 K under magnetic field 5 T, which indicates that the change magnetic transition of In15 alloy is more obvious and is expected to have better magnetocaloric properties. The phase transition temperatures of each alloy at low magnetic field from Fig. 2.20 are shown in Table 2.6.

Figure 2.21a shows the magnetic entropy curve of In14 alloy under the applied magnetic field of $0 \leq H \leq 50$ kOe. As the temperature increases, In14 possesses the first-order structural phase transition with the magnetic entropy change reaches the maximum positive value (negative MCE), and then the magnetic entropy change decreases rapidly. The magnetic entropy becomes negative when the second-order magnetic phase change occurs (conventional MCE). With the increase of magnetic

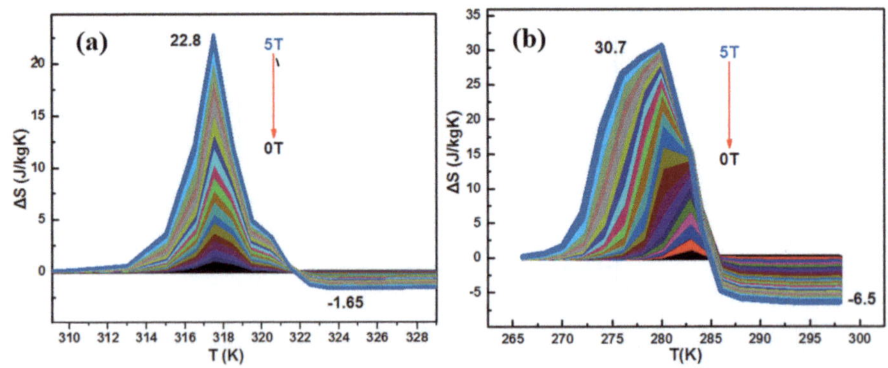

Fig. 2.21 Magnetic entropy change of Ni-Mn-In alloy. **a** In14, **b** In15

field, the absolute value of magnetic entropy change increases, reaching 317.5 J/kg K at 22.8 K and −1.65 K at 325.5 K.

Figure 2.21b shows the magnetic entropy change curve of In15 alloy. Similar to the In14, the magnetic entropy becomes positive when the first-order structural phase transition occurs in the alloy and then decreases rapidly. The magnetic entropy becomes negative when the second-order magnetic phase transition occurs. Besides, with increase of external magnetic field, the peak temperature of positive entropy change shifts to low temperature, and the effect of external magnetic field on the phase transition of In15 alloy is more obvious. The In15 alloy reaches the maximum positive magnetic entropy change of 30.7 J/kg K at 280 K and negative magnetic entropy change of −6.45 J/kg K at 294 K. Since positive and negative magnetic entropy changes exist in an adjacent temperature range, In14 and In15 can absorb and release heat after applying magnetic field at different temperatures, which makes it more flexible in widening the working temperature range as a magnetic refrigeration material.

2.7.2.2 Ni–Mn–In–Co Alloy

Co doping will greatly increase the magnetization of ternary Ni–Mn–In alloys and affect the variable magnetic transition and magnetocaloric properties [35]. Figure 2.22 shows the *M–T* curves of annealed $Ni_{44.5}1Mn_{36.35}In_{13.94}Co_{5.20}$ alloys under 500 Oe and 50 kOe. The increase in magnetization is due to the magnetic transition from paramagnetic or antiferromagnetic martensite to ferromagnetic austenite during heating. The maximum magnetization of austenite in the annealed alloy is 126 emu/g at 50 kOe, and the difference of magnetization between austenite and martensite is 121 emu/g, which is the highest magnetization in the Ni–Mn-based alloys [36–38]. Table 2.7 summarizes the MT temperatures and Curie point. At a

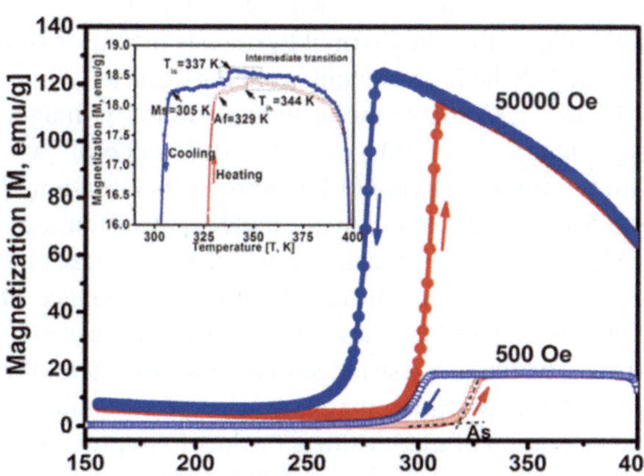

Fig. 2.22 *M–T* curves of Ni-Mn-In-Co alloy under 200 Oe and 50 magnetic fields after heat treatment **a** In14; **b** In15; **c** In16

Table 2.7 Ni–Mn–In–Co martensite phase transition temperature (A_s, A_f, M_s, M_f), austenitic curie temperature (T_c) and magnetically induced phase transition velocity of the alloy

Magnetic field (Oe)	Phase transition temperature (K)						Transition temperature change rate (K/10 kOe)	
	M_s	M_f	A_s	A_f	M_p	A_p	$\Delta M_p/\Delta H$	$\Delta A_p/\Delta H$
500	305	293	319	329	300	324	4.6	3.6
50,000	280	270	300	309	277	306		

low field of 500 Oe, the starting temperature of the inverse martensitic transformation is A_s 319 K and the end temperature is A_f 329; during cooling, the martensite starting temperature is 305 K, and the end temperature of martensite is 293 K.

Moreover, beyond the MT temperature, a small bulge appears above 377 K on the cooling curve and above 344 K on the heating curve (arrows in Fig. 2.22), representing a weak first-order intermediate transformation, similar to Ni–Mn–Ga alloys with composition close to stoichiometry. Therefore, the coarse grains produced by the heat treatment process of annealing affect the internal stress and internal interface, which changes the magnetoelastic coupling state that induces the intermediate phase transition in the alloy. Figure 2.23 shows the magnetization curves and magnetocaloric effect of Ni–Mn–In–Co alloy after heat treatment.

2.7.3 *Ni–Mn–Sn–Fe Alloy*

A summary of the magnetocaloric properties of Ni–Mn–Sn–Fe alloys [39–42] is presented. The *M*–*T* curve of the $Ni_{50-x}Fe_xMn_{38}Sn_{12}$ alloy after 24 h heat treatment at a temperature 1173 K is shown in Fig. 2.24a. With the increase of Fe doping content, the *MT* temperature decreases and the saturation magnetization of austenite M_A increases. The post-annealing hysteresis (A_P–M_P) statistics at 50 kOe magnetic field indicate that the thermal hysteresis increases significantly when $x \geq 4.2$, which is related to the high content secondary phase.

Figure 2.25 shows the austenite saturation magnetization M_A, martensite saturation magnetization M_M and the difference of two-phase magnetization $\Delta M_{A\text{-}M}$ with the variation curve of Fe atomic concentration. With the increase of Fe content ($\leq$ 5.5 at.%), the maximum M_A and M_M of austenite at 50 kOe increases. The increase of M_A is obviously more than that of the two-phase magnetization difference, which is beneficial to the magnetic refrigeration performance. On the other hand, the saturation magnetization of Fe4.2 and Fe5.5 alloy martensite also increases significantly under 50 kOe, which is a resistance to the improvement of magnetic refrigeration. How to improve the saturation magnetization of austenite without increasing the saturation magnetization of martensite is the key to improve the magnetic entropy change. The study shows that Fe doping has little effect on Curie point. Higher magnetic entropy change can be realized by reducing the Sn content without changing the Fe content.

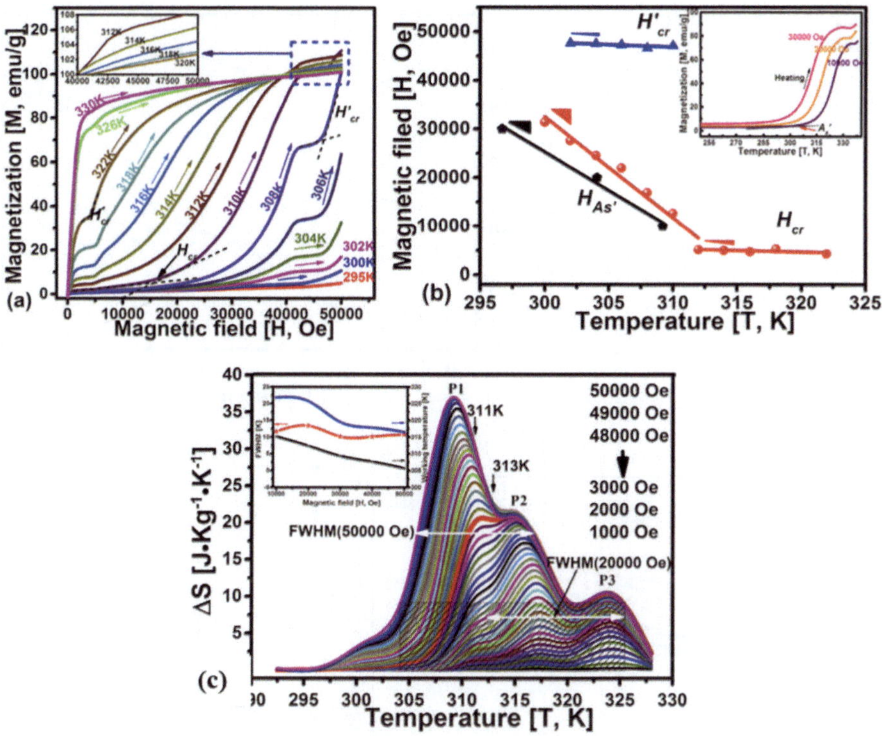

Fig. 2.23 Magnetization curve and magnetocaloric effect of Ni–Mn–In–Co alloy. **a** Isothermal magnetization curve, **b** meta-magnetic transition temperatures at various magnetic fields, **c** magnetic entropy change

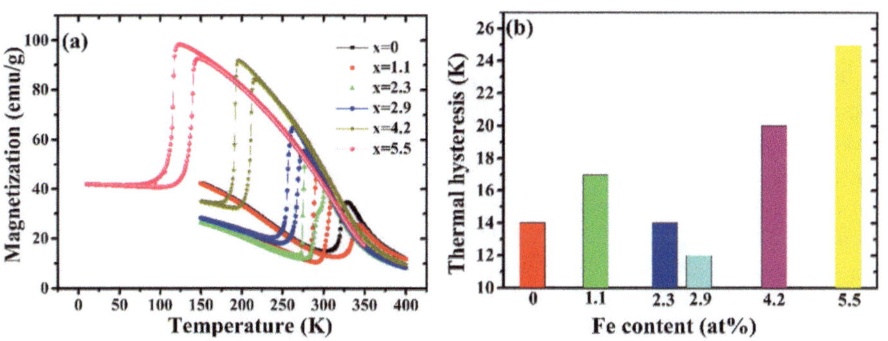

Fig. 2.24 Magnetization curves and hysteresis in $Ni_{50-x}Fe_xMn_{38}Sn_{12}$ alloy. **a** M–T curves, **b** hysteresis at a field of 50 kOe

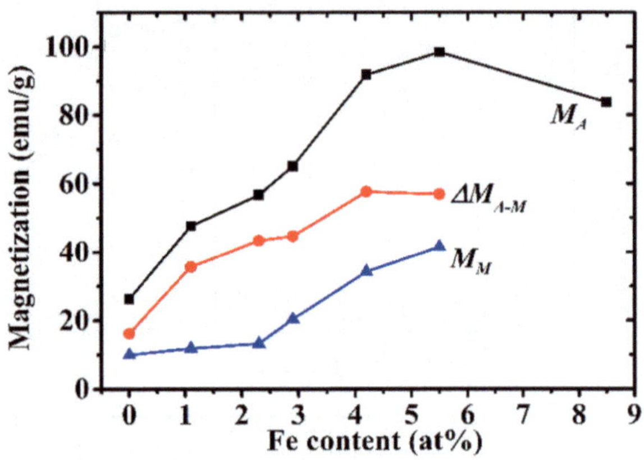

Fig. 2.25 Magnetization difference of $Ni_{50-x}Fe_xMn_{38}Sn_{12}$ alloy at 50 kOe

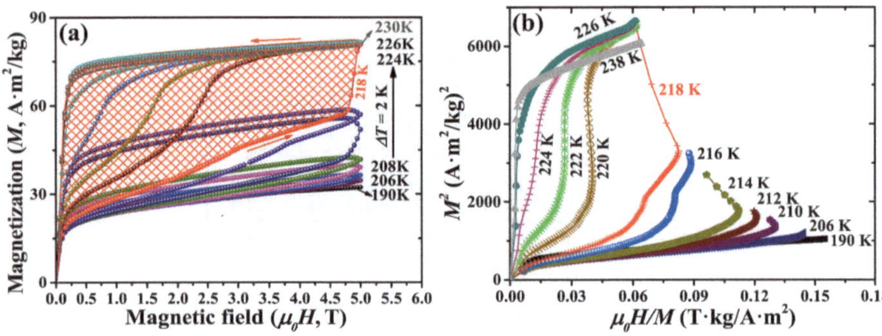

Fig. 2.26 Isothermal magnetization and Arrott plots of Fe4.2 alloy. **a** M–H curve during cooling, **b** Arrott plots

The isothermal magnetization curves of Fe4.2 alloy at different temperatures are shown in Fig. 2.26a. For eliminating the influence of historical process on magnetic phase change, the sample was cooled to 180 K, 1 K/min and heated to the temperature to be tested, and then demagnetization experiment was carried out. At 190 K, the alloy shows a paramagnetic state, and the magnetization curve of the magnetic demagnetization process coincides. There is little hysteresis phenomenon, that is, no magnetotropic martensitic transformation at this temperature. The saturation magnetization at this temperature is 32 emu/g, which is consistent with the result of M–T curve heating process at 5 T. At 218 K, hysteresis is produced in the process of magnetization and demagnetization. However, this variable magnetic transition behavior disappears during demagnetization and remains ferromagnetic. This is related to the low Zeeman energy provided by the magnetic field which is not enough to overcome the large thermal hysteresis during the phase transition. At temperature A_f, the magnetization of the austenitic posture increases sharply in the low field but reaches saturation quickly with the increase of the magnetic field, accompanied by a small amount of hysteresis. Figure 2.26b is Arrott corresponding curve. The curve shows an

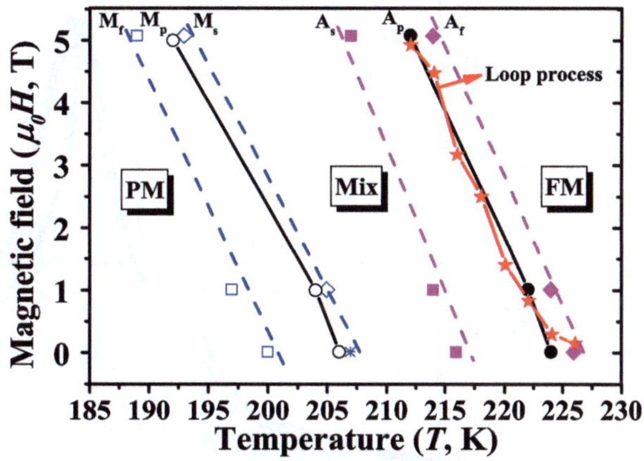

Fig. 2.27 Phase diagram of magnetostructural transition from *MT* and *MH* curves in Fe4.2 alloy

obvious "S" shape with significant negative slope, implying obvious first-order *MT*. In addition, the alloy shows magnetic transition near the structural transformation temperatures range.

Figure 2.27 is the magnetostructure phase diagram of the Fe4.2 alloy, where the critical magnetic field is determined by half of the austenite saturated magnetic field (~ 40 A m^2/kg). A_p, M_p represents the austenite and martensite peak temperature, respectively. So-called loop method is used to reach a complete martensite state before each test *M–H* cycle. In this method, the magnetic-induced inverse *MT* from weak magnetic martensite to ferromagnetic austenite does not occur, indicating no state of coexistence of ferromagnetic phase and weak magnetic phase during isothermal transformation. In other words, the magnetic entropy change obtained by *M–H* calculation measured using loop method is reproducible.

The isothermal magnetic entropy change of Fe4.2 alloy near the MT temperature is shown in Fig. 2.28. In the ΔS_m curves, there is a nonlinear relationship between the maximum value and the applied magnetic field: the magnetic field increases and the maximum value shifts to the low-temperature region. The onset point of the peak curve corresponds to the critical magnetic field where the magnetic phase transition occurs at that temperature. So, the presence of ΔS_m is due to the meta-magnetic transition. The maximum value ΔS_m at 50 kOe is 33.8 J kg^{-1} K^{-1} with a working temperature window 216.5–223.5 K. The magnetic entropy change is 12.2 J kg^{-1} K^{-1} and ΔT_{FWHM} 8 K under a magnetic field 20 kOe, which is accessible via permanent magnets.

Figure 2.28 shows the curves of the Fe4.2 alloy in ΔS_m–H martensitic transformation temperature range. To reveal the relationship between magnetic entropy change and variable magnetic transition, the critical magnetic field of variable magnetic transition is listed in the illustration. At $A_s < T < A_f$ temperature range, there is no platform for the magnetic transition process from weak magnetic martensite to ferromagnetic austenite, which indicates that the magnetic entropy change under different magnetic fields is attributed to the variable magnetic transition. The starting critical magnetic field H_{c1} is about 2 kOe in the whole martensitic temperature range (212–216 K),

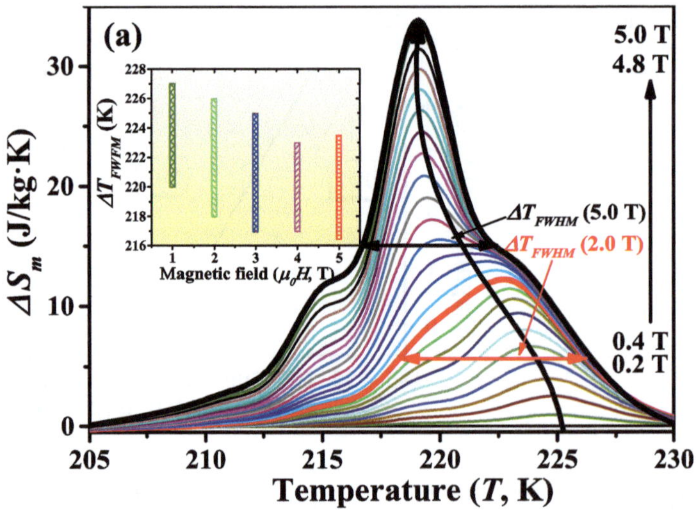

Fig. 2.28 Magnetic entropy change (ΔS_m) of Fe4.2 alloy

indicating that magnetically induced inverse martensitic transformation is relatively easy. The termination critical magnetic field H_{c2} changes to the low magnetic field region with the increase of temperature, and the magnetic field reaches saturation when the temperature is higher than 218 K. The magnetic entropy change increases rapidly from the beginning of the H_{c1} to reach saturation at the H_{c2}, and the results are shown in Fig. 2.29. Specifically, the magnetic field reaches saturation H_{c2} rapidly at 218 K, resulting in a large magnetization difference between 218 and 220 K. The maximum magnetic entropy change is reached at 219 K. At temperature < 218 K, the magnetization of austenite does not reach saturation, which leads to the increase of magnetic entropy with the increase of magnetic field. This indicates that only part of MT occurs under a magnetic field 50 kOe.

2.8 Conclusions

Bulk polycrystalline MSMAs ingots are usually produced by elemental melting and mold casting. As the martensitic transformation temperature and alloy properties are sensitive to composition segregation, post-fabrication annealing is often required. Single crystals are more difficult to prepared than cast ingots. High temperature with long time annealing is also necessary for single crystals since the severe composition segregation along the growth direction.

MSMAs are brittle at low temperature but are ductile at temperatures higher than the chemical ordering point, where the high-temperature phase shows a partially ordered state. Good plasticity is demonstrated in polycrystalline Ni–Mn–Ga alloy.

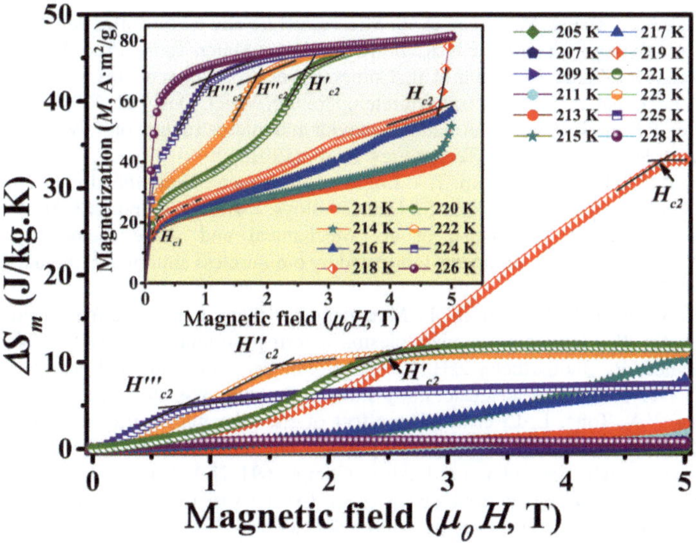

Fig. 2.29 ΔS_{m}–H curve of Fe4.2 alloy in the temperature range of martensitic transformation

The Ni–Mn–Ga alloy can be shaped by hot extrusion and superplastic deformation at such high temperatures.

MSMAs exhibit multifunctional properties such as MFIS, MCE and so on. The first-order martensitic transformation and second-order magnetic transition coexist. As the martensitic transformation temperature is sensitive to composition while the magnetic transition not, various magnetostructural coupling states can be realized, which provide strategies for adjusting the properties of the MSMAs.

Meta-magnetic transition behavior exists in Ni–Mn–X (X = Sn, In, Sb) alloys, which renders them magnetic shape memory, giant magnetocaloric effect and so on. Doping by a fourth element or even fifth element, such as Fe, Co, is summarized. The doping of Co element enhances the magnetization greatly, while other elements such as Fe may induce the formation of secondary phases at grain boundaries or in grains, which alter the mechanical and functional properties.

References

1. K. Ullakko, J.K. Huang, C. Kantner et al., Large magnetic-field-induced strains in Ni_2MnGa single crystals. Appl. Phys. Lett. **69**(13), 1966–1968 (1996)
2. R. KAINUMA, Y. IMANO, W. ITO, et al. Metamagnetic shape memory effect in a Heusler-type $Ni_{43}Co_7Mn_39Sn_{11}$ polycrystalline alloy. Applied Physics Letters, **88**(19), 192513 (2006)
3. L.S. Wei, X.X. Zhang, J. Liu et al., Orientation dependent cyclic stability of the elastocaloric effect in textured Ni-Mn-Ga alloys. AIP Adv. **8**(5), 055312 (2018)
4. L.S. Wei, X.X. Zhang, M.F. Qian et al., Introducing equiaxed grains and texture into Ni-Mn-Ga alloys by hot extrusion for superplasticity. Mater. Des. **112**, 339–344 (2016)

5. L.S. Wei, X.X. Zhang, M.F. Qian et al., Microstructure and texture after deformation-induced grain growth in polycrystalline $Ni_{48}Mn_{30}Ga_{22}$ alloys. Mater. Today-Proc. **2**, 863–866 (2015)
6. L.S. Wei, X.X. Zhang, M.F. Qian et al., Compressive deformation of polycrystalline Ni-Mn-Ga alloys near chemical ordering transition temperature. Mater. Des. **142**, 329–339 (2018)
7. L.S. Wei, Research on hot deformation behavior and microstructure of polycrystalline Ni-Mn-Ga alloys, Harbin Institute of Technology, Harbin (2013)
8. D.L. Schlagel, R.W. McCallum, T.A. Lograsso, Influence of solidification microstructure on the magnetic properties of Ni-Mn-Sn Heusler alloys. J. Alloy Compd. **463**, 38–46 (2008)
9. K. Ito, W. Ito, R.Y. Umetsu et al., Mechanical and shape memory properties of $Ni_{43}Co_{7}Mn_{39}Sn_{11}$ alloy compacts fabricated by pressureless sintering. Scripta Mater. **63**(12), 1236–1239 (2010)
10. K. Ito, W. Ito, R.Y. Umetsu et al., Metamagnetic shape memory effect in polycrystalline NiCoMnSn alloy fabricated by spark plasma sintering. Scripta Mater. **61**(5), 504–507 (2009)
11. R. Ahamed, R. Ghomashchi, Z.H. Xie et al., Powder processing and characterisation of a quinary Ni-Mn-Co-Sn-Cu Heusler alloy. Powder Technol. **324**, 69–75 (2018)
12. F. Chen, Y.X. Tong, L. Li et al., The effect of step-like martensitic transformation on the magnetic entropy change of $Ni_{40.6}Co_{8.5}Mn_{40.9}Sn_{10}$ unidirectional crystal grown with the Bridgman-Stockbarger technique. J. Alloy. Compd. **691**, 269–274 (2017)
13. D.Y. Cong, Y.D. Wang, P. Zetterstrom et al., Crystal structures and textures of hot forged $Ni_{48}Mn_{30}Ga_{22}$ alloy investigated by neutron diffraction technique. Mater. Sci. Technol. **21**(12), 1412–1416 (2005)
14. D.Y. Cong, Y.D. Wang, R.L. Peng et al., Crystal structures and textures in the hot-forged Ni-Mn-Ga shape memory alloys. Metall. Mater. Trans. A **37A**(5), 1397–1403 (2006)
15. Y.Q. Ma, S.Y. Yang, Y. Liu et al., The ductility and shape-memory properties of Ni-Mn-Co-Ga high-temperature shape-memory alloys. Acta Mater. **57**(11), 3232–3241 (2009)
16. T.G. Langdon, Superplasticity in metals and ceramics—an examination of flow mechanisms. Creep Fract. Eng. Mater. Struct. 295–304 (1993)
17. G. Erdelyi, H. Mehrer, A.W. Imre et al., Self-diffusion in $Ni_{2}MnGa$. Intermetallics **15**(8), 1078–1083 (2007)
18. K. Edalati, Z. Horita, Correlations between hardness and atomic bond parameters of pure metals and semi-metals after processing by high-pressure torsion. Scripta Mater. **64**(2), 161–164 (2011)
19. Y.V.R.K. Prasad, K.P. Rao, S. Sasidhara, Hot working guide: compendium of professing maps. ASM Int. 1–24 (1997)
20. R.D. Doherty, D.A. Hughes, F.J. Humphreys et al., Current issues in recrystallization: a review. Mater. Sci. Eng. A **238**(2), 219–274 (1997)
21. I. Aaltio, O. Soderberg, Y.L. Ge et al., Twin boundary nucleation and motion in Ni-Mn-Ga magnetic shape memory material with a low twinning stress. Scripta Mater. **62**(1), 9–12 (2010)
22. A. Sozinov, N. Lanska, A. Soroka et al., Highly mobile type II twin boundary in Ni-Mn-Ga five-layered martensite. Appl. Phys. Lett. **99**(12) (2011)
23. M. Potschke, S. Weiss, U. Gaitzsch et al., Magnetically resettable 0.16% free strain in polycrystalline Ni-Mn-Ga plates. Scripta Mater. **63**(4), 383–386 (2010)
24. U. Gaitzsch, J. Romberg, M. Potschke et al., Stable magnetic-field-induced strain above 1% in polycrystalline Ni-Mn-Ga. Scripta Mater. **65**(8), 679–682 (2011)
25. U. Gaitzsch, M. Pötschke, S. Roth et al., Mechanical training of polycrystalline 7M $Ni_{50}Mn_{30}Ga_{20}$ magnetic shape memory alloy. Scripta Mater. **57**(6), 493–495 (2007)
26. X.H. Wan, Research on the martensitic transformation and magnetocaloric effect in Ni-Mn-Ga micro-particles, Harbin Institute of Technology, Harbin (2016)
27. J.F. Duan, P. Huang, H. Zhang et al., Magnetic entropy changes of NiMnGa alloys both on the heating and cooling processes. J. Alloy. Compd. **441**(1–2), 29–32 (2007)
28. Z.B. Li, Y.D. Zhang, C.F. Sanchez-Valdes et al., Giant magnetocaloric effect in melt-spun Ni-Mn-Ga ribbons with magneto-multistructural transformation. Appl. Phys. Lett. **104**(4) (2014)
29. J. Marcos, L. Manosa, A. Planes et al., Multiscale origin of the magnetocaloric effect in Ni-Mn-Ga shape-memory alloys. Phys. Rev. B **68**(9) (2003)

30. L. Tocado, E. Palacios, R. Burriel, Entropy determinations and magnetocaloric parameters in systems with first-order transitions: study of MnAs. J. Appl. Phys. **105**(9) (2009)
31. V.A. L'Vov, E.V. Gomonaj, V.A. Chernenko, A phenomenological model of ferromagnetic martensite. J. Phys. Condens. Matt. **10**(21), 4587–4596 (1998)
32. Y. Long, Z.Y. Zhang, D. Wen et al., Phase transition processes and magnetocaloric effects in the Heusler alloys NiMnGa with concurrence of magnetic and structural phase transition. J. Appl. Phys. **98**(4) (2005)
33. N.V.R. Rao, R. Gopalan, V. Chandrasekaran et al., Microstructure, magnetic properties and magnetocaloric effect in melt-spun Ni-Mn-Ga ribbons. J. Alloy. Compd. **478**(1–2), 59–62 (2009)
34. R.Z. Su, Research on the magnetocaloric effect on Ni-Mn-In and Ni-Mn-Sn shape memory bulk alloys and microwires, Harbin Institute of Technology, Harbin (2015)
35. X.X. Zhang, M.F. Qian, S.P. Miao et al., Enhanced magnetic entropy change and working temperature interval in Ni-Mn-In-Co alloys. J. Alloy. Compd. **656**, 154–158 (2016)
36. W. Ito, K. Ito, R.Y. Umetsu et al., Kinetic arrest of martensitic transformation in the NiCoMnIn metamagnetic shape memory alloy. Appl. Phys. Lett. **92**(2), 021908 (2008)
37. R.Y. Umetsu, K. Ito, W. Ito et al., Kinetic arrest behavior in martensitic transformation of NiCoMnSn metamagnetic shape memory alloy. J. Alloy. Compd. **509**(5), 1389–1393 (2011)
38. R. Kainuma, Y. Imano, W. Ito et al., Metamagnetic shape memory effect in a Heusler-type $Ni_{43}Co_7Mn_{39}Sn_{11}$ polycrystalline alloy. Appl. Phys. Lett. **88**(19), 192513 (2006)
39. H.H. Zhang, M.F. Qian, X.X. Zhang et al., Martensite transformation and magnetic properties of Fe-doped Ni-Mn-Sn alloys with dual phases. J. Alloy. Compd. **689**, 481–488 (2016)
40. H.H. Zhang, X.X. Zhang, M.F. Qian et al., Enhanced magnetocaloric effects of Ni-Fe-Mn-Sn alloys involving strong metamagnetic behavior. J. Alloy. Compd. **715**, 206–213 (2017)
41. H.H. Zhang, X.X. Zhang, M.F. Qian et al., Effect of partial metamagnetic and magnetic transition coupling on the magnetocaloric effect of Ni-Mn-Sn-Fe alloy. Intermetallics **105**, 124–129 (2019)
42. H.H. Zhang, Investigation on the regulation of magnetostructural transition and magnetocaloric effects in Ni-Mn-Sn-Fe(Co) alloys, Harbin Institute of Technology, Harbin (2019)

Chapter 3
Preparation and Properties of Magnetic Shape Memory Alloy Foams

3.1 Introduction

Foam is a light porous mass of fine bubbles formed in or on the surface of a material. The introduction of pores in a material modifies the structural and functional properties. In structural materials, the foam has lightweight, high energy adsorption and low modulus, while for functional foams, the introduction of pores improves certain functional properties such as sound, light, electricity and magnetism. For example, polycrystalline Ni–Mn–Ga foam shows reduced constraints for grain boundary motion and thus improved magnetic-field-induced strain (MFIS) [1–4]. The porous Ni–Mn–Ga alloy can be cycled >30 million times without breaking (high mechanical stability [5]). In addition, when foams are used as a magnetocaloric material, the pores allow liquid to flow inside, acting like a micropump (fluid flows in and out due to deformation of the alloy). This renders foams' efficient heat exchange capacity since their high specific surface area [6, 7].

The preparation methods of foams include gas-, liquid- and solid-phase methods. As the conventional molten metal foaming method is not easy to control element volatilization and surface oxidation, it is difficult to be used in magnetic shape memory alloy foams. The feasible methods for preparing MSMA foams include template replication and powder metallurgy. This chapter describes the preparation methods, microstructure, martensitic transformation behavior and MFIS of MSMA foams. The relationship between porous architecture and FMIS is detailed and discussed.

X. Zhang and M. Qian, *Magnetic Shape Memory Alloys*,
https://doi.org/10.1007/978-981-16-6336-9_3

3.2 Preparation Methods of MSMA Foam

3.2.1 Template Replication

The template replication method is to use the space holder as a template, fill the interspace between space holders and finally remove the template material to get foam. The advantage of template replication is that the shape, size, content and distribution of the pores can be feasibly controlled, showing a strong designability.

The requirements for template materials include: (1) The melting point of the template material is higher than that of the foam alloy; (2) no excessive interfacial reaction between the template and metal; (3) only applicable to open-pore foams, i.e., sintering between the templates is required for complete removal of the template material; and (4) a different solubility of the template and metal in order to selectively remove the template in the foam.

3.2.1.1 Preparation of Ni–Mn–Ga Foam

Sodium aluminate ($NaAlO_2$) is the first template material used to prepare Ni–Mn–Ga foam because it has a high melting point (~1650 °C) and good chemical stability and is soluble in hot water or acid. The process of preparing $NaAlO_2$ powder with a certain particle size includes: (1) pressing $NaAlO_2$ fine powders at a pressure >125 MPa and sintering at 1500 °C for >1 h in air; (2) mechanically crushing and grinding the sintered flakes; and (3) sieving to get different particle sizes. Because $NaAlO_2$ has strong hygroscopicity, it has to be stored under dry conditions; otherwise, its chemical properties (such as solubility in water and acid) change. Figure 3.1a, b represents SEM images showing the irregular shape of $NaAlO_2$ powders with the size of 250–355 and 75–90 μm.

When preparing a template preform with particles of different sizes, careful selection of the $NaAlO_2$ particle size is required. According to the close-packed model of

Fig. 3.1 Morphology of sodium aluminate ($NaAlO_2$) particles prepared by crushing and sieving. **a** 250–355 μm and **b** 75–90 μm

a crystal, the small particle should be 0.1–0.2 times the large one to ensure that the small particles fully occupy the gap of the large particles. For example, if the large particles have a diameter 355–500 μm, the small ones should be 53–75 μm; the ratio of the large and small particles should be 73:27. The mixing of the two particles can be carried out in a liquid (such as acetone for $NaAlO_2$) via a so-called layer-by-layer method [3].

After mixed powders are obtained, proper sintering is usually required. For $NaAlO_2$, sintering at 1550 °C for 3 h is helpful to obtain a strong preform with necks between powders. In this way, open-pore foams can be obtained by template replication. On the one hand, the strength of the preform is increased to avoid cracking during the infiltration process. On the other hand, the powders are interconnected, which is beneficial for subsequent dissolution and removal.

The sintered preform can be placed in an alumina crucible and then place an alloy ingot on top of the preform. The crucible is evacuated (pressure $<10^{-3}$ Pa to avoid oxidation) and heated above melting of the alloy and finally applies the gas pressure to drive the molten alloy infiltrating into the preform. During the subsequent cooling process, heat treatment can be performed to induce composition homogenization and chemical ordering.

In order to fully remove $NaAlO_2$ in the preform without excessive dissolution of the foam alloy, it is necessary to find a solution that can dissolve $NaAlO_2$ but not or slightly dissolve the alloy. Figure 3.2a shows the corrosion rate of Ni–Mn–Ga alloy by different acid solutions. It can be seen that sulfuric acid hardly dissolves Ni–Mn–Ga alloy, while hydrochloric acid dissolves the alloy at a higher rate. To this end, 34% H_2SO_4 solution can be used to dissolve and remove $NaAlO_2$. When the porosity of the foam needs to be increased, 10% HCl solution can be used to dissolve the alloy around the pores. At the same time, when $NaAlO_2$ is fully surrounded by metal, 10% HCl solution can be used to dissolve the metal so as to expose $NaAlO_2$ and facilitate the acid solution. This is especially important for the complete removal of $NaAlO_2$ in the double-pore foams because small-sized $NaAlO_2$ particles are usually wrapped in metal.

The mass loss in 34% H_2SO_4 solution showing the dissolution of $NaAlO_2$ in double-pore foam is shown in Fig. 3.2b. After dissolving for 2000 min, only 87% $NaAlO_2$ is removed, indicating that it is difficult to completely remove $NaAlO_2$ using only sulfuric acid solution. So, an applicable process of removing $NaAlO_2$ in double-pore foam is: using 34% H_2SO_4 solution to remove large $NaAlO_2$; then using 10% HCl solution to dissolve the small-sized $NaAlO_2$ or the metal around the large pores and expose the small-sized $NaAlO_2$; and finally using 34% H_2SO_4 solution to dissolve the exposed $NaAlO_2$. Figure 3.2c is the time-dependent porosity curve when foam sample was immersed in 34% H_2SO_4 solution for 645 min (corresponding to the rapid dissolution stage of Fig. 3.2b), and then transferred to 10% HCl solution for 1140 min. The final porosity reaches 55%, which is higher than the actual content of $NaAlO_2$ in the material (45%), indicating that $NaAlO_2$ is completely removed. Figures 3.3 and 3.4 show the pore morphology of Ni–Mn–Ga foams with single and double pores.

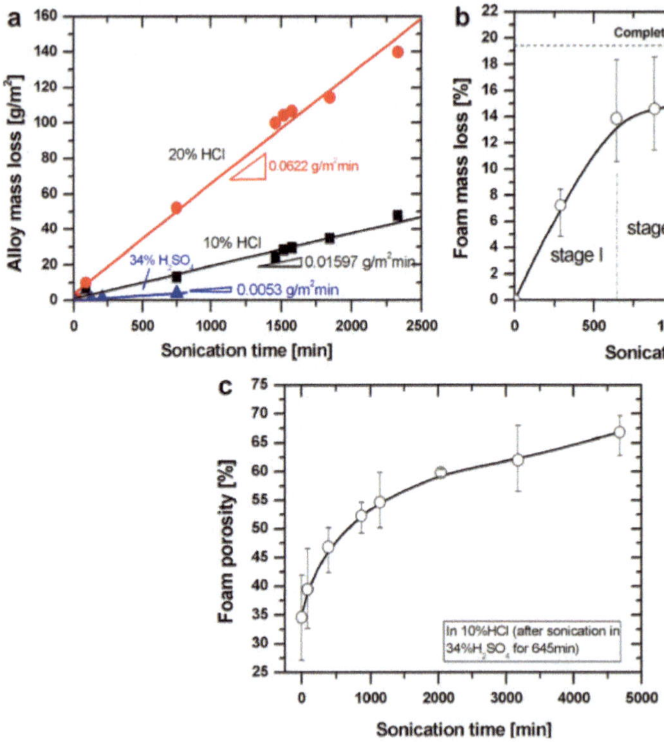

Fig. 3.2 Dissolution of space holder in Ni–Mn–Ga foam. **a** Ni–Mn–Ga bulk alloy corrosion rate curve in 10% HCl, 20% HCl and 34% H_2SO_4 solution; **b** weight loss curve of double-pore foam in 34% H_2SO_4 solution; **c** the weight loss curve of double-pore foam transferred to 10% HCl solution after soaking in 34% H_2SO_4 solution for 645 min [3]

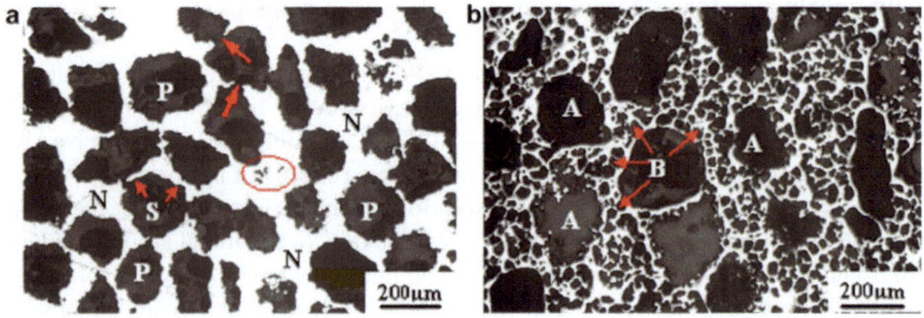

Fig. 3.3 Ni–Mn–Ga foams with various architectures [3]. **a** Single pore and **b** dual pore

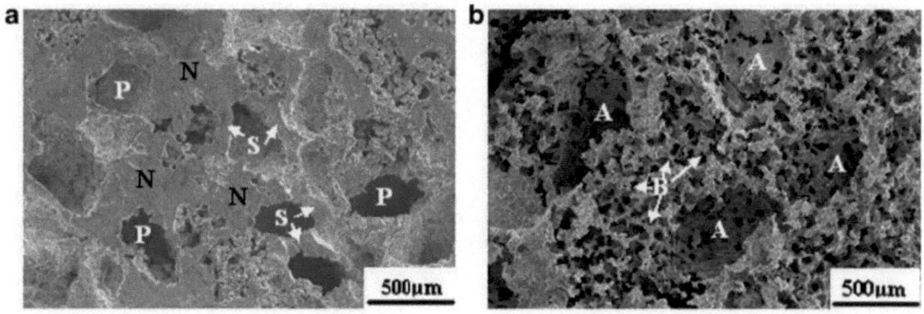

Fig. 3.4 SEM micrographs showing 3-D architecture of Ni–Mn–Ga foams [3]. **a** Single pore and **b** dual pore

3.2.1.2 Preparation of Ni–Mn–In–Co Foam

For double-pore Ni–Mn–In–Co foam prepared with $NaAlO_2$ space holder, 1% HCl and 10% HCl are proved suitable for the removal of $NaAlO_2$. Among these two acids, 1% HCl is used to react with $NaAlO_2$, while 10% HCl to etch Ni–Mn–In–Co alloy to expose $NaAlO_2$ wrapped in the alloy.

Figure 3.5 shows the pore structure of single-pore Ni–Mn–In–Co foam with different porosities. At low porosity, the foam contains many fine dendritic alloys inside the pores. This is produced by the $NaAlO_2$ particles which are not dense enough. As the corrosion time prolongs and the porosity increases, the dendritic alloy gradually dissolves, showing relatively clean pores. However, some cracks appear in the nodes and struts.

Figure 3.6 shows the morphology of double-pore Ni–Mn–In–Co foam with different porosities. The highest porosity of the sample is 77%, which is higher than the largest porosity of single-weight foam of 73%. The pore structure is close to the designed structure, with many small pores distributing in the large struts and nodes. However, in the foam with porosity 77%, many cracks exist. As a result, the foam structure is very loose, and the pores are almost completely connected.

3.2.2 Powder Metallurgy

Powder metallurgy is a method of preparing foam by mixing and sintering alloy powder (with pore forming agent). Considering that most magnetic memory alloys are brittle, the raw powders can be prepared by mechanical crushing and sieving [4]. During sintering, sufficient pressure is usually necessary. A high temperature is also necessary for sintering. Taking Ni–Mn–Ga alloy as an example, it has good plastic deformation ability in the B2 phase region (above ~800 °C, below the melting point). When Ni–Mn–Ga powders are sintered at 1000, 1050 and 1080 °C for 2 h with pressure 70 MPa, the sintering condition is poor at 1000 °C, only a small amount

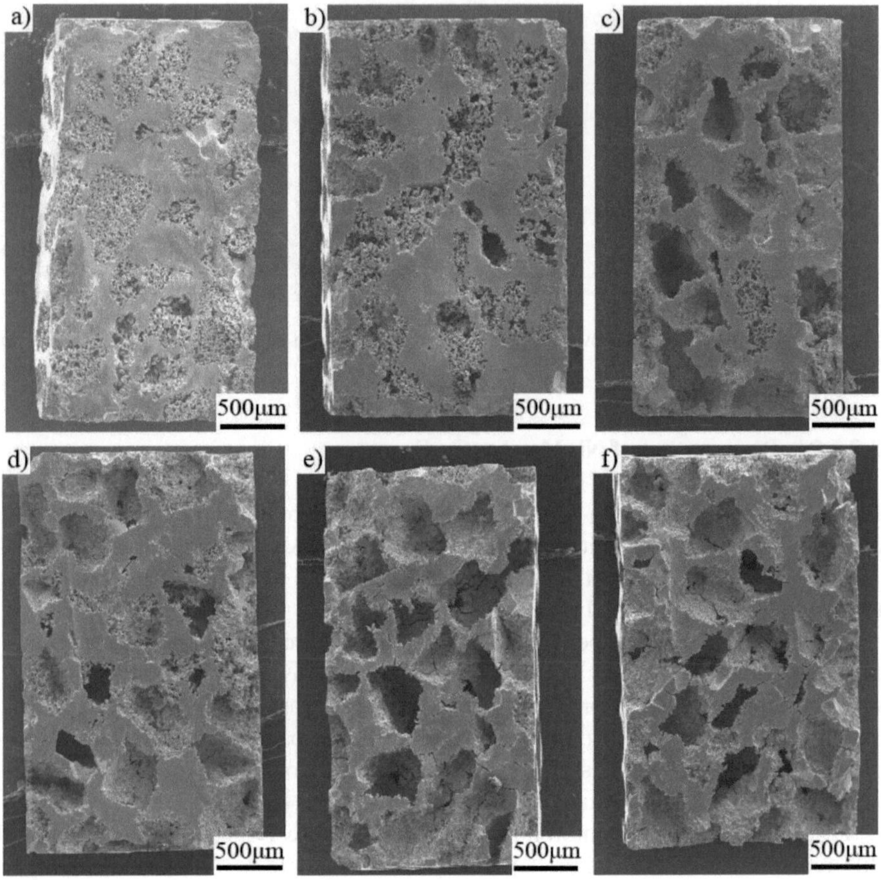

Fig. 3.5 SEM photographs showing pore architecture of single-pore Ni–Mn–In–Co foams with different porosities. **a** 50%; **b** 52%; **c** 58%; **d** 61%; **e** 65%; **f** 67%

of boundary between particles at 1050 °C and fully densified at 1080 °C. Therefore, 1080 °C is an ideal sintering temperature of Ni–Mn–Ga alloy particles.

Based on the above results, Ni–Mn–Ga foam was prepared by PM by sintering at 1080 °C for 2 h. The foam is prepared using Ni–Mn–Ga powders of size 250–355 μm and $NaAlO_2$ particles with size 75–120 μm. The prepared foam is shown in Fig. 3.7. Compared with the foam prepared by the template replication method, the foam prepared by PM has thicker nodes and struts. Some small-sized pores also exist in the walls due to the existence of space holder. During high-temperature sintering, the alloy powder undergoes plastic deformation, leaving an irregular pore shape. This is different from the foam prepared by template replication, which produces regular pores similar to the space holders.

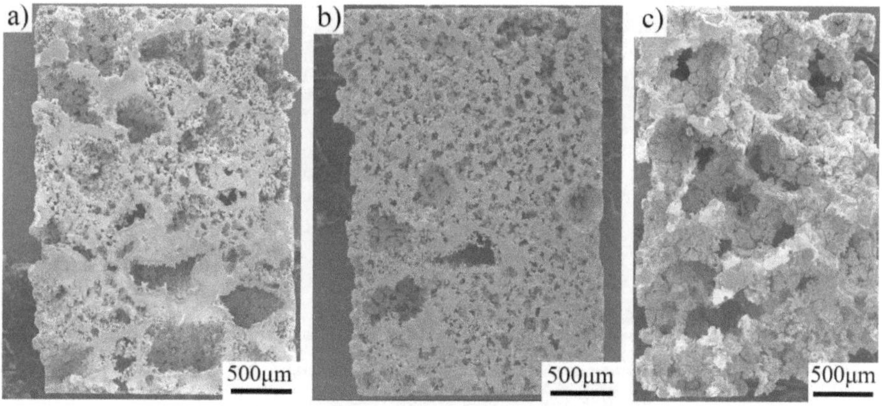

Fig. 3.6 SEM images of double-pore Ni–Mn–In–Co foam with different porosities. **a** 60%; **b** 67%; **c** 77%

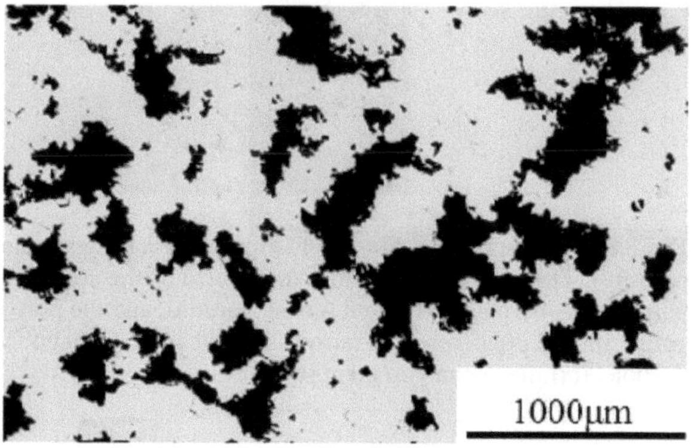

Fig. 3.7 Optical microscope of Ni–Mn–Ga foam showing two-dimensional pore structure in the foam transverse section

3.3 Magnetic-Field-Induced Strain (MFIS) of Ni–Mn–Ga Foam

3.3.1 Single-Pore Foam

In 2007, Boonyongmaneerat et al. [5] prepared a single-pore Ni–Mn–Ga foam by template replication and first reported MFIS in foams. The composition of the master alloy is $Ni_{50.6}Mn_{28.0}Ga_{21.4}$, and the space holder is $NaAlO_2$ (particle size 355–500 μm). Figure 3.8 represents the optical micrograph showing nodes and struts of single-pore Ni–Mn–Ga foam. Figure 3.8a shows the pore structure immediately after the space holder is removed (porosity 55%). Further corrosion results in a foam with a porosity of 76%, as shown in Fig. 3.8b. It can be seen that some fine pores are corroded and broken, with reduced size of the nodes. After successive corrosion to create a porosity of 76% (Fig. 3.8c), bamboo grains are formed in the struts; i.e.,

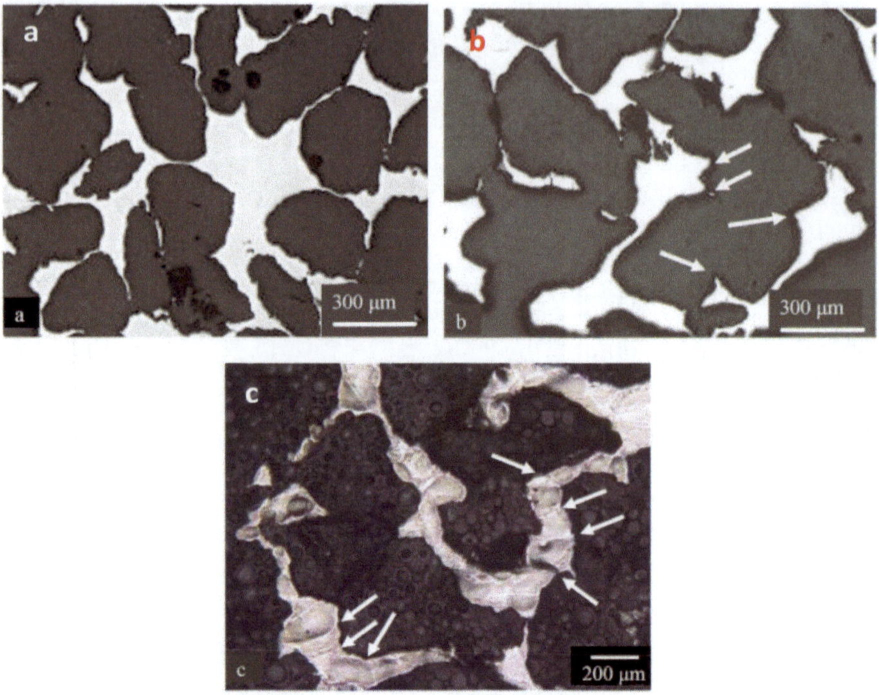

Fig. 3.8 Single-pore Ni–Mn–Ga foam prepared by template replication [5]. **a** Alloy A (55% porosity). The pore size is similar to the size of the space holder; **b** alloy B, porosity 76%, the pore edge is partially dissolved (arrow portion), and the pore size is larger than the space holder size; **c** alloy B after corrosion (arrows show that the grain boundary is perpendicular to the edge of the hole, forming a bamboo grain structure)

the grain spans the entire width of the struts. In this case, the twin in the struts has smaller constraints caused by grain boundary.

Figure 3.9 shows a schematic diagram of a rotating magnetic field for MFIS test developed by Müllner [2] at Boise State University, USA. The two ends of the sample (1) are placed between the sample holder (3) and a moving head (2); the sample holder (3) is placed in the tube (4); (5) is the cover plate of the sample holder; the sample is stretched on the horizontal plane. The displacement Δx along the vertical direction is produced by a quartz tube (6) and the horizontal turning structure (7), which can be measured by an extensometer. Cool/hot air (5) is blown on the sample surface. The temperature of the sample is measured by a thermocouple (9). The dotted line in the figure indicates the rotation axis of the magnetic field (the direction of the magnetic field is along the horizontal direction ΔZ).

Figure 3.10 shows the strain change of a foam sample with a porosity of 76% when the magnetic field rotates 360°. It can be seen that the sample elongates and shortens twice during 360° rotation, and each extension and shortening are defined as one magnetomechanical cycle. During 360° rotation, the sample undergoes two

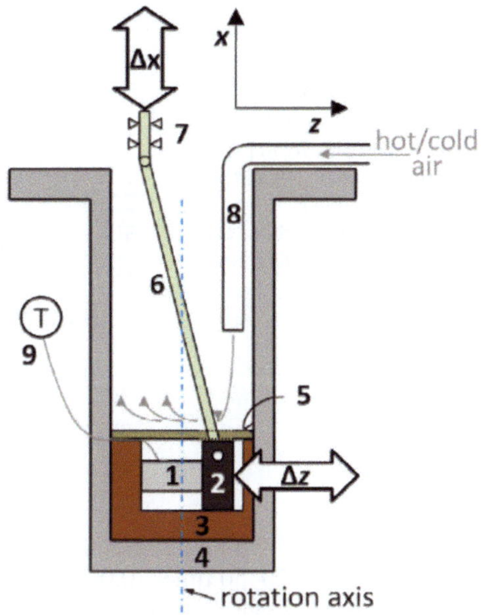

Fig. 3.9 Schematic diagram of magnetic-field-induced strain (MFIS) under a rotating magnetic field measurement device [2]

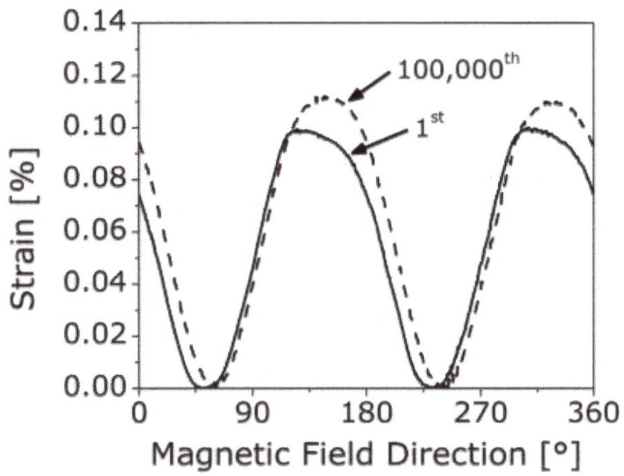

Fig. 3.10 Magnetic-field-induced strain (MFIS) of Ni–Mn–Ga foam B (implemented for the first cycle, dashed line for the 100,000th cycle) [5]

magnetomechanical cycles. The plastic strain generated in the first cycle was 0.097%; after 105 magnetomechanical cycles, the plastic strain increased to 0.115%.

The relationship between the plastic strain and number of magnetic–mechanical cycles is shown in Fig. 3.11. The initial strain of sample A_F (porosity 55%) is 0.003%, which increases to 0.064% after undergoing a field cooling treatment and decreases to 0.010% after 40 cycles. For foam B (porosity 76%), the initial strain is 0.097%, which is increased to 0.110% during the subsequent 1000 cycles and finally changes between 0.080 and 0.115%. The sample B_F strain is ~0.040% at 40 cycles and slowly

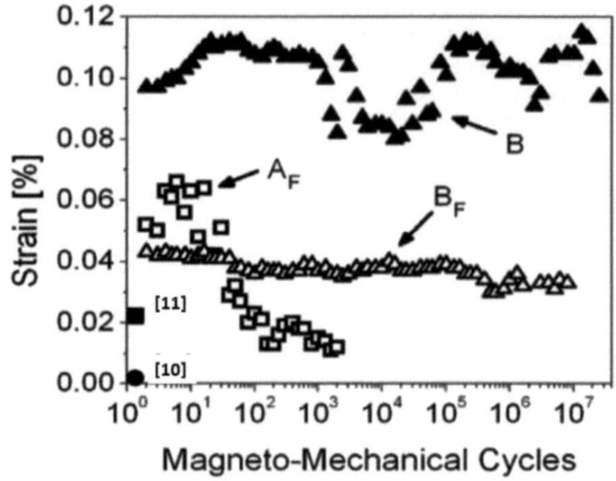

Fig. 3.11 Relationship between the magnetic-field-induced strain of the Ni–Mn–Ga foam A_F (porosity 55%), B and B_F (porosity 76%) [5]. The fine-grained [10] and coarse-grained [11] Ni–Mn–Ga alloy ribbons are also marked in the diagram

decreases to 0.030% in the subsequent 500,000 cycles. The above-mentioned strain (0.04–0.10%) is 20–50 times (0.002%) in fine-grained Ni–Mn–Ga alloy ribbon [8] and 2–4 times the coarse-grained Ni–Mn–Ga ribbon (0.025%) [9]. This indicates that a large number of the free surface in the foam have less restrictions on the movement of twin boundaries.

Since the two ends of the sample are fixed at the two ends of the sample stage, one end can move freely in the radial direction. When the crystal orientation is not proper, the resistance of the twin boundary movement is very small. For the bamboo grains, the nodes are connected to each other to form a three-dimensional structure. This three-dimensional structure also constrains the movement of the twin boundaries. Although the constraint is smaller than the polycrystalline bulk material, it is larger than the single crystal, thereby reducing the strain that foams can produce. If there are no nodes connected between the edges of the bamboo grains (like a short rod or microwire assembly), the constraint is smaller and the resulting strain that can be generated is greater.

Figure 3.12a shows the MFIS of the single-pore Ni–Mn–Ga foam [3]. The strain gradually changes from 0.24% (78 cycles) at the beginning to 0.18% (506 cycles) and then remains 0.18–0.19% in 20,378 cycles. The decrease in strain at the beginning may be due to the breakage of the pore nodes and struts during the cycle. In addition, the formation of twin dislocations and their restrictions to the movement of the twin boundary may also be the reason for the decrease of MFIS. The training effect induced by multiple magnetomechanical cycles also has a positive effect on the increase in strain; however, in single-pore foam, this positive effect is affected by the pinning effect caused by fracture or dislocation initiation in struts and nodes.

Figure 3.12b shows the change of strain during the 360° rotation of the magnetic field. Here, the strains at 68 and 20,378 cycles are shown. In the first cycle (0–180°), two strain peaks are generated at 60° and 115°. For 68 cycles, the strains are 0.20% and 0.24%, respectively; in the second cycle (181–360°), the peak strain

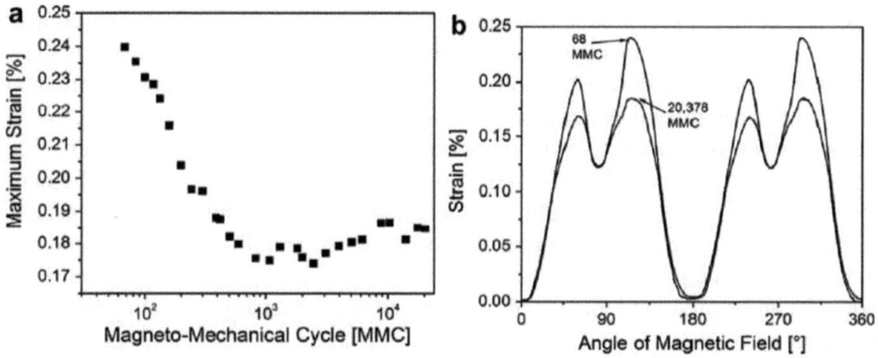

Fig. 3.12 Magnetic-field-induced strain of single-pore Ni–Mn–Ga foam (porosity 56.6%) [3]. **a** Variation of MFIS with the number of cycles; **b** variation of MFIS with magnetic field rotation angle (rotation 68th and 20,378th)

appears at 240 and 300° corresponding to the first cycle. In a single crystal, usually, only one strain peak appears in each cycle. Here, the two peaks that appeared in each cycle reflect the polycrystalline characteristics of the alloy. That is, grains of different orientations produce strain in different directions. So, the contribution to the measured average strain in the z-direction is different. The two peaks that appear in each cycle may correspond to the two main twin variants in the alloy. These two twin variants produce strain in their respective directions under the action of the magnetic field. However, due to the constraints, some of the variants cannot fully exert their respective strain. In short, what is tested here is the average strain in the presence of multiple variants of polycrystalline alloys. Finally, compared with the 506th cycle, the strain at the 20,378th cycle is reduced by 0.16 and 0.18%, but the shape of the peak does not change significantly. Only at 105°, the second peak (135°/315°) has a shoulder. This indicates that a third twin variant may be induced during the cycle.

3.3.2 Dual-Pore Foam

Here, the template replication method is also used to prepare the double-pore Ni–Mn–Ga foam. The parent alloy ingots used are $Ni_{52.0}Mn_{24.4}Ga_{23.6}$ and $Ni_{52.3}Mn_{23.9}Ga_{23.8}$. The space holder is sodium aluminate particles with sizes 500–600 μm and 75–90 μm, respectively, and the mass ratio of large/small space holder particles is 73:27. The prepared foam structure is shown in Fig. 3.12. The pore prism walls of the large pores also contain fine pores formed by fine space holders with twins in these fine nodes and struts. From the magnetization curve, the austenite starting and finishing temperatures are 30 °C and 43 °C, respectively, and the martensite starting and ending temperatures are 35 °C and 24 °C, respectively. The Curie point of the foam is 88 °C. In addition, the saturation magnetization of the foam is 73 $Am^2\ kg^{-1}$.

Zhang et al. [3] prepared single- and double-pore Ni–Mn–Ga foams by template replication and multi-step etching with acids. The morphology of the foams is shown in Figs. 3.3 and 3.4. Figure 3.13 shows the twin morphology in the double-pore foam. The thickness of the twin wafer layer in the as-cast state is 1–5 μm with twins covering the small struts but not the thick nodes. In the alloy after chemical ordering annealing, the twin wafer layer width is 5–10 μm, which spans the whole width of the nodes and struts, which can increase the foam magnetocrystalline anisotropy. Because the free surface can absorb dislocations and promote the mobility of twin boundaries, the mobility of coarse twins covering the nodes and struts is better than that of fine twins. It can be predicted that the alloy after chemical ordering annealing has higher MFIS properties.

Figure 3.14 shows the MFIS of the double-pore foam (porosity 66.8%) [3]. Figure 3.14a shows the relationship between the maximum strain and the number of magnetic–mechanical cycles. The state of the sample includes the ordered processing

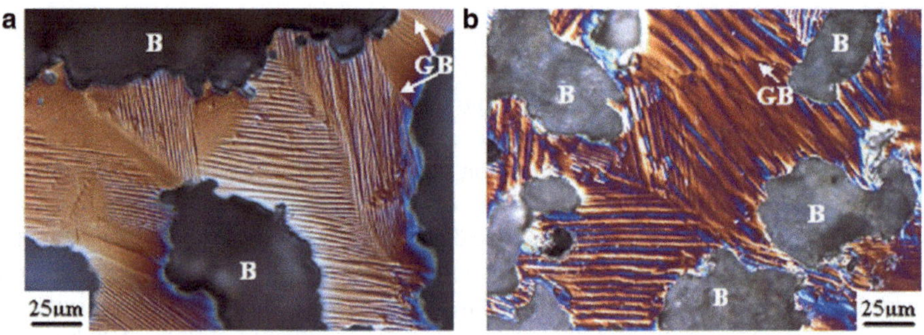

Fig. 3.13 Twin crystal morphology in double-pore Ni–Mn–Ga foams [3]. **a** As-cast state; **b** chemical ordering annealing state. Bamboo grain boundary (GB) is highlighted in **b**

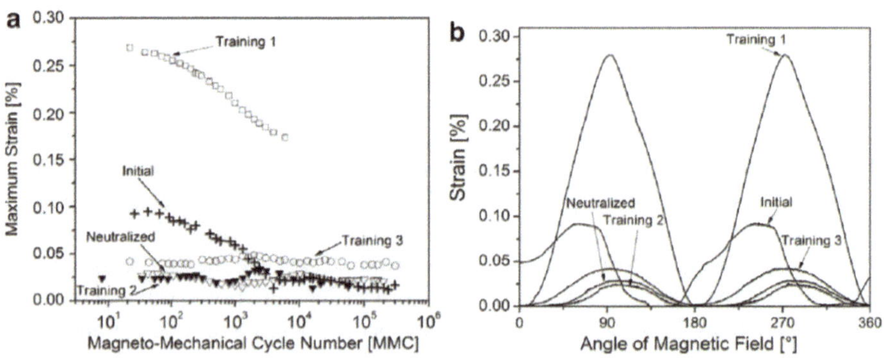

Fig. 3.14 Magnetic-field-induced strain of double-pore foam (porosity 66.8%) [3]. **a** The relationship between the magnetic-field-induced strain of states (ordered state, first training, neutral state, second and third training states); **b** the relationship between magnetic receptive strain and magnetic field rotation angle in five states

state, the neutral state and the training state for different times. The initial strain of the alloy in the ordered state is 0.10%, which is reduced to 0.014% after 60,000 cycles, and no strain reduction occurs after 300,000 cycles under this strain. After the first thermal–magnetic training, the strain of the alloy increases significantly to 0.28% and then decreases to 0.17% after 6000 cycles. The maximum strain of 0.28% is more than twice that of single-pore alloys (54% porosity strain is 0.01%, and 76% porosity strain is 0.12%); it is also slightly larger than the maximum strain of 0.24% of single-pore pores in Fig. 3.12. These strain differences reflect the complex effects of grain orientation, size, grain boundaries and pore edges/nodes on MFIS. In polycrystalline alloys, it is usually necessary to introduce texture to obtain considerable MFIS, while foams can obtain higher strains (strain values similar to Terfenol-D 0.2%) by conventional casting.

The alloy after thermal–magnetic training is heated to 150 °C (above *Mf*), and then the temperature is reduced at zero field; during this process, the preferred orientation of the twin variant obtained by the original training is eliminated (neutral state), resulting in the strain of the alloy. It is reduced to 0.027% and further to 0.014% during the subsequent 1300 cycles and remains 0.020–0.027% in 200,000 cycles. This shows that thermomagnetic training is favorable for the improved MFIS properties of as-cast or ordered heat-treated foams. After the alloy was subjected to a second thermomagnetic training, the strain does not increase significantly. After the third training, the strain increases slightly to 0.04% and remains basically unchanged at 151,000 times. The different effects of this training on strain may be due to defects in the alloy, that is, material damage caused by crack initiation and propagation of struts/nodes during 805,000 cycles. In addition, it can be seen that the strain reduction of the double-pore foam is smaller than that of the single-pore foam, indicating that the double-pore foam has higher mechanical stability. The reasons are: (a) In the double-pore foam, twins traverse the entire nodes and struts, which renders easier movement of twin boundaries, thus reducing the tendency of twin interaction and crack initiation; (b) the crack growth rate of the small pore struts/nodes in the double-pore foam alloy has a growth rate less than the single-pore foam because cracks may be stopped by when they encounter the free surfaces.

Figure 3.14b shows the relationship between MFIS and magnetic field angle. The alloy in the chemical ordered state has a wide strain peak (0.1%) between 20–130° and 160–310°. The peak shape is not symmetrical, indicating that the strain peak is actually a combination of many small peaks, corresponding to the multi-variant state. After the first thermal–magnetic training, an approximately symmetrical narrow single peak (0.28%) appeared at 90/270°, showing the uniform twin structure and twin boundary movement state after training. Subsequent neutralization heat treatment and another thermomagnetic cycle no longer change the shape of the strain peak.

Here, we can roughly discuss the correlation between pore structure, grain size and magnetic-field-induced strain. The edge of the hole can be regarded as a microwire, in which the twin crystals traverse the entire width. In this case, the edge of the hole can produce a larger strain. When the twin boundary ends at the grain boundary, the twin boundary is greatly restricted and may not move under the action of the magnetic field. Therefore, some of the grains in the actual foam contribute to the magnetic-field-induced strain, while the rest do not. In addition, the orientation of the struts and nodes is different, and the contribution to the strain is also different. Assuming that the grain size in the alloy is equal to the pore size in the single-pore foam or the size of the macro-pores in the double-pore alloy, many twins must terminate at grain boundaries in the single-pore foam. These twins are difficult to move; however, for double-pore alloys, the same twins are divided into many small twins by fine pores. The twin boundaries terminate at the free surface, and the mobility of the twin boundaries is significantly improved, corresponding to the larger MFIS.

The MFIS obtained is smaller than the theoretical values. In thick nodes or pore edges, stable 120° grains and multivariable structures within the grains may be formed, which is not favorable for reducing the resistance of twin boundary movement. If finer pores can be introduced into the coarse pores and nodes, the specific surface area can be further increased, and the size of the pores/nodes can be reduced, which induces the formation of bamboo grains and thus further reduction of the twin boundary motion resistance.

Another foam sample with dual pores is used for the measurement of MFIS as well. The size of the sample is 2 × 3 × 6 mm. According to the martensitic transformation temperature of the alloy, the MFIS is measured at 16 °C (in the martensite state) first. The rotating magnetic field is 0.97 T. The results are shown in Fig. 3.15 [2]. At the beginning, the alloy showed a strain of 2.1%, which is almost 20 times the strain of the single-pore alloy; during the subsequent 2000 magnetic–mechanical cycles, the strain increases to 3.4% and stabilizes 15,000 cycles, followed by 75,000 cycles. Finally, the strain moderately reduces to 2.0% and stable circulation reaches 161,000 times. After the sample is removed and reinstalled, the strain is reduced to 0.5%, which may be caused by the chaotically arranged twins introduced during the reloading process. Then, the sample is removed from the sample for a second time and placed in a vibrating sample magnetometer for thermomagnetic training (by raising the sample from room temperature to 150 °C under a 20 kOe magnetic

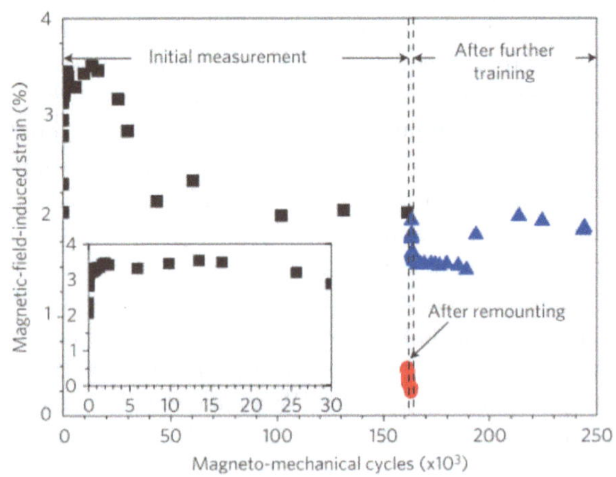

Fig. 3.15 Magnetic-field-induced strain properties of the Ni–Mn–Ga foam with dual pores at 16 °C [2]

field and lowering it to room temperature again). After reloading on the sample holder, the sample produced a strain of 1.5–1.9% in 90,000 cycles, indicating that the magnetocaloric training is effective.

Thermomagnetic training shows that chaotically arranged twins rearrange during training and can restore the MFIS capacity. If the sample is subjected to a magnetic–thermal cycle in a magnetic field during the cooling process, the martensite formed may be induced by the magnetic field and more importantly form a favorable orientation that significantly improves the MFIS performance of the alloy. Subsequent measurement results prove this effect, as shown in Fig. 3.16 [2]. It can be seen that the alloy can produce a constant strain of 1.4% at 16 °C. After reaching 35–41 °C, the strain quickly decreases to near zero after being transformed into austenite. In the subsequent cooling process, the strain rapidly increased to 2.2% at 22–23 °C (close to *Mf*). We know that after thermal cycling under this magnetic field, a preferentially oriented martensite variant can be produced in the sample. After the temperature of the sample is reduced to −100 °C under zero field, the alloy will undergo an intermediate martensitic transformation, which may disturb the regular martensitic deformation structure originally produced. During the heating process, the strain of

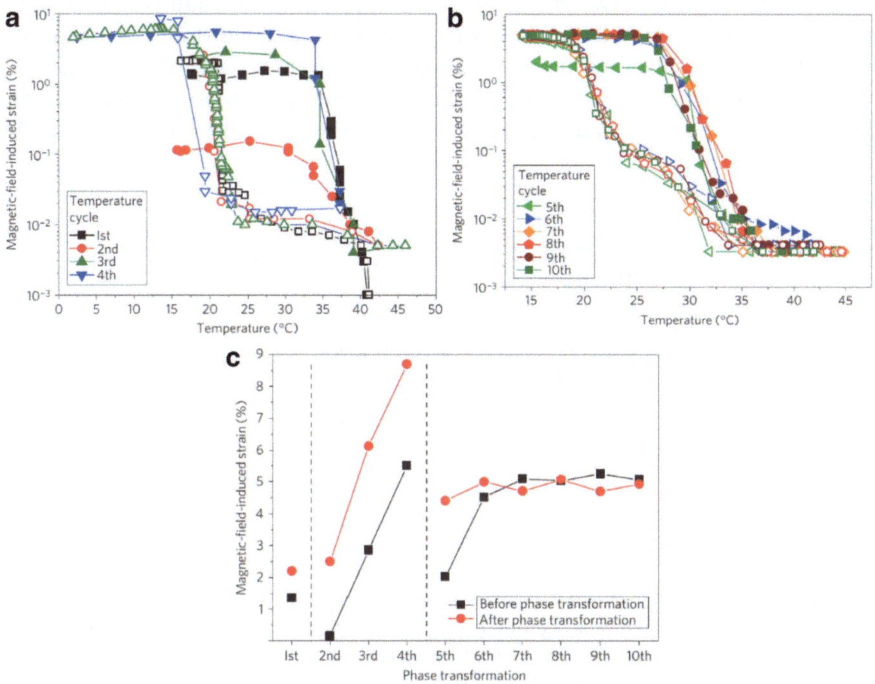

Fig. 3.16 Magnetic-field-induced strain properties of dual-pore Ni–Mn–Ga foam during the martensitic transformation [2]. **a** 1–4 cycles; **b** 5–10 cycles (solid symbol for heating process and hollow symbol for cooling process); **c** maximum MFIS value during heating and cooling processes (two dotted lines in the figure are the sampling and reloading, respectively)

the sample was greatly reduced to 0.2%, indicating that the internal regular variants disappear and it is difficult to produce large strains. However, during the cooling process for a second time, the sample recovers 2.5% of the strain. During the third and fourth thermal cycles, the sample can produce a maximum strain of 8.7%.

In addition, during the first 5 cycles, the strain before the phase transition is significantly smaller than the strain after the phase transition. The strain value at the 6th cycle is also slightly increased, and it is basically stable at 4.4–5.1% in the range of 10 times. These phenomena may indicate that the sample has undergone effective magneto-thermomechanical training during the first 5 cycles: During the martensite phase transformation under the magnetic field, the generated martensite twins are affected by the external magnetic field and produce the *c*-axis preferential variants parallel to the direction of the magnetic field. After a few cycles, only variants with high twin boundary mobility remain, thus giving the alloy higher MFIS properties.

However, the above-mentioned magnetomechanical training effect still cannot explain why the sample can produce up to 8.7% MFIS after the fourth cycle. Figure 3.17 shows four different austenite unit cells (upper) and corresponding martensite unit cells (lower, easy magnetization axis *z*-axis parallel to the direction of the external magnetic field) [2]. The strain obtained in the *z*-direction here is related to the angle α between the *c*-axis and the *z*-axis of the unit cell. In addition, the crystal orientation in the nodes/struts is arbitrary, and the strain obtained by an isolated single-crystal alloy can be calculated by the average $\cos\alpha$ of α between 0 and $\pi/4$, so that the resulting strain is 73% theoretical strain of the single crystal, where the theoretical strain is calculated using $1 - c/a$ (*a* and *c* are the longest and shortest

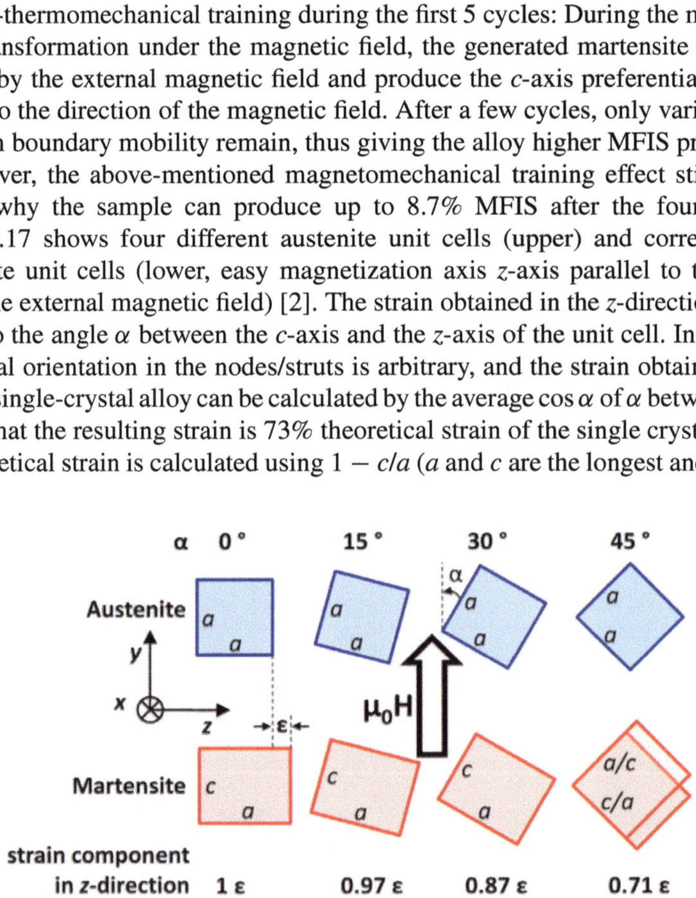

Fig. 3.17 Relationship between the magnetic-field-induced strain and the orientation of the variants in the nodes/struts during thermal cycling [2]. α is the angle between the *c*-axis crystallographic orientation of the sample and the easy magnetization direction *z*. The possible average strain in the *z*-direction is determined by the average value of the crystallographic theoretical strain ε and cos (α) (where α is all Euler angles between 0 and $\pi/4$)

axes of the martensite unit cell, respectively). For 14M martensite, $c/a = 0.90$, untextured isolated single crystal pore edges can produce 7.3% strain ; however, the present foam produces 8.7% strain, implying that texture is produced during solidification.

The research here also shows that porous materials can be trained by thermal cycling (i.e., thermal–magnetic cycling) under a magnetic field. Considering that the sample is fixed at both ends, for dense materials, according to the principle of Saint-Venant, the material about 1 mm away from the two fixed ends is limited by the fixed end and cannot move freely; that is, this part of the alloy will not contribute to the total strains. For porous materials, this stress zone is estimated to be only about 20 μm; that is, a higher fraction of the material can contribute to the total strain.

In short, despite the intrinsic brittleness of the Ni–Mn–Ga alloy, the dual-pore porous material can be steadily cycled more than one million times under the strain conditions generated. In addition, this alloy can be prepared by economical casting methods, which shows promising application potential for low-cost, high-performance active sensors and actuators.

3.3.3 *Magnetocaloric Effect (MCE)*

The research on the magnetocaloric properties of porous magnetic shape memory alloys is still scarce. Sasso et al. [12] studied the magnetocaloric properties of $Ni_{54.8}Mn_{20.2}Ga_{25.0}$ foam (porosity 44–58%). The phase transition temperature decreases with the applied magnetic field at a rate of 0.8 K T^{-1}, and eventually, an entropy change of 2.5 J kg^{-1} K^{-1} occurs.

3.4 Magnetostructural Transition and Magnetocaloric Effect of Ni–Mn–In–Co Foam

The foam has a large specific surface area and high heat transfer efficiency and is very suitable as a solid refrigeration material. After introducing pores, the uncoordinated deformation due to volume change during the first-order phase transition is alleviated. So, it can improve cycle life, phase transition resistance, transition hysteresis and refrigeration efficiency. Here, based on the analysis of phase transformation characteristics of foams with different porosities and pore structures, a phenomenological model of phase transformation of foams is established, and the magnetocaloric properties of single-pore alloys are tested to obtain a large cooling capacity (RC) [7].

3.4.1 Single-Pore Ni–Mn–In–Co Foam

$Ni_{45}Mn_{37.3}In_{12.7}Co_5$ alloy ingot is used to study the relationship between magnetization and temperature, as shown in Fig. 3.18a. The sample is tested at 200 Oe and 50 kOe. The martensite phase is paramagnetic, while the austenite phase is ferromagnetic. So, the applied magnetic field can stabilize the austenite phase; that is, the magnetic field shifts the phase transition temperature to a low temperature. It can be seen that the bulk Ni–Mn–In–Co has a large hysteresis, about 13 K at 200 Oe and about 24 K at 50 kOe. The hysteresis in the high magnetic field is greater than that in the low magnetic field. The tests of single-pore foams with 50, 61, 67 and 73% porosities show that the hysteresis is significantly reduced, and the phase transition width is larger than that of dense bulk samples. The results of the 73% porosity foam test are shown in Fig. 3.18b. The phase transition temperatures of alloys with different porosities are listed in Tables 3.1 and 3.2, where M_s, M_f, A_s and A_f correspond to the martensitic transformation starting, finishing temperatures, austenite starting and finishing temperatures, respectively. T_0 represents the temperature at which the Gibbs free energy of the austenite phase and the martensite phase is equal. In thermoelastic martensite, $T_0 = (M_s + A_f)/2$ is generally considered; T_M and T_A are the average temperature of martensitic transformation and the average temperature of austenite transformation, respectively, $T_M = (M_s + M_f)/2$, $T_A = (A_s + A_f)/2$;

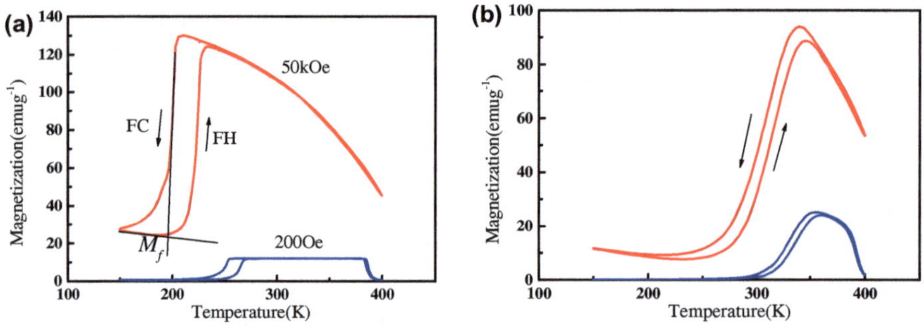

Fig. 3.18 Magnetization and temperature dependence of bulk and porous $Ni_{45}Mn_{37.3}In_{12.7}Co_5$ alloys. **a** Dense bulk alloy; **b** porous alloy

Table 3.1 Transformation temperatures of single-pore $Ni_{45}Mn_{37.3}In_{12.7}Co_5$ foam at 200 Oe (unit K)

Sample	M_s	M_f	A_s	A_f	T_0	T_M	T_A	T_{hys}	$M_s - M_f$	$A_f - A_s$
Bulk sample	254	248	259	268	261	251	264	13	6	9
50%	342	306	317	351	347	324	334	10	36	34
61%	345	304	314	350	348	325	332	8	41	36
67%	346	307	313	352	349	327	333	6	39	39
73%	348	311	318	353	351	330	336	6	37	35

Table 3.2 Transformation temperatures of single-pore $Ni_{45}Mn_{37.3}In_{12.7}Co_5$ foam at 50 kOe (unit K)

Sample	M_s	M_f	A_s	A_f	T_0	T_M	T_A	T_{hys}	$M_s - M_f$	$A_f - A_s$
Bulk sample	203	195	218	228	216	199	223	24	8	10
50%	328	265	284	335	332	297	310	13	63	51
61%	325	266	279	334	330	296	307	11	59	55
67%	328	269	279	336	332	299	308	9	59	57
73%	334	274	285	340	337	304	313	9	60	55

T_{hys} represents the phase transition lag, $T_{hys} = (A_f - M_s + A_s - M_f)/2$. $M_s - M_f$ and $A_f - A_s$, respectively, indicate the width of the phase transition between the positive phase transition and the reverse phase transition.

3.4.2 *Dual-Pore Ni–Mn–In–Co Foam*

3.4.2.1 Martensitic Transformation Characteristics

The M–T curves of Ni–Mn–In–Co foams with a porosity of 60, 67 and 77% are shown in Fig. 3.19. It can be seen that the double-pore foam also has the characteristics of large phase transition width and small hysteresis. The phase transition temperature determined by the tangent method is shown in Tables 3.3 and 3.4.

Comparing the phase transition hysteresis with the bulk alloy and single heavy pore alloy, it is found that the phase transition hysteresis of the double-pore alloy is larger than that of the single-pore alloy and smaller than that of the bulk sample. Obviously, compared with the single-pore alloy, the deformation caused by the phase change volume change of the double-pore alloy is easier to coordinate and the accumulated elastic energy becomes smaller, so the phase change resistance and hysteresis should be smaller. According to the phase transition characteristics of Ni–Mn–In–Co foam, the phenomenological model of phase transition is established.

3.4.2.2 Phenomenological Interpretation of Martensitic Transformation Behavior of Foam

The thermodynamic process of thermoelastic martensitic transformation can be expressed by the following formula [13–15]:

$$\Delta G_{chem}{}^{M-P} - \Delta G_{elast}{}^{M-P} + E_{irr}{}^{M-P} = 0 \tag{3.1}$$

where $\Delta G_{chem}{}^{M-P}$ is chemical-free energy difference (a.u.) between the martensite and parent phases during MT, $\Delta G_{elast}{}^{M-P}$ is elastic energy accumulated during the

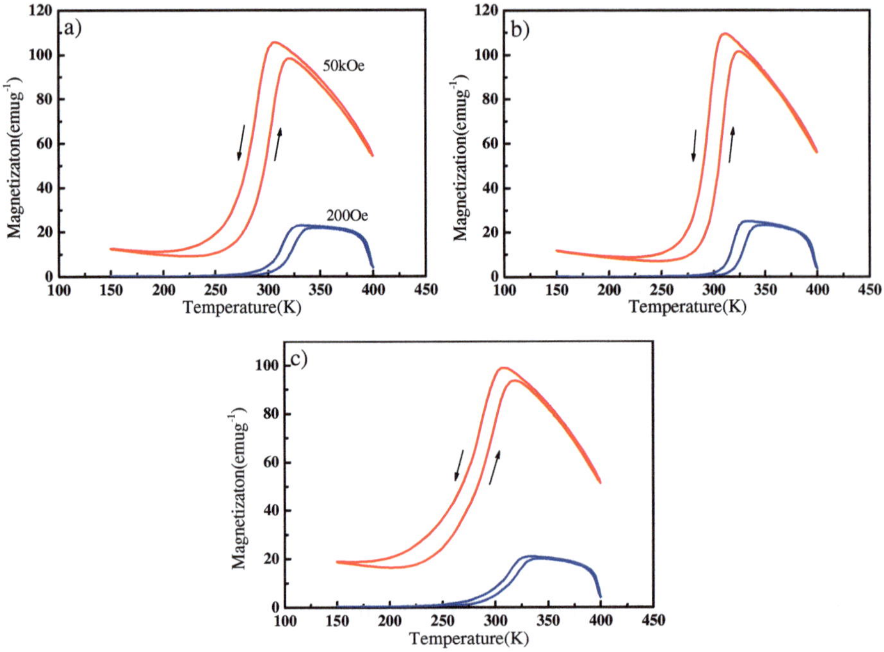

Fig. 3.19 Magnetization curves of double-porosity Ni–Mn–In–Co foams with different porosities. **a** 60%; **b** 67%; **c** 77%

Table 3.3 Phase transition point of double-pore foam under 200 Oe (K)

Sample (%)	M_s	M_f	A_s	A_f	T_0	T_M	T_A	T_{hys}	$M_s - M_f$	$A_f - A_s$
60	322	301	311	333	327.5	311.5	322	10.5	21	22
67	325	310	319	337	331	317.5	328	10.5	15	18
77	324	291	298	332	328	307.5	315	7.5	33	34

Table 3.4 Phase transition point of double-pore foam under 50 kOe (K)

Sample	M_s	M_f	A_s	A_f	T_0	T_M	T_A	T_{hys}	$M_s - M_f$	$A_f - A_s$
0.6	300	266	284	314	307	283	299	16	34	30
0.67	305	279	295	318	311.5	292	306.5	14.5	26	23
0.77	302	252	264	312	307	277	288	11	50	48

phase transition (a.u.) and $E_{irr}{}^{M-P}$ is irreversible dissipated energy (a.u.) generated during MT.

Thermoelastic martensite changes in structure and volume during MT. Elastic strain energy due to deformation mismatch may accumulate at the phase interface. When the stored elastic strain is high enough, dislocations can be induced or cracks can be formed, with accumulated elastic strain energy released and the interface stress relaxed. As the reverse phase transition occurs, the elastic energy accumulated is partially dissipated. At this time, the hysteresis occurs. The resistance of phase interface motion is also a source of hysteresis. The hysteresis mainly comes from the frictional resistance and elastic energy dissipation.

The effect of the pore structure on the martensitic transformation behavior is studied by comparing the sample hysteresis. We first consider the influence of elastic energy loss on the phase transition hysteresis in porous Ni–Mn–In–Co alloy. In bulk Ni–Mn–In–Co alloys, martensite reliefs are formed on the surface and some cracks are generated, indicating that the elastic strain energy accumulated by the phase transition is relieved by plastic deformation or crack initiation. Therefore, the bulk alloy can consume the stored elastic energy by crack initiation, resulting in the reduction of the driving force and increase of hysteresis. For foams, the deformation during MT is easier to coordinate due to the existence of pores. In addition, the elastic energy produced by the small-sized struts/nodes can be stored in the thick struts/nodes, resulting in smaller hysteresis in foams.

Figure 3.20 shows micrographs of a single-pore Ni–Mn–In–Co foam with a porosity of 50% after multiple phase transformations. It can be seen that the sample has more cracks after two MT cycles; the density of cracks keeps almost intact after

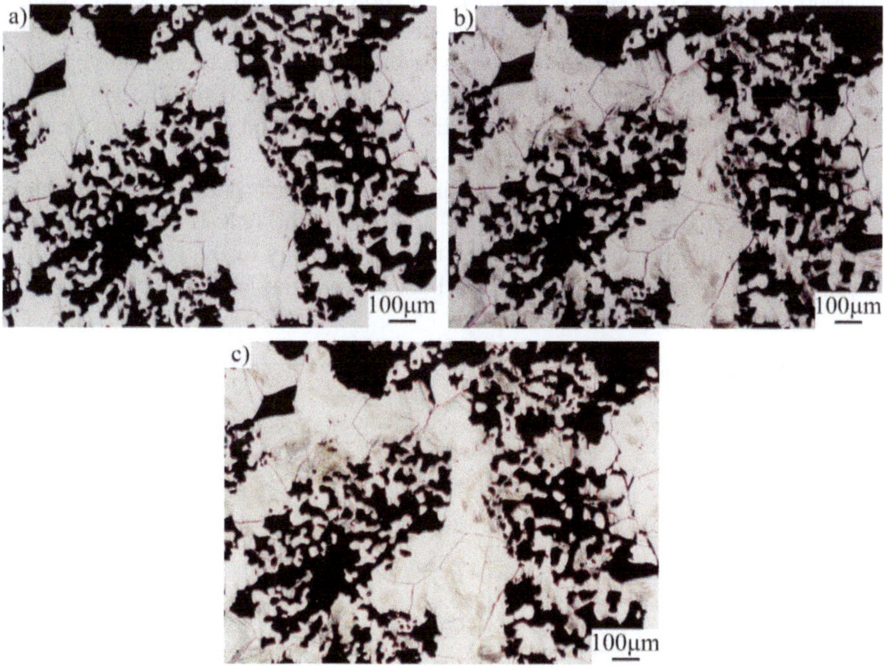

Fig. 3.20 Optical images of single-pore Ni–Mn–In–Co foam with 50% porosity after multiple phase transformations. **a** Initial state; **b** after two MT cycles; **c** after nine MT cycles

nine cycles. This shows that, when the porosity of the alloy is small, crack initiation is also necessary to relax the elastic strain energy. After the crack has been formed, the deformation coordination is improved when the phase transition of the foam occurs again. So, there is no need to continue to relax elastic strain energy by initiating cracks. Therefore, when the porosity of the sample is increased, the deformation coordination during phase transformation is enhanced, the elastic energy loss is smaller, and thus the hysteresis is smaller.

The effect of the pore structure on the frictional resistance of phase interface movement is considered. In Cu–Al–Ni microwires with bamboo grains with different diameters [15], the hysteresis increases with decreasing wire diameter when the wire diameter is 10–100 μm. By analyzing the effects of interface energy, surface energy, elastic strain energy, heat transfer process and phase interface friction work on the hysteresis, it is found that the hysteresis has a power exponential relationship with the wire diameter (about −0.5), while M_s and grain size also exhibit a power exponent close to −0.5 in many alloys and compounds. This means that the power exponent relationship is the same, but they have different physical backgrounds. The result implies that the superelastic hysteresis has a size effect. When the sample diameter is less than 100 μm and the specific surface area is large, increased hysteresis may occur.

For the Ni–Mn–In–Co foam, after grain growth, the struts and nodes exhibit bamboo grains. As the porosity increases, the pore struts and nodes become thinner and the free surface pins the phase interface, resulting in larger hysteresis. In order to verify this point of view, a sample with double-pore foam is eroded into particles, thereby eliminating the impact of elastic energy loss on hysteresis. The M–T tests of the particles are shown in Fig. 3.21, and the phase transition temperatures are listed in Table 3.5.

The hysteresis statistics of the bulk, single-pore, double-pore alloy and particles obtained by corrosion are shown in Fig. 3.22. Here, the bulk is regarded as foam with zero porosity. It can be considered that the porosity of the particles is 100%. It can be seen that the hysteresis of the bulk alloy is the largest; the hysteresis of

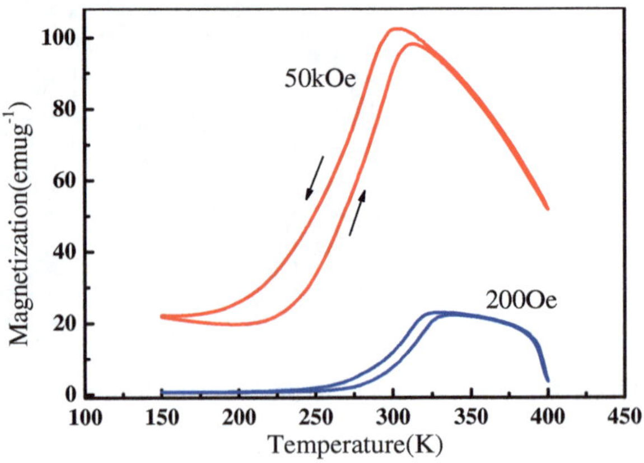

Fig. 3.21 Magnetization curve of Ni–Mn–In–Co particles produced from the foam by acid erosion

Table 3.5 Phase transition point (K) of Ni–Mn–In–Co particles produced by foam by acid erosion

Field	M_s	M_f	A_s	A_f	T_0	T_M	T_A	T_{hys}	$M_s - M_f$	$A_f - A_s$
200 Oe	321	283	292	329	325	302	310.5	8.5	38	37
50 kOe	298	232	246	309	303.5	265	277.5	12.5	66	63

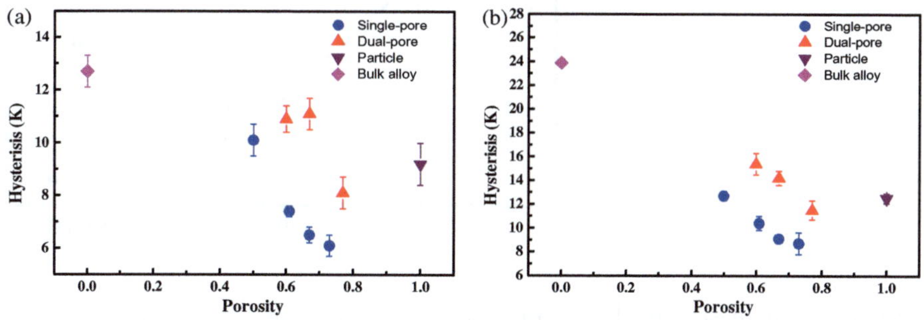

Fig. 3.22 Hysteresis of Ni–Mn–In–Co bulk alloy, foams and particle under different magnetic fields. **a** 200 Oe; **b** 50 kOe

the double-pore alloy is greater than that of the single-pore alloy; for the single-pore foam, the hysteresis decreases with increasing porosity; for the double-pore foam, the hysteresis first increases with the porosity. When the sample is eventually corroded into particle, the hysteresis increases again.

First, the hysteresis characteristics of the double-pore alloy with 77% porosity and particles are analyzed. It can be seen that the double-porosity alloy with a porosity of 77% has some cracks, which can be regarded as separate particles. The diameter of the particles is <100 μm. Compared with the particles formed after corrosion, the samples with a porosity of 77% have only large particles. When the particle size in the sample is large, the specific surface area is small, and the pinning of the free surface to the phase interface is small. So the hysteresis is small. After continuous corrosion of the sample, the particle size decreases and the specific surface area increases. The pinning effect of the free surface on the phase interface is enhanced, resulting in increased hysteresis.

Compared with the single-pore alloy, the double-pore alloy has smaller pore struts and nodes and larger specific surface area. The free surface has a stronger pinning effect on the phase interface. So, the hysteresis of the double-pore alloy is greater than that of the single-pore foam.

Considering the influence of elastic energy dissipation and friction resistance of phase interface on the hysteresis, it can be seen that the single-pore alloy has larger pore struts and nodes (equivalent diameter >100 μm), and the size effect of martensitic transformation is weak. The increase of the porosity means that the deformation is easily coordinated, and the hysteresis decreases with the increase of the porosity. On the other hand, the struts and nodes in the double-pore foam are <100 μm, and thus, the size effect of MT needs to be considered: (1) When

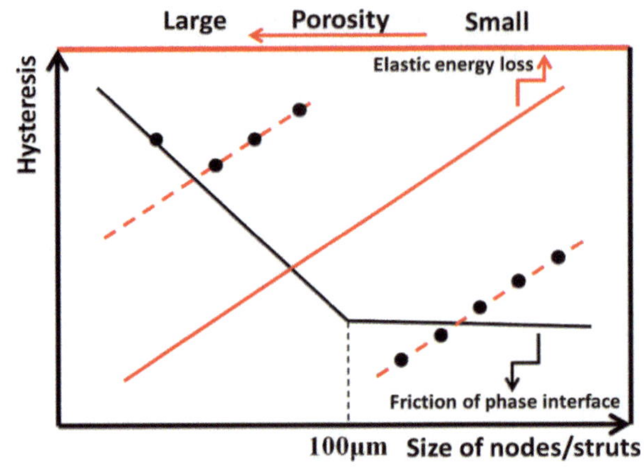

Fig. 3.23 Schematic diagram of phenomenological interpretation of hysteresis origin of Ni–Mn–In–Co foam

the porosity increases, the deformation is easy to coordinate, and the loss of elastic energy decreases with the increase of the porosity; (2) as the pore struts and nodes become thinner, the pinning effect of the free surface is enhanced, and the resistance to phase interface movement increases. At this time, the dependence of hysteresis on porosity depends on the dominant factor. When the elastic energy loss is the dominant factor, the hysteresis decreases with increasing porosity; when the phase interface friction loss dominates, the hysteresis is determined by the foam porosity. The hysteresis of martensitic transformation in foams is shown in Fig. 3.23. The solid circles represent the experimental results. The straight line just acts as a guide for the eye.

3.4.3 Magnetocaloric Properties of Ni–Mn–In–Co Foam

In bulk Ni–Mn–In–Co alloys, the meta-magnetic behavior induces giant MCE. In the foams, the introduction of pores can effectively improve the cycle life and reduce hysteresis; the increased specific surface area induced by high porosity is also favorable to the heat exchange between the working material and flowing agent. Here, only the magnetocaloric effect of the single-pore foam is analyzed.

The M–H curve test of the single-pore foams with 50 and 67% porosity is shown in Fig. 3.24. For simplification, only selected isothermal magnetization curves are shown in Fig. 3.24a. When the sample is in the martensite phase, its magnetization behavior is consistent with the paramagnet; when it is austenite, the magnetization behavior is ferromagnet. At 313 K, mixed phases of austenite and martensite exist. The alloy shows hysteresis during the first-order phase transition. The area enclosed by the isothermal magnetization curve is the magnetic hysteresis.

Extract the magnetization curve from the M–H curve using the Maxwell equation, and the adiabatic entropy change can be obtained:

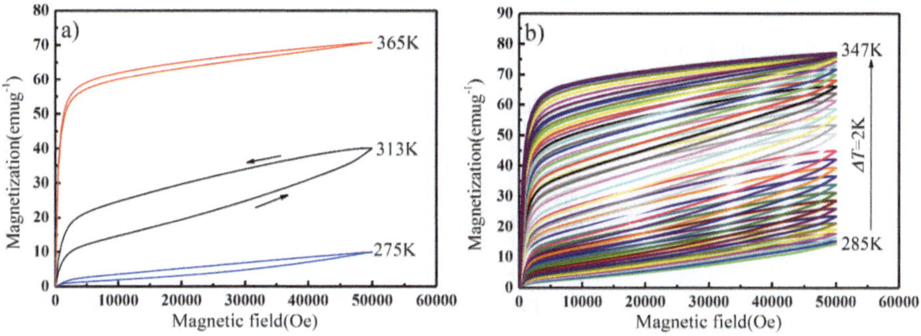

Fig. 3.24 Isothermal magnetization curves of single-porosity Ni–Mn–In–Co foam with different porosities. **a** 50%; **b** 67%

$$\Delta S = \sum_i \frac{M_{i+1} - M_i}{T_{i+1} - T_i} \Delta H_i \tag{3.2}$$

where ΔS—total entropy change (J kg^{-1} K^{-1}) of the material change phase;

H—intensity of the applied magnetic field (T);

M—magnetization of the material at temperature T and magnetic field H (emu g^{-1}).

The magnetic entropy changes of the samples with a porosity of 50 and 67% calculated according to the Maxwell formula are shown in Fig. 3.25. The magnetic entropy curve of the sample shows the characteristics of low peak height and huge peak width, and its half-height width (i.e., working interval) is large. The relationship between the magnetic hysteresis of the sample and temperature can be obtained by calculating the area enclosed by the M–H curve, as shown in Fig. 3.26. The magnetic hysteresis of the sample with a porosity of 67% is slightly smaller than that of the sample with a porosity of 50%. The magnetic hysteresis reaches a maximum at about

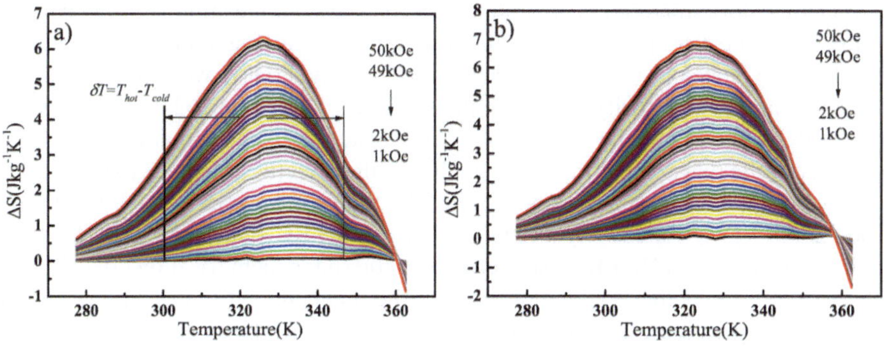

Fig. 3.25 Magnetic entropy change of Ni–Mn–In–Co foam with single porosity with different porosities. **a** 50%; **b** 67%

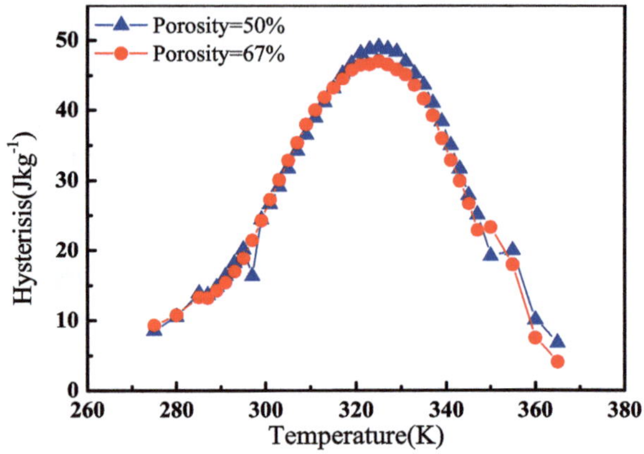

Fig. 3.26 Magnetic hysteresis of Ni–Mn–In–Co foam with single porosity with different porosities

325 K, which means that the meta-magnetic behavior (magnetic-field-induced phase transition) is most intense near this temperature. The temperature and the T_A value of the sample (330–335 K) are relatively close.

The average magnetic hysteresis of the sample can be calculated by:

$$\overline{\text{Hys}} = \frac{\int_{T_{\text{cold}}}^{T_{\text{hot}}} \text{Hys} * \text{d}T}{T_{\text{hot}} - T_{\text{cold}}} \tag{3.3}$$

where T_{cold}—temperature of the low-temperature heat source (K);

T_{hot}—temperature (K) of the high-temperature heat source;

Hys—magnetic hysteresis at temperature T (J kg^{-1}).

$T_{\text{hot}} - T_{\text{cold}} = \Delta T$ is usually estimated by the half-height width of the magnetic entropy curve. The sample with a porosity of 50% has an average magnetic hysteresis of 40.6 J kg^{-1}, and the sample with a porosity of 67% has an average magnetic hysteresis of 39.1 J kg^{-1}. There is not much difference between the two.

In addition to the isothermal entropy change, the refrigeration capacity (RC) of a magnetocaloric material is also important. Its physical meaning is the maximum value of the heat transfer between the low-temperature heat source and the high-temperature heat source when the magnetocaloric material is working. The foams with 50% and 67% porosity have RC values of 231.3 J kg^{-1} and 262.7 J kg^{-1}, respectively. RC value minus the average magnetic hysteresis is the actual heat exchange capacity of the magnetocaloric material, that is, the effective RC value (RC_{eff}). After calculation, the samples with a porosity of 50 and 67% have RC_{eff} of 190.7 J kg^{-1} and 223.6 J kg^{-1}, respectively. Figure 3.27 shows the RC value and phase transition point of the common Ni–Mn-based ferromagnetic shape memory alloy [16–25]. The abscissa represents the peak temperature of the magnetic entropy (i.e., the optimal working temperature). The porosities of 50 and 67% in the figure are single-pore Ni–Mn–In–Co foams. The rest of the data are from the literature. The solid symbols represent RC_{eff}, and the open symbols represent RC. It can be seen that the RC value

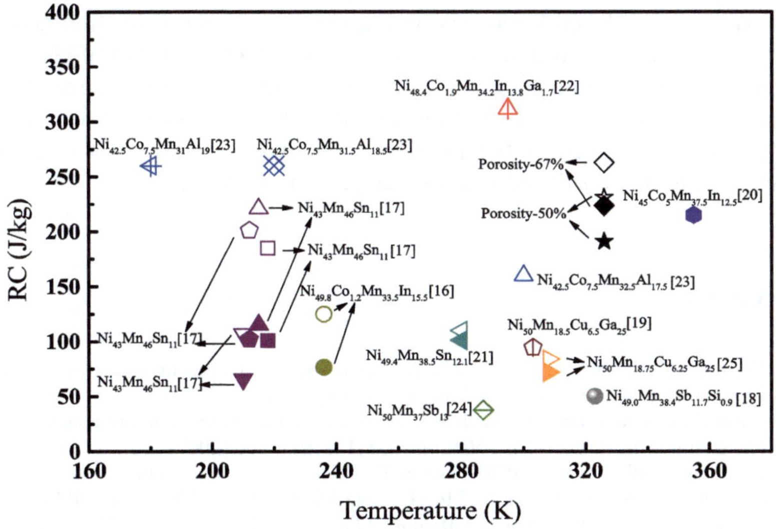

Fig. 3.27 RC value of Ni–Mn-based Heusler alloy [16–25]

of the foam is at a higher position, which is comparable to other Ni–Mn–In(Sn)-based alloys. Although the peak magnetic entropy of the foam is only 7–8 J kg^{-1} K^{-1}, which is lower than the 20–30 J kg^{-1} K^{-1} of the bulk alloy, the operating temperature range of the foam is as high as 46 K, which makes the RC value of the foam and bulk alloys almost equivalent. In addition, the phase transition hysteresis of the foam is reduced, which has an RC$_{eff}$ value greater than that of the bulk alloy, and its phase transition point is near room temperature, which meets the requirements for room-temperature magnetic refrigeration. On the whole, porous materials have their unique advantages in the field of magnetic refrigeration and may become one of the development trends of magnetocaloric materials in the future.

3.5 Conclusions

Magnetic shape memory alloy foams can be prepared by template replication and powder metallurgy. The former may produce foams with designed pore size, porosity, shape and distribution. Single- and dual-pore foams are prepared with tailored pore architecture and composition in MSMAs.

Bulk Ni–Mn–Ga foam shows superior MFIS due to the significantly reduced constraint of grain boundary on twin boundary. The strain as high as 8.7% is achieved in dual-pore Ni–Mn–Ga foam during cycling across MT temperature under a magnetic field of 0.97 T.

Foams exhibit reduced hysteresis during martensitic transformation and better cyclic life. The reduced hysteresis is related to the lower constraint of phase transformation boundary in thin nodes and struts. This is favorable for the reversibility of magnetocaloric effect and refrigeration capacity. The foam exhibits lower magnetic entropy change compared to its bulk alloy counterpart; however, it showed a high RC value due to the broadened working temperature interval.

References

1. M. Chmielus, C. Witherspoon, R.C. Wimpory et al., Magnetic-field-induced recovery strain in polycrystalline Ni-Mn-Ga foam. J. Appl. Phys. **108**(12), 123526 (2010)
2. M. Chmielus, X.X. Zhang, C. Witherspoon et al., Giant magnetic-field-induced strains in polycrystalline Ni-Mn-Ga foams. Nat. Mater. **8**(11), 863–866 (2009)
3. X.X. Zhang, C. Witherspoon, P. Müllner et al., Effect of pore architecture on magnetic-field-induced strain in polycrystalline Ni-Mn-Ga. Acta Mater. **59**(5), 2229–2239 (2011)
4. H.T. Liu, *Superelasticity and Shape Memory Effect in Ni-Mn-Ga Foams* (Harbin Institute of Technology, Harbin, 2013)
5. Y. Boonyongmaneerat, M. Chmielus, D.C. Dunand et al., Increasing magnetoplasticity in polycrystalline Ni-Mn-Ga by reducing internal constraints through porosity. Phys. Rev. Lett. **99**(24), 247201 (2007)
6. K.A. Gschneidner, V.K. Pecharsky, A.O. Tsokol, Recent developments in magnetocaloric materials. Rep. Prog. Phys. **68**(6), 1479–1539 (2005)
7. J. Gu, *Preparation and Magnetocaloric Effect of Ni-Mn-In-Co Foams* (Harbin Institute of Technology, Harbin, 2017)
8. P. Lazpita, G. Rojo, J. Gutierrez et al., Correlation between magnetization and deformation in a NiMnGa shape memory alloy polycrystalline ribbon. Sens. Lett. **5**(1), 65–68 (2007)
9. S.H. Guo, Y.H. Zhang, B.Y. Quan et al., The effect of doped elements on the martensitic transformation in Ni-Mn-Ga magnetic shape memory alloy. Smart Mater. Struct. **14**(5), S236–S238 (2005)
10. Y.J. Wu, *Shape Memory Alloy and Its Application* (China University of Science and Technology Press, Hefei, 1993), pp. 66–70, 82–91
11. P.J. Brown, J. Crangle, T. Kanomata et al., The crystal structure and phase transitions of the magnetic shape memory compound Ni_2MnGa. J. Phys. Condens. Matter **14**(43), 10159–10171 (2002)
12. C.P. Sasso, P.Q. Zheng, V. Basso et al., Enhanced field induced martensitic phase transition and magnetocaloric effect in $Ni_{55}Mn_{20}Ga_{25}$ metallic foams. Intermetallics **19**(7), 952–956 (2011)
13. P.J. Shamberger, F.S. Ohuchi, Hysteresis of the martensitic phase transition in magnetocaloric-effect Ni-Mn-Sn alloys. Phys. Rev. B **79**(14) (2009)
14. R.F. Hamilton, H. Sehitoglu, Y. Chumlyakov et al., Stress dependence of the hysteresis in single crystal NiTi alloys. Acta Mater. **52**(11), 3383–3402 (2004)
15. Y. Chen, C.A. Schuh, Size effects in shape memory alloy microwires. Acta Mater. **59**(2), 537–553 (2011)
16. L. Huang, D.Y. Cong, L. Ma et al., Large reversible magnetocaloric effect in a Ni-Co-Mn-In magnetic shape memory alloy. Appl. Phys. Lett. **108**(3), 32405 (2016)
17. Y. Zhang, L.L. Zhang, Q. Zheng et al., Enhanced magnetic refrigeration properties in Mn-rich Ni-Mn-Sn ribbons by optimal annealing. Sci. Rep. **5**, 11010 (2015)
18. R.C. Zhang, M.F. Qian, X.X. Zhang et al., Magnetocaloric effect with low magnetic hysteresis loss in ferromagnetic Ni-Mn-Sb-Si alloys. J. Magn. Magn. Mater. **428**, 464–468 (2017)
19. S.K. Sarkar, P.D. Sarita Babu et al., Giant magnetocaloric effect from reverse martensitic transformation in Ni-Mn-Ga-Cu ferromagnetic shape memory alloys. J. Alloys Compd. **670**, 281–288 (2016)

20. D. Bourgault, J. Tillier, P. Courtois et al., Large inverse magnetocaloric effect in $Ni_{45}Co_5Mn_{37.5}In_{12.5}$ single crystal above 300 K. Appl. Phys. Lett. **96**(13), 13250113 (2010)
21. P. Czaja, R. Chulist, M.J. Szczerba et al., Magnetostructural transition and magnetocaloric effect in highly textured Ni-Mn-Sn alloy. J. Appl. Phys. **119**(16), 165102 (2016)
22. T. Paramanik, I. Das, Near room temperature giant magnetocaloric effect and giant negative magnetoresistance in Co, Ga substituted Ni-Mn-In Heusler alloy. J. Alloy. Compd. **654**, 399–403 (2016)
23. Y. Kim, W.B. Han, H.S. Kim et al., Phase transitions and magnetocaloric effect of $Ni_{1.7}Co_{0.3}Mn_{1+x}Al_{1-x}$ Heusler alloys. J. Alloys Compd. **557**, 265–269 (2013)
24. J. Du, Q. Zheng, W.J. Ren et al., Magnetocaloric effect and magnetic-field-induced shape recovery effect at room temperature in ferromagnetic Heusler alloy Ni-Mn-Sb. J. Phys. D Appl. Phys. **40**(18), 5523–5526 (2007)
25. S. Stadler, M. Khan, J. Mitchell et al., Magnetocaloric properties of $Ni_2Mn_{1-x}Cu_xGa$. Appl. Phys. Lett. **88**(19) (2006)

Chapter 4
Preparation and Heat Treatment of Magnetic Shape Memory Alloy Microwires

4.1 Introduction

Spinning method is a process in which the tip of a metal roller wheel (copper wheel) with high rotating speed contacts a thin-layer molten metal, and then the molten metal is rapidly cooled by the heat exchange between the molten liquid, copper wheel and inert gas environment and finally turns into a one-dimensional microwire. Reasonably adjusting and optimizing the rotating speed of copper wheel, the feed rate of molten metal, heating power, temperature and the viscosity of molten alloy are necessary to obtain high quality microwires. In addition, adopting reasonable heat treatments is the key to obtain excellent properties of the alloy. Therefore, it is of great significance to study the effects of different heat treatment processes on microstructure, martensitic transformation and magnetic properties of one-dimensional microwire. This chapter takes ternary Ni–Mn–Ga alloy as an example to introduce its preparation and heat treatment techniques [1–4].

4.2 Wire Preparation Technology

4.2.1 Optimization of Processing Parameters for Melt Spinning Process

4.2.1.1 Design of Alloy Composition

The phase transition temperature of Ni–Mn–Ga–X alloy strongly depends on the alloy composition [5] and affects the structure and properties of the martensite. Therefore, the design and regulation of the alloy composition are extremely important. Taking Ni–Mn–Ga as an example, Mn elements in the alloy are easy to volatilize at high temperature, which affects the properties of the alloy. During the preparation

X. Zhang and M. Qian, *Magnetic Shape Memory Alloys*,
https://doi.org/10.1007/978-981-16-6336-9_4

of Ni–Mn–Ga microwire, the loss of elements occurs in the arc melting, spinning, as well as high temperature heat treatment process. Therefore, the composition should be optimized and controlled in microwire preparation process. In this section, the nominal components of the microwires are designed as Ni: 50.6 at.%, Mn: 28.0 at.% and Ga: 21.4 at.% [4]. In the design process, 2 wt.% of Mn elements preparation NMG1 and NMG2 were added to compare. The addition of excess Mn element was to compensate the loss of Mn element in the subsequent preparation and heat treatment process. The composition of different microwires is shown in Table 4.1. Through the composition analysis on the surface and internal components of NMG2 microwires, it is found that the elements in the prepared microwires are equal to the nominal components, and the Mn elements are reduced by 0.7 at.% in comparison with the design composition, while the nominal components still have a margin of 0.8 at.%, which can guarantee the content of Mn elements in the subsequent heat treatment process.

Figure 4.1a shows the XRD results of the two microwires at room temperature. Both microwires contain austenite and Martensite phases at room temperature. However, the peak of martensite is very weak in NMG1 microwires due to the content of martensite components is low, as indicated by the black arrow in the figure. As the content of Mn increases, the phase transition temperature of NMG2 microwire

Table 4.1 The composition design for microwire

Alloy elements	Chemical composition, at.%			
	Nominal composition	Designed composition	Avergage composition for NMG1	Avergage composition for NMG2
Ni	50.6	49.5	49.9 ± 0.3	50.6 ± 0.4
Mn	28.0	29.5	28.5 ± 0.5	28.8 ± 0.3
Ga	21.4	20.9	21.6 ± 0.5	20.6 ± 0.4

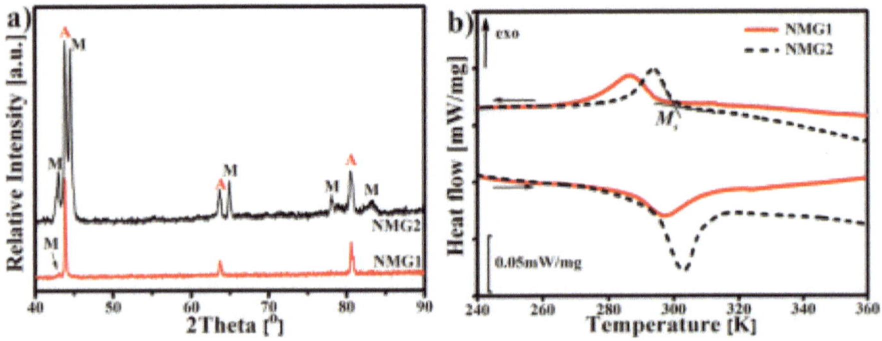

Fig. 4.1 **a** XRD patterns of microwires at room temperature; **b** the comparison of DSC phase transition temperature for NMG1 and NMG2

increases, so the proportion of martensite peak at room temperature increases greatly. Figure 4.1b shows the DSC thermal analysis results of the two microwires. The results show that the phase transition temperature of NMG2 is indeed slightly higher than that of NMG1, but both are around room temperature. The phase transition temperature nearing room temperature has some advantages, for example, the SE needs to be obtained in austenite state, and the resistance of stress-induced martensite increases with the increase of temperature; SME acquisition needs to be carried out in martensite state. In addition, room-temperature magnetic refrigeration has greater practical value.

As mentioned earlier, the difference of the transformation temperature will lead to the change of martensite type, so the low temperature XRD analysis was performed on the two microwires, and the results are shown in Fig. 4.2. It is found that the martensite of these two microwires is both 7 M martensite with monoclinic non-uniform structure at 263 K [6]. The lattice parameters of the two microwires are obtained by full spectrum fitting, as shown in Fig. 4.2. Due to the difference in composition, the lattice parameters of the two microwires are slightly different.

To sum up, the composition of the NMG1 and NMG2 microwires with 2 wt.% excessive Mn addition is less different from nominal composition, and the martensitic transformation and phase composition of the two microwires are consistent. It is considered that the deviation of composition is within the allowable range of the test, which can guide the subsequent test. Hereinafter, all the alloy microwires were

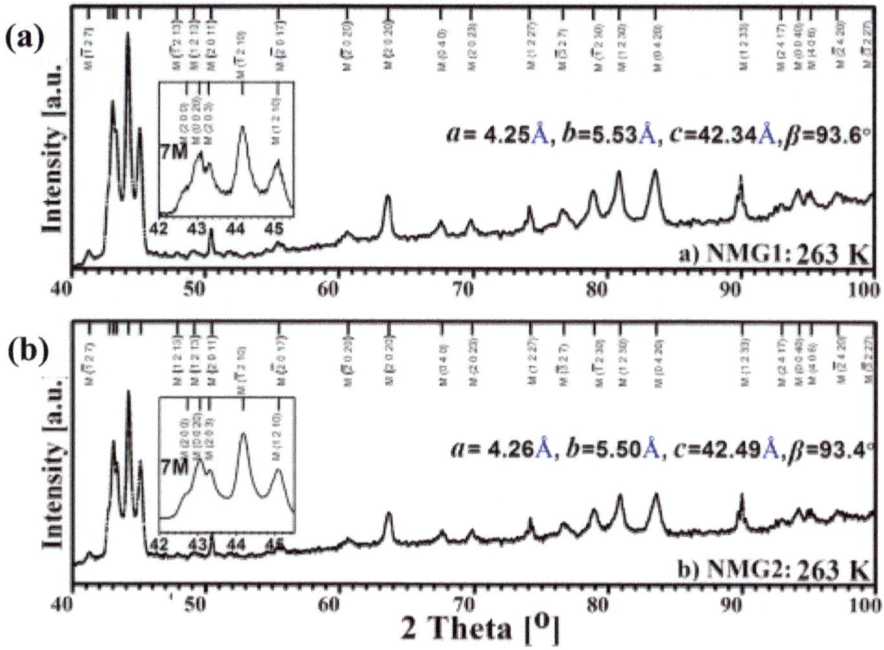

Fig. 4.2 Low temperature (263 K) XRD phase analysis: **a** NMG1; **b** NMG2

prepared by adding 2 wt.% Mn elements to the nominal design composition. Such doping rules are well verified in the preparation of Ni–Mn–Ga–(Cu, Fe) alloys.

4.2.1.2 Optimization of Preparation Process Parameters

In the spinning process, Rayleigh waves tend to form resulting in the inhomogeneity of microwire diameter [7]. In addition, the roundness of the microwire is also an important parameter affecting its properties. In this section, taking the preparation process of Ni–Mn–Ga as an example, the influence of the preparation parameters on the diameter uniformity and roundness of microwire is analyzed [3].

Schematic diagram of microwire preparation by spinning method is shown in Fig. 4.3, and the main process is as follows: (1) The alloy ingot is firstly cut into a cylinder with a height of 8–10 mm by electric firewire, then cleaned in acetone solution, ultrasonically, to remove the surface oil stain, and finally put into the drying box at 393 K for 2 h; (2) The clean and dried ingot was put into the boron nitride crucible. The cavity of crucible was firstly vacuumed to $0.5–5 \times 10^{-3}$ Pa and then filled with high purity argon. Repeating the above process 3 ~ 4 times and finally making the cavity be filled with 35–65 MPa high purity argon; (3) Start the metal roller and set the copper wheel rotation speed (V_w). In the groping process, the rotation speed interval is 13—30 m/s; (4) Turn on the induction heating device. Adjust the heating power until the upper end of the alloy solution forms a spherical molten pool; (5) control the feeding rate (V_m) of crucible and make the crucible move towards the roller; (6) As the copper wheel contacts the alloy bath, the microwire is prepared.

The rotation speed of copper wheel (V_w) is an important parameter that affects the formation of Rayleigh wave and the roundness of microwire [8, 9]. Research

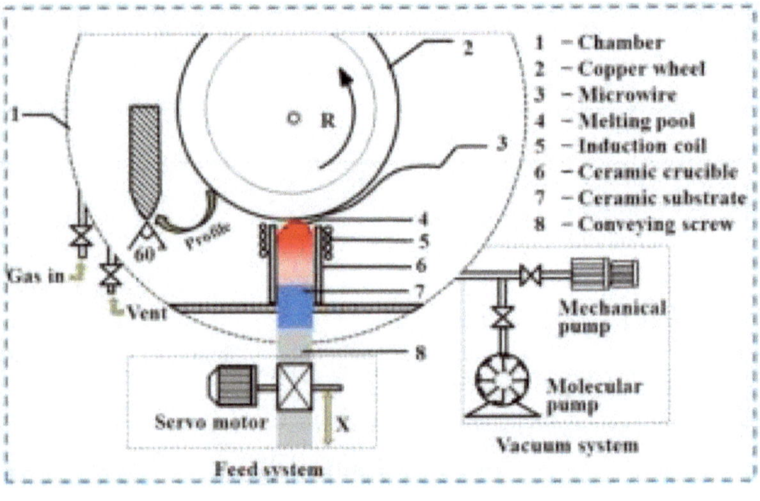

Fig. 4.3 Schematic diagram of microwire preparation by spinning method

shows that when V_w is very low, the alloy liquid has enough time to transfer heat to the copper wheel tip where the alloy nucleates and solidifies, thus inheriting the shape of the copper wheel tip and forming a plane on one side of the microwire. On the other hand, due to the high surface energy of the molten liquid, Ni–Mn–Ga alloy tends to shrink the surface area and spherify during solidification, eventually forming Rayleigh wave and circular free solidification surface. When V_w is higher, the liquid alloy sends out by rotating copper wheel and does not have enough time to nucleate on the copper wheel tip but homogeneously nucleates in the atmosphere, so that it forms a continuous microwire with circular cross-section under the action of the surface energy. In the preparation of the Co-based amorphous alloy wire process, Wang [8] obtained the microwire with a homogeneous circular cross-section under the condition of high V_w.

In addition, the viscosity of molten alloy is another important parameter, which is mainly determined by heating power and the temperature of alloy. When the viscosity is low, only a few alloy melts can be carried by the copper wheel, so the small-diameter microwires will be formed. The viscosity of molten alloy also affects its surface energy and its wettability with copper wheel. Theoretically, continuous Ni–Mn–Ga microwires with uniform diameter and circular cross-section can be prepared by reasonably adjusting and optimizing the preparation process parameters. However, liquid alloys tend to form intermittent spherical products rather than continuous microwires at high temperature and high *Vw* conditions, because they usually have high surface energy and low viscosity [10], which is different from the results reported by Wang et al. [8]. Therefore, it is considered that the homogeneous and continuous microwires may also depend on the properties of the alloy itself.

In this section, the characteristics of the products prepared by different *Vw*, *Vm* and heating powers are compared and studied, and the results are shown in Fig. 4.4. Under the condition of high heating power ~23 kW and high feeding speed *Vw* ~30 m/s, the alloy melt solidifies into microspheres with diameters ranging from 50 to 300 μm as increasing Vm from 40 to 120. As the heating power is reduced to ~14 kW, however, a broadband product is formed due to the increase in the viscosity of the molten alloy, as shown in (b). Thus, the heating power is adjusted to medium power ~20 kW while the Vw is reduced to ~13 m/s. In this condition, the microwire has obvious rayleigh wave, which may be related to the vibration of the copper wheel at low speed rotation, as shown in (c). In this study, by comparing and analyzing the relationship between the various preparation parameters, finally, the heating power is set as ~20 kW, Vw is ~24 m/s, and the Vm is in 60–90 μm/s, the macroscopic morphologies of the microwires prepared this parameter combinations are shown in Fig. 4.5. It can be seen that the continuity of microwire is good, the Rayleigh wave phenomenon is weakened and the diameter is uniform, as shown in Fig. 4.4d. Comparing the insets in (c) and (d) reveals that the uniformity of microwire diameter has been greatly improved in the range of 300 μm.

However, given the alloy cannot form obtain continuous microwires with circular cross-section at high *Vw*, the cross-section of the microwires in this study is all "*D*" shape, as shown in the inset of Fig. 4.5. Also, the roundness of the microwires in different preparation stages is different as well, which will be described in detail

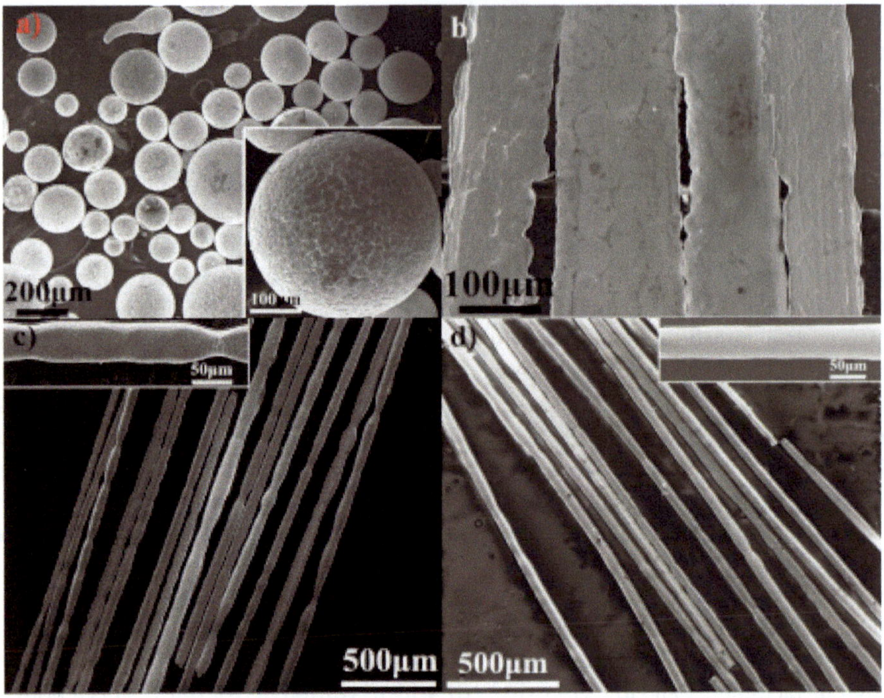

Fig. 4.4 Influence of processing parameters on the morphology of prepared products **a** ~23 kW, ~30 m/s; **b** ~14 kW, ~30 m/s; **c** ~20 kW, ~13 m/s; **d** ~20 kW, ~24 m/s

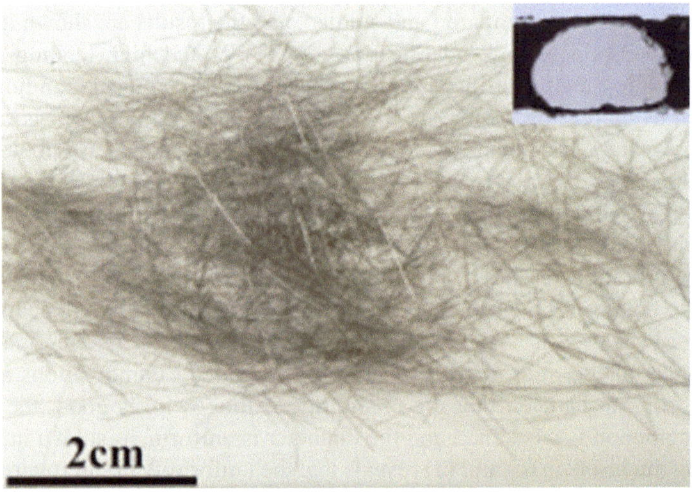

Fig. 4.5 Macro morphology of Ni–Mn–Ga microwire prepared by spinning method after parameter optimization

later. To sum up, the process parameters for preparing a continuous and uniform Ni–Mn–Ga microwire with, "*D*" shape cross-section are the heating power of ~20 kW, Vw of ~24, and Vm of 60–90 μm/s.

4.2.2 *Taylor Drawing Method*

The manufacturing techniques of micro-nano microwire include mechanical drawing, template synthesis, phase separation spinning, electrostatic spinning, etc. Taylor method, also known as Taylor-Ulitovsky method, is a process in which equal metals are put into a glass tube, heated, melted, and as coated glass is softened, and then mechanically drawn to form glass-coated metal microwires. The method can be used to prepare continuous microwires up to 10,000 m. The melting point of the metal needs to match the softening temperature of the coated glass. According to the melting point of ferromagnetic shape memory alloy, borosilicate glass is generally used. Due to the high solidification rate of internal micron filaments, this method can be used to prepare alloys with uniform composition and fine structure, or the alloys need to avoid the second phase precipitation. In some cases, amorphous microwires can be obtained.

4.3 Microstructure Evolution of Ni–Mn–Ga–X Alloy Microwires During Melt-extraction

4.3.1 *Microstructure and Phase Composition of Prepared Microwires*

After composition design and preparation process parameters optimization, a variety of microwires can be prepared according to different purposes, as shown in Table 4.2. This section analyzes the microstructure and phase structure of the prepared microwire [4].

4.3.1.1 Ni–Mn–Ga Microwire

Figure 4.6 shows the TEM morphology of as-prepared Ni–Mn–Ga microwire and corresponding electron diffraction patterns [4]. It can be seen from Fig. 4.6a that the content of martensite in NMG1 microwire is very low at room temperature. Therefore, no martensite is found in the micro area of TEM. It can be seen that the as-prepared NMG1 microwire presents obvious characteristics of the lining strain at room temperature, which is consistent with the microstructure of L21 austenite obtained by Bennett et al. [11]. The superlattice dots pointed by white arrow in the

Table 4.2 Phase and composition of as-prepared Ni–Mn–Ga–X microwires

No	Phase composition at room temperature	Phase composition at low temperature	Chemical composition (at.%)
NMG1	A + less 7 M	7 M	$Ni_{49.9\pm0.3}Mn_{28.5\pm0.5}Ga_{21.6\pm0.5}$
NMG2	A + 7 M	7 M	$Ni_{50.6\pm0.4}Mn_{28.8\pm0.3}Ga_{20.6\pm0.4}$
NMG3	A	A	$Ni_{50.3\pm0.2}Mn_{26.9\pm0.3}Ga_{22.8\pm0.3}$
NMG4	7 M	7 M	$Ni_{50.5\pm0.3}Mn_{29.1\pm0.4}Ga_{20.4\pm0.4}$
NMGC1	7 M	7 M	$Ni_{49.4\pm0.3}Mn_{26.2\pm0.2}Ga_{20.7\pm0.3}Cu_{3.7\pm0.1}$
NMGF1	A	–	$Ni_{50.3}Mn_{25.1}Ga_{23.2}Fe_{1.4}$
NMGF2	A	–	$Ni_{50.3}Mn_{25.3}Ga_{22.0}Fe_{2.4}$
NMGF3	A	–	$Ni_{50.5}Mn_{24.8}Ga_{21.5}Fe_{3.2}$
NMGF4	A	–	$Ni_{50.2}Mn_{25.1}Ga_{20.4}Fe_{4.3}$
NMGF5	A	7 M	$Ni_{50.0}Mn_{25.0}Ga_{19.6}Fe_{5.5}$
NMGF6	7 M	7 M	$Ni_{50.0}Mn_{25.2}Ga_{18.4}Fe_{6.5}$

inset of Fig. 4.6a also verifies its ordered structure. In order to study the characteristics of the martensite in NMG1 microwire, TEM analysis was carried out after low temperature treatment, as shown in Fig. 4.6b, the martensite in NMG1 microwire presents a thin but not clear twin laminar structure, containing multiple twin crystals. The SAED spots show that there are six small spots between the two main spots (7 M martensite), which is consistent with the XRD results.

Figure 4.6c shows the TEM morphology of NMG2 microwire at room temperature. The microstructure of the microwire is composed of austenite and martensite, where the martensite presents twin lamellae structure and the austenite contains a large number of dislocations. The formation of the defects is related to the rapid solidification process. Moreover, the phase interface between martensite and austenite is clear and obvious. The corresponding SAED confirmed that it is 7 M Martensite, which is consistent with the XRD results. Figure 4.6d shows the TEM morphology of a grain close to the outer surface of NMG3 microwire, showing that the microwire has austenite structure at room temperature. It has a tweety morphology similar to that shown in Fig. 4.6a and possesses superlattice spots. In addition, subgrain boundaries can be observed, which is due to the existence of high defect density induced by rapid solidification in prepared microwires.

By XRD analysis and subsequent full-spectrum fitting, crystal structure information of the prepared microwires (NMG1, NMG2 and NMG3) at room temperature and low temperature is obtained, as shown in Table 4.3. At low temperature, both NMG1 and NMG2 is composed of 7 M martensite with monoclinic structure and the β of NMG1 and NMG2 is 93.6° and 93.4°, respectively, which are consistent with the crystal structure of 7 M non-public structure martensite, as reported by Righi et al. [6]. According to the lattice parameters, the average volume (V_a) of each constituent atom can be calculated, and the results are shown in Table 4.3. For a given

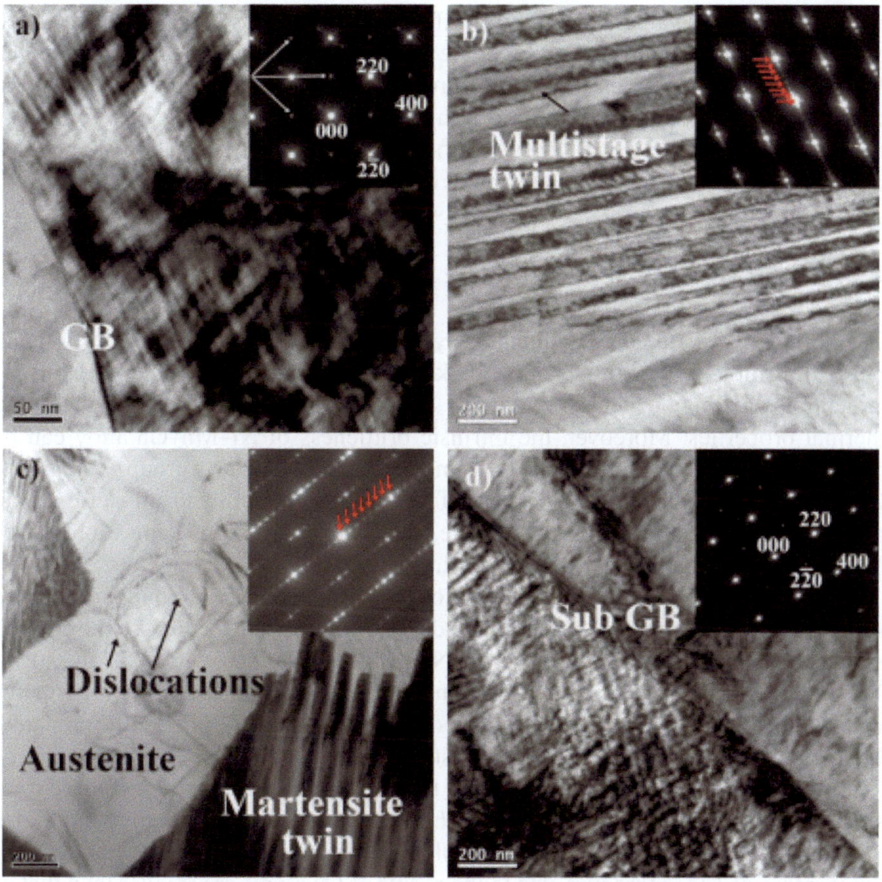

Fig. 4.6 TEM morphologies and SAED spots of as-prepared microwires. **a** NMG1 (room temperature); **b** NMG1 (low temperature); **c** NMG2; **d** NMG3

Table 4.3 Crystal structure and phase composition of as-prepared microwires

Microwires	Crystal structure	Lattice constant						
		a(Å)	b(Å)	c(Å)	α(°)	β(°)	γ(°)	V_a(Å^3)
NMG1	Room temperature A	5.83	–	–	90	90	90	12.38
	Low temperature 7 M	4.25	5.53	42.34	90	93.6	90	12.41
NMG2	High temperature A	5.84	–	–	90	90	90	12.45
	Low temperature 7 M	4.26	5.50	42.49	90	93.4	90	12.42
NMG3	Room temperature A	5.83	–	–	90	90	90	12.38
	Low temperature A	5.83	–	–	90	90	90	12.38

microwire, the V_a values are slightly different in different states, which is related to the shear distortion of the lattice in the phase transition process. Meanwhile, the V_a of different microwires shows a different variation trend in the phase transition process, which is due to the difference in the internal stress and defects formed in the rapid solidification for different microwire processes. The variation should also be related to the difference in the content of Ni, Mn and Ga atoms.

4.3.1.2 Ni–Mn–Ga–Cu Microwires

The doping of Cu element simultaneously adjust the magnetic phase transition and phase transition of Ni–Mn–Ga alloy, making the alloy reach the state of magnetic-structural coupling, which is conducive to the acquisition of high magnetic and thermal properties. Moreover, the intrinsic brittleness of Ni–Mn–Ga alloy can be reduced by adding the fourth group of elements. Among the fourth group of elements, Cu has attracted much attention because of its low price, its ability to efficiently adjust phase transition temperature and promote MCE. In addition, replacing Ga by Cu in Ni–Mn–Ga alloy can also reduce the cost.

Considering the loss of Mn in the ingot preparation, microwire preparation and subsequent heat treatment, 2wt. % Mn is added to compensate the loss, so the composition of the microwire is designed as Ni50Mn25Ga25-Xcux (x = 3.8at. %) [4, 12]. Using the optimized processing parameters of spinning process, the microwires with diameter of ~20–80 m and length of ~30–150 mm are efficiently prepared. Meanwhile, all of them show metallic luster.

In ferromagnetic shape memory alloy, the temperature of martensitic phase transition is increased with the increase of *E*/*A* and the decrease of lattice monocyte volume [13]. Therefore, Cu($3d^{10}4s^1$) replacing Ga($4S^24P^1$) will inevitably lead to the increase of *E*/*A*. Moreover, due to the radius of Cu atoms are smaller than that of Ga atoms, the cell will shrinkage so that phase transition temperature will increase. The reason why Cu atoms replacing Ga atoms makes T_c decrease is that the presence of Cu atoms affects the first neighbour atomic spacing of Mn–Mn. The first neighbour atomic spacing of Mn–Mn has the best magnetic energy in the standard atomic measurement ratio of Ni_2MnGa. Increasing or decreasing the atomic spacing of Mn–Mn will reduce the alloy's magnetic energy [14], thus reducing the value of T_c. Therefore, it is possible to obtain the microwire with the magnetic structure coupling state near room temperature via Cu atoms replacing Ga atoms. In addition, it has been reported that Cu can replace Ga to improve plasticity [15, 16]. Similarly, in order to further improve the magnetic properties of the microwire, chemically ordered heat treatment was carried out. The SEM surface morphology of NMGC1 microwire before and after the ordering is shown in Fig. 4.7. The inset shows that the microwire diameter is uniform similar to the as-prepared Ni–Mn–Ga microwire, Fig. 4.7a shows that the surface of NMGC1 microwire presents a cellular crystal morphology, and there are obvious traces of plastic deformation on the surface (black arrow), indicating that the doping of Cu improves the plasticity of the microwire. The size of cellular gains on the surface of as-prepared microwire is approximately

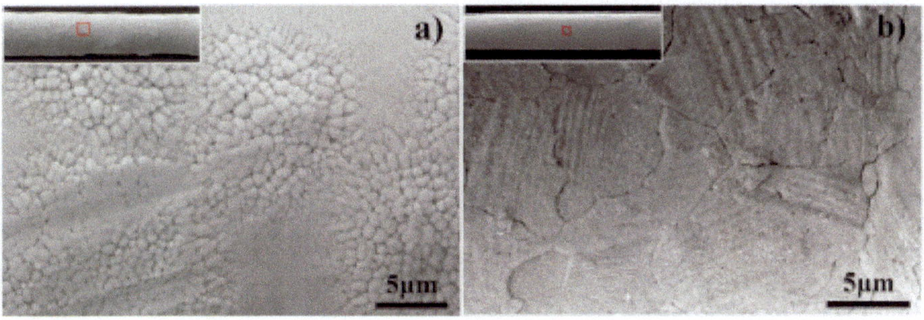

Fig. 4.7 SEM morphology of NMGC1 microwire before and after ordered heat treatment. **a** NMGC1; **b** NMGC1C

0.5–3 m. However, the actual size cannot be distinguished. The surface morphology of the ordered microwire is shown in Fig. 4.7b. Due to the element loss at grain boundary, the grain boundary is visible and the grain size ranges from 1 to 15 m. At room temperature, the microstructure of the microwire is martensite with twin laminates spanning grains.

The microwires before and after the NMGC alloy ingots and ordered heat treatment were analyzed by XRD at room temperature, to study the structural characteristics of the microwires. The results are shown in Fig. 4.8. The grey spectral line at the lower part of figure is the standard spectral line of 7 M martensite with a non-public structure in Ni–Mn–Ga alloy [6]. After calibration, it is found that, except for a slight shift in peak position, NMGC alloy ingot and microwires before and after ordered heat treatment are composed of 7 M martensite with a non-public structure at room temperature.

Fig. 4.8 XRD patterns of NMGC ingot, as-prepared and ordered microwires at room temperature

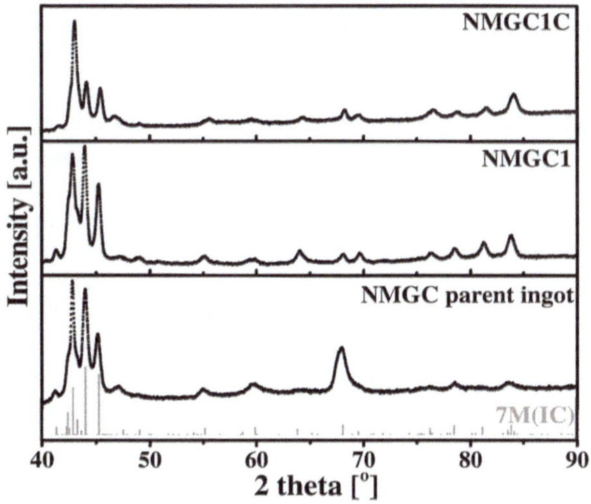

Table 4.4 Microwire structure information of NMGC ingot, and microwires before and after ordering treatment

Materials	Phase	Lattice constant						
		a(Å)	b(Å)	c(Å)	α(°)	β(°)	γ(°)	V_a(Å^3)
Parent ingot	Room temperature 7 M	4.28	5.49	42.23	90	92.5	90	12.39
NMGC1	Room temperature 7 M	4.25	5.52	42.22	90	93.3	90	12.36
NMGC1C	Room temperature 7 M	4.25	5.52	42.10	90	93.5	90	12.32

After the XRD spectrum is fitted with the full spectrum, the lattice parameter information is obtained, as shown in Table 4.4. The lattice constant of ingot is $a = 4.28$ Å, $b = 5.49$ Å, $c = 42.23$ Å, $\beta = 92.5°$, so $V_a = 12.39$ Å^3; The lattice constant of NMGC1 microwire is $a = 4.25$ Å, $b = 5.52$ Å, $c = 42.22$ Å, $\beta = 93.3°$, so $V_a = 12.36$ Å^3; The lattice constant of NMGC1C microwire is $a = 4.25$ Å, $b = 5.52$ Å, $c = 42.10$ Å, $\beta = 93.5°$, so $V_a = 12.32$ Å^3; compared with the 7 M of Ni–Mn–Ga microwire of prepared state and the ordered structure, the doping of Cu has very little influence on the structural parameters of the microwire.

4.3.1.3 Ni–Mn–Ga–Fe Microwires

The doping of Fe element will affect the crystal structure, phase transition temperature, Curie temperature and intrinsic brittleness of Ni–Mn–Ga alloy. The microstructure of $Ni_{50}Mn_{25}Ga_{25\text{-}x}Fe_x$ ($x = 1, 2, 3, 4, 5, 6$) was studied by doping different contents of Fe [17]. The above alloys are respectively labeled as NMGF1-6, and the true composition is shown in Table 4.2. Figure 4.9 shows the transmission electron micrograph of the austenite structure of the prepared NMGF4 microwire and the corresponding selection electron diffraction pattern. (a) displays the austenitic "crested" pattern of NMGF4 microwire, which can be interpreted as softening and phonic branching in the martensitic pre-phase transition lattice in the microwire, and this structure also exists in Ni–Fe–Ga, Ni–Mn–Ga and Ni–Mn–Fe–Ga alloys [18–20]. The existence of austenite at room temperature indicates that the MT initial temperature (M_s) of the microwire is lower than room temperature. The electron diffraction results of area A shows that the austenite possesses cubic symmetry. The spot shows that the structure of the microwire is an orderly body-centered cubic L21 structure, as shown in Fig. 4.9a, the ordering degree of the as-prepared microwire is analyzed by HRTEM techniques. Figure 4.9b is a high-resolution image of the microwire, and corresponding FFT (Fig. 4.9c) shows the long-range order arrangement of atoms. In addition, certain dislocation lines were observed in some areas of the as-prepared NMGF4 microwire, as shown in Fig. 4.9d. Figure 4.9e is SADE patterns, showing that the spots of the body-centered cubic structure were elongated along the perpendicular direction to form satellite-like spots [18]. HRTEM observation in Fig. 4.9d showed that the microstructure of the microwire was composed of fine long micro-twins at the nanometer scale. The microstructure of micro-twins

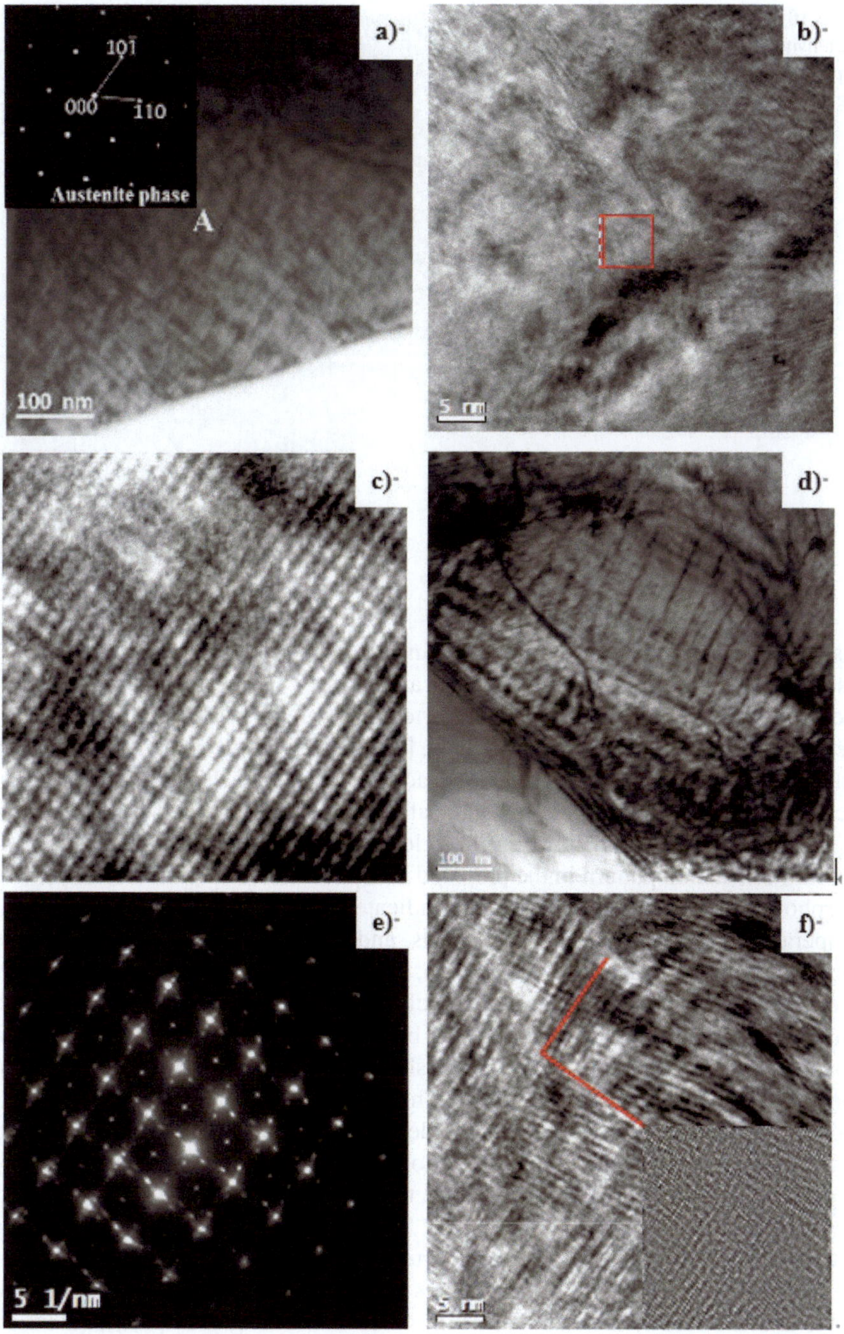

Fig. 4.9 TEM image of austenite structure in as-prepared NMGF4 microwire. **a** Austenitic phase and diffraction; **b** HRTEM images; **c** IFFT image; **d** Twins; **e** Twin partial diffraction; **f** HRTEM; **g** IFFT images

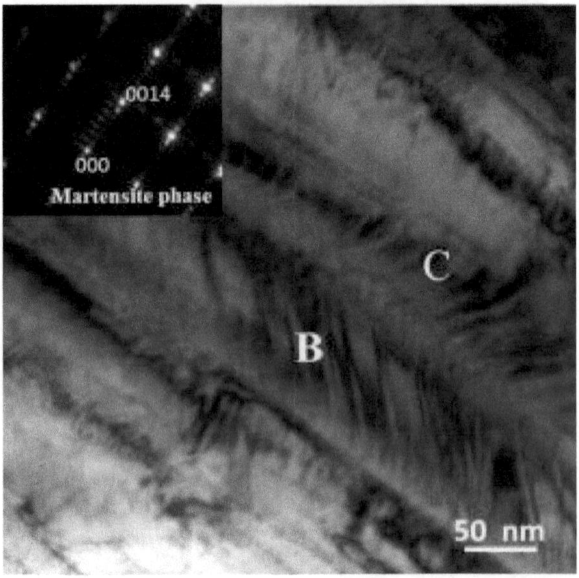

Fig. 4.10 TEM image of martensite structure of as-prepared NMGF5 microwire

has a meshed structure and the nano-twins interpenetrate to a certain extent. The inset shows the ordered arrangement of atoms in two directions perpendicular to each other by means of inverted and inverted leaf transformations. This is similar to the phenomenon observed by Sarkar et al. [21] in TiNi alloy, which is called chaotic glassy state, namely glassy martensite. In addition, there are also superlattice spots in the diffraction pattern, indicating that the alloy has a highly ordered crystal structure.

Figure 4.10 shows the TEM bright-field image and the corresponding selection electron diffraction pattern in the prepared NMGF5 microwire martensite state. The morphologies of the bright field images indicate that the martensite variants at room temperature are composed of small sheets, and two variants in different directions constitute a good adaptive configuration, with a typical "human" shape, and the interface between the variants is not very clear. The SADE patterns in the inset show that there are also weak diffraction spots between every two main diffraction spots along $<001>_M$ direction. Measurements show that every six weak diffraction spots constitute a group, as marked in figure. The distance between the strong diffraction spots in the direction of $(220)_{L21}$ is divided into 7 equal parts on average, and the strong diffraction spots meet the diffraction conditions of the cubic crystal system. According to the Ni–Mn–Ga alloy phase transformation research of classification, the martensite belongs to 7 M martensite. Therefore, the composition of microwire has seven layers of modulation cubic martensite structure [19, 20].

The as-prepared NMGF1-6 microwires were analyzed by XRD at room temperature, and the results are shown in Fig. 4.11. Calibrating the spectra shows that only austenite diffraction peaks existed in NMGF1-4 at room temperature, so it is a highly ordered cubic L21 structure, which also indicates that the initial phase transition temperature of these alloys is lower than room temperature. With the increase

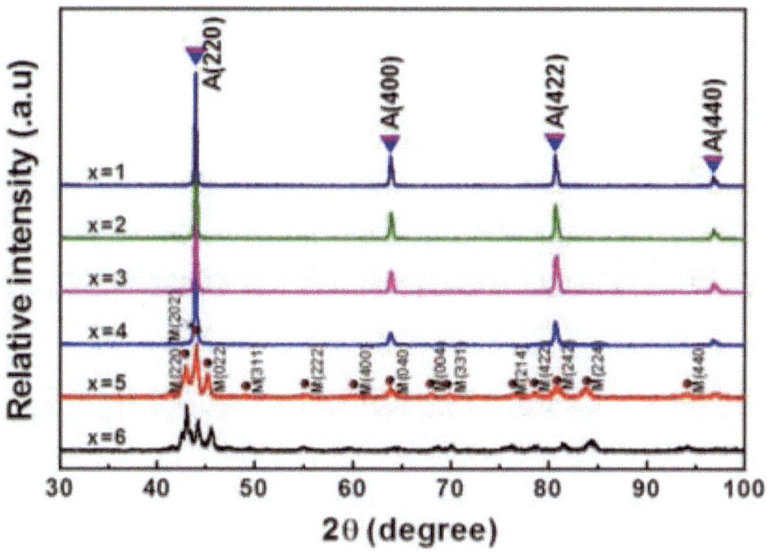

Fig. 4.11 XRD pattern of as-prepared $Ni_{50}Mn_{25}Ga_{25-x}Fe_x$ microwire

of Fe content, the parent phase diffraction peak decreases and the martensite diffraction peak increases. Meanwhile, some Bragg peaks split due to the low symmetry of martensite, the main peak (220) split into (220), (202) and (022), and the parent phase (422) split into (214), (422), (242) and (224). These two kinds of microwires have basically the same crystal structure, which conforms to the characteristic of 7 M martensite diffraction peak [22, 23]. The phase analysis results are consistent with TEM analysis. The lattice constants and the volume of the cells are listed in Table 4.5.

Figure 4.12a and b show the change of microwire lattice constant and monocyte volume with Fe content. The lattice constants of the austenitic phase basically remain unchanged. As for the martensite phase, as $x > 5$, the lattice constant a increases with

Table 4.5 Lattice constants and cel volume of $Ni_{50}Mn_{25}Ga_{25-x}Fe_x$ ($x = 1$–6) microwires

Fe (at.%)	phase	Lattice constant				Cell volume (Å^3)
		a (Å)	*b* (Å)	*c* (Å)	c/a	
1	A	5.829	–	–	–	198.0
2	A	5.826	–	–	–	197.7
3	A	5.826	–	–	–	197.7
4	A	5.828	–	–	–	198.0
5	A	5.824	–	–	—	197.6
	M	6.123	5.804	5.546	0.9057	197.1
6	M	6.159	5.755	5.507	0.8941	195.2

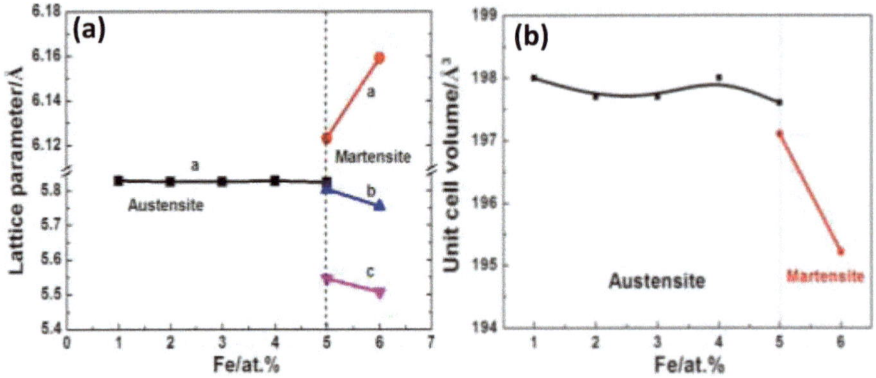

Fig. 4.12 The change of the crystal lattice constant and monocyte volume of as-prepared microwire with Fe content at room temperature. **a** lattice parameters; **b** cell volume

the increase of Fe content, while both *b* and *c* decrease. Heusler alloy L21 structure can be regarded as composed of four face-centered cubic (FCC) lattices that pass through each other [24]. At $x < 4$, Ni atoms occupy the A and C bits, Ga atoms occupy the D bits, Mn and Fe atoms will occupy the B bits randomly. Because the atomic scattering amplitude of Mn and Fe atoms is very close, it is difficult to confirm the arrangement of Mn and Fe atoms by X-ray diffraction analysis. When $x > 5$, as the Fe content increases, excess Mn and Fe atoms will replace the positions of Ga atoms. Due to the atomic radius of Fe (1.27 Å) and Mn (1.32 Å) are smaller than that of Ga (1.40 Å), the single-cell volume of martensite phase is shrinking. The lattice shrinkage enhances the stability of martensite phase, so the phase transition can carry out at a higher temperature.

4.3.2 Characteristics of Martensitic Phase Transition in As-Prepared Microwire

4.3.2.1 Ni–Mn–Ga Microwire

Martensitic transformation is one of the most important characteristics of Ni–Mn–Ga ferromagnetic shape memory alloys. There are several methods to measure the phase transition temperature depending on the signal collected. Figure 4.1b shows the DSC curve of as-prepared microwire in the process of heating and cooling, which reflects the change of heat flow in the phase transition process. The phase transition temperature can be obtained by the baseline tangent method, as shown at M_s point. When the ferromagnetic shape memory alloy is heated to the ferromagnetic paramagnetic transition temperature, the magnetic transition occurs. However, magnetic transformation is a weak first-order transformation (FOT) or second-order transformation (SOT), and the enthalpy change generated by the phase transition process is very small, so it is not easy to detect by DSC curve. Therefore, according to the magnetization characteristics of different magnetic states, enthalpy can be detected by measuring the change curve (M–T) of magnetization intensity with temperature by VSM. In addition, due to the different structures of ferromagnetic martensite and austenite, martensite has stronger magnetic crystal anisotropy, so it is more difficult for martensite to be magnetized under a low magnetic field. Therefore, the M–T curve under a low magnetic field can also measure the martensitic phase transition temperature of the alloy. However, due to the difference of detection signal, the temperature of phase transition measured by the two methods has a little deviation. Figure 4.13 shows the M–T curve of the as-prepared microwire, and the phase transition temperature is also obtained by the tangent method. It can be seen from the figure that there is an unusually sharp peak in the M–T curve near T_c. Literature [25] reports that for rapidly solidifying materials, such sharp peaks originate from the anisotropy of stress in the material, while the anisotropy of stress comes from the internal stress and strong magnetoelasticity formed during the rapid solidification process. This situation will be improved after stress relief annealing. In addition, this situation is also related to the domain interface movement during the transition from paramagnetic state to ferromagnetic state.

According to DSC and M–T curves of prepared microwires, phase transition information of different microwires in martensitic transformation process is summarized, and the results are shown in Table 4.6. First of all, due to the difference in the rate of rise and drop and the difference in the signal acquisition during the two methods, the phase transition width of DSC results ($|A_s–A_f|$ or $|M_f–M_s|$) is wider than that of VSM results, but the temperature range of phase transition agrees well. M_p and A_p are defined as the peak temperature of rise and drop in DSC curve, and the peak temperature difference $|A_p–M_p|$ is defined as phase transition hysteresis in thermal cycling process. For M-T curves, $|A_s–M_f|$ is defined as phase transition hysteresis.

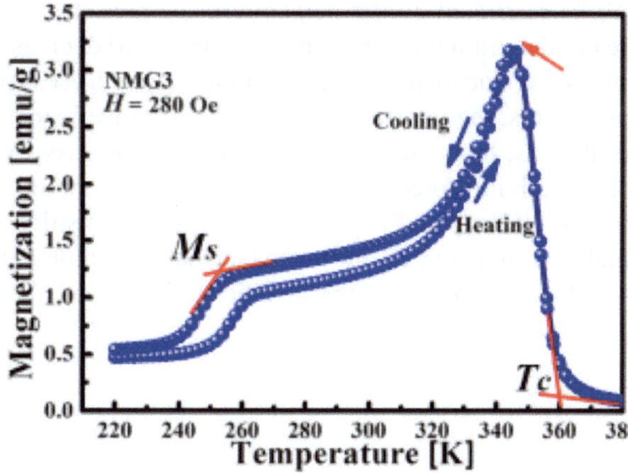

Fig. 4.13 Typical M–T curve of Ni–Mn–Ga microwire

Table 4.6 Comparison of microwire martensitic phase transition temperatures

Wires	Method	A_S	A_f	M_S	Mf	T_C	Hysteresis	Q_{cooling}	Q_{heating}
		K						J/g	
NMG1	DSC	288.6	309.6	295.1	274.7	–	12.2	−4.5	4.9
	VSM	294.9	304.6	297.7	289.6	360.4	5.3	–	–
NMG2	DSC	294.7	309.7	299.8	285.6	–	9.4	−4.3	4.8
	VSM	299.3	305.1	302.3	294.2	360.5	5.1	–	–
NMG3	DSC	245.7	263.4	251.7	237.4	–	11.4	−3.3	3.6
	VSM	249.7	262.4	253.3	239.4	359.5	10.3	–	–
NMG4	DSC	352.5	364.9	360.4	343.2	–	8.0	−6.1	6.7
	VSM	355.2	368.1	359.1	350.1	366.2	5.1	–	–

The acquisition method of T_c is shown in Fig. 4.13. Q_{cooling} and Q_{heating} represent phase change enthalpy in cooling and heating processes, respectively. According to the comparative analysis of phase change enthalpy value and phase change temperature, it is found that the four microwires all belong to the second martensite range [5]. NMG3, due to its low phase transition temperature, has a lower enthalpy than the average level of type II martensite, so it is closer to type I martensite. NMG4 is closer to the third martensite due to its high phase transition temperature and high enthalpy. For Mn-rich Ni50Mn25 + xGax alloys, T_c basically does not change with composition [26]. It can be seen from Table 4.6 that the T_c of as-prepared microwires has little difference, which is around 360 K.

Phase transition hysteresis is one of the main characteristics of first-order phase transition, which is mainly related to the resistance during phase transition. The phase transition resistance is mainly due to the irreversible elastic strain energy and friction losses such as defects and interface movement. The specific surface area of the alloy can be greatly increased by preparing the bulk polycrystalline structure into microwire. Moreover, due to the grains in the alloy have more free-surface, the crystal lattice volume change during the martensitic transformation of polycrystalline alloy will be restricted by the neighbouring grains. On the contrary, as the free surface area of the grain increases, the resistance will decrease. However, in the as-prepared microwire, due to that the rapid solidification process produces greater internal stress and defects, the interface movement resistance is increased. In addition, because of the defects, the stored elastic strain energy is easy to dissipate and cannot play the role of driving force of inverse phase transition, so the phase transition hysteresis increases. Compared with NMG1, the phase transition hysteresis of NMG2 is slightly smaller, which may be due to the fact that the diameter of NMG2 microwire is smaller than that of NMG2, i.e. the specific surface area is larger.

4.3.2.2 Ni–Mn–Ga–Fe Microwire

This section introduces the martensitic phase transition characteristics of Ni–Mn–Ga–Fe microwires [17, 27, 28]. Figure 4.14 shows the DSC curves of the as-prepared microwire. It can be seen from the figure that there is only one endothermic peak and one exothermic peak respectively in the heating and cooling process. The heat absorption peak corresponds to the inverse martensitic transformation of the alloy from the low temperature martensite phase to the high temperature austenite phase, while the exothermic peak corresponds to the martensitic transformation of the alloy from the austenite phase to the martensite phase, indicating that the MT of the microwires in this group is all one-step thermoelastic MT. Moreover, with the increase of Fe content, the phase transition peak strength increases. Although the alloy composition changes little, the changing range of temperature is significant, which indicates that the MT temperature of this kind of microwires is very sensitive to the alloy composition. The values of phase transformation temperature, phase change hysteresis and phase change enthalpy are listed in Table 4.7.

Since there is a part of elastic strain energy which can be released in thermoelastic martensite and become the driving force of reversing change, the driving force of reversing change $\Delta G^{M\to\gamma}$ is composed of two parts: Chemical driving force $\Delta G_C^{M\to\gamma}$ and mechanical driving force $\Delta G_M^{M\to\gamma}$, which can be expressed as $\Delta G_C^{M\to\gamma} = \Delta G_C^{M\to\gamma} + \Delta G_M^{M\to\gamma}$.

When $\Delta G_M^{M\to\gamma} = 0$, $T_0 = T_0'$, that is, the critical temperature at which cooling and heating are equal, and this is the general martensitic phase transition; when $\Delta G_M^{M\to\gamma} > 0$, $\Delta G^{M\to\gamma} > \Delta G_C^{M\to\gamma}$, $T_0 > T_0'$, the phase transition is thermoelastic equilibrium between T_0 and $T_0{}'$, thus martensite increases with temperature, or, decreases with temperature. According to the degree of critical temperature drop, Wayman classified thermoelastic martensitic phase transitions into two categories

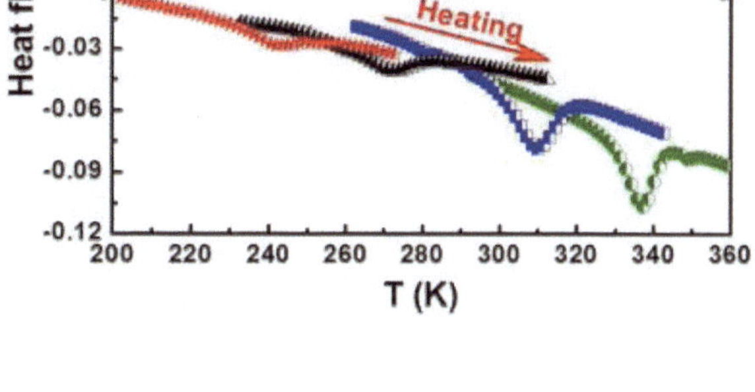

Fig. 4.14 DSC curves of $Ni_{50}Mn_{25}Ga_{25\text{-}x}Fe_x$ (x = 1–6) microwire

Table 4.7 Phase transition temperature of as-prepared microwire

Fe	A_s(K)	A_f(K)	A_{pv}K)	M_s(K)	Mf(K)	M_p(K)	T_0	T_0	A_p-M_p (K)	$Q_{Cooling}$ (J/g)	$Q_{Heating}$ (J/g)
$x = 1$	–	–	–	–	–	–			–	–	–
$x = 2$	–	–	–	–	–	–			–	–	–
$x = 3$	232.7	249.7	242.4	242.5	2.2	231.7	246.1	227.5	10.7	1.990	1.193
$x = 4$	253.5	281.2	272.5	272.6	245.8	262.7	276.9	249.7	9.8	2.05	2.101
$x = 5$	297.5	318.8	310.3	309.4	291	301.1	314.1	294.3	9.2	4.600	4.523
$x = 6$	329.4	343	337.2	334.3	320.5	328.3	338.7	325	8.9	4.462	4.400

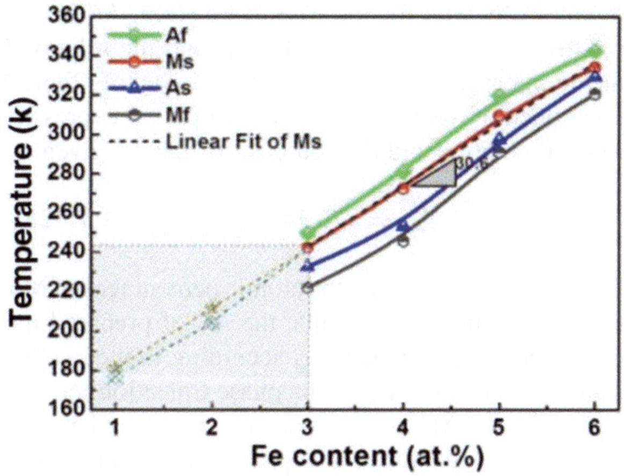

Fig. 4.15 Influence of Fe content on the MT temperature of $Ni_{50}Mn_{25}Ga_{25\text{-}X}Fe_x$ microwire

[29]: (1) as $M_s < T_0'$, $A_f > A_s > T_0 > T_0' > M_s > M_f$; (2) as $M_s > T_0'$, $A_f > T_0 > M_s > A_s > T_0' > M_f$.

Figure 4.15 shows the effect of Fe content on the MT temperature of $Ni_{50}Mn_{25}Ga_{25\text{-}X}Fe_x$ microwire. From the change of phase transition temperature in the figure, it can be seen that different Fe contents have the characteristics of Eq. 4.1. According to the change of phase transition temperature shown in the figure, the relationship $A_f > T_0 > M_s > A_s > T_0' > M_f$ holds for different Fe contents. Therefore, the single MT of this alloy belongs to the 2nd type MT, as classified by Wayman, and Fe doping doesn't change the type of MT.

In addition, under the condition of keeping the relative atomic percentage of Ni and Ga unchanged and replacing Mn with Fe, the temperature of positive and inverse martensite phase transitions of the alloy (M_s, M_f, A_s, A_f) increases linearly with the increase of Fe content. When the content of Fe increases from 3 to 6%, the phase transition temperature increases by 91.8 K. In the as-prepared $Ni_{50}Mn_{25}Ga_{25\text{-}X}Fe_x$ ($x = 1 \sim 6$) microwire, the increased rate of phase transition temperature is:

$$\frac{\Delta T}{\mathrm{Fe}(at.\%)} = \left|\frac{T_6 - T_3}{\mathrm{Fe}_6 - \mathrm{Fe}_3}\right| = \left|\frac{334.3 - 242.5}{6 - 3}\right| = 30.6K/at.\% \tag{4.1}$$

where,

ΔT/Fe (at.%)	the increasing rate of phase transition temperature;
Fe_6	the atomic percentage of Fe is 6 at.%;
Fe_3	the atomic percentage of Fe is 3 at.%;
T_6	the MT temperature of the alloy at 6 at.% Fe content;
T_3	the MT temperature of the alloy at 3 at.% Fe content.

In other words, for the as-prepared $Ni_{50}Mn_{25}Ga_{25\text{-}X}Fe_x$ ($x = 3$–6) microwire, when Fe element replaces Ga element, the MT temperature increases by about 30.6 K with every 1 at. % increase of Fe element, thus the MT temperature of the alloy changes dramatically with the increase of Fe atom. At the same time, by analyzing the true microwire composition (Table 4.2) and the relation between MT temperature and the composition of Ni–Mn–Ga-Fe microwire (Table 4.7), which is similar to three-component Ni–Mn–Ga alloy, the empirical expression for the relationship between MT temperature and the composition can be given:

$$T_{M_s} = 973.14 - 26.44X_{M_n} - 7.32X_{G_a} + 25.84X_{F_e} \quad (4.2)$$

where $X_{\mathrm{Mn}}, X_{\mathrm{Ga}}, X_{\mathrm{Fe}}$ are the atomic percentages of Mn, Ga and Fe in the alloy respectively. Through this formula, the M_s of prepared quaternary alloy can be roughly estimated. At the same time, according to the coefficient of each element, Fe addition can obviously adjust the phase transition temperature. According to Eqs. (4.1) and (4.2), the MT temperature for $x = 1$ and $x = 2$ were respectively calculated, which have been added to Fig. 4.15. The results calculated by the two formulas are slightly different.

The influence of alloy composition on phase transition temperature can be measured by the ratio of electron concentration *e/a*, where e is the number of valence electrons in the alloy, *a* is the number of atoms [30, 31]. Based on the periodic table of elements, the electron number of Heusler alloy is calculated by the usual method of energy band calculation. The valence electron configuration of Ni, Mn, Ga and Fe atoms is $3d^84s^2$, $3d^54s^2$, $4s^24p^1$ and $3d^64s^2$, respectively. Therefore, the valence electrons of Ni, Mn, Ga and Fe atom are 10, 7, 3 and 8 respectively. Therefore, the electron concentrations of the different alloys can be calculated by the following formula:

$$\frac{e}{a} = \frac{X_{Ni} \times 10 + X_{Mn} \times 7 + X_{Ga} \times 3 + X_{Fe} \times 8}{100}. \quad (4.3)$$

The variation of phase transition temperature of $Ni_{50}Mn_{25}Ga_{25\text{-}x}Fe_x$ microwire with the electron concentration is plotted in Fig. 4.6. It can be seen that with the increase of electron concentration, the characteristic temperature of microwire phase transition increases monotonically. By fitting the relation between MT temperature (M_s) and electron concentration, the empirical formula for the relationship between phase transition temperature and valence electron concentration of Fe-doped prepared microwire is obtained and can be expressed as:

$$M_s(\mathrm{K}) = 681(e/a) - 4993 \quad (4.4)$$

Compared with the empirical formula of ternary Ni–Mn–Ga alloy ($M_s(\mathrm{K}) =$ 702.5(*e/a*)-5067 [32]), the change rate of phase transition temperature with electron concentration is slightly lower with regard to Fe doped microwire. According to Eq. (4.4), the MT temperature for $x = 1$ and $x = 2$ was respectively calculated, which were then added to Fig. 4.16. It can be seen that the values calculated by Eq. (4.4) are close to those calculated by Eqs. (4.1) and (4.2).

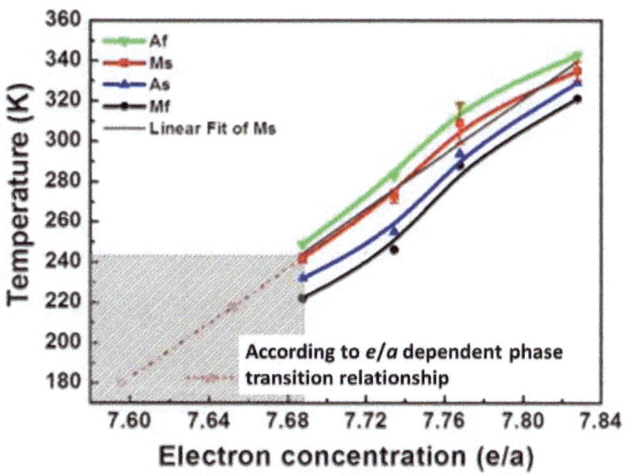

Fig. 4.16 Effect of electron concentration (*E*/*A*) on the MT temperature of $Ni_{50}Mn_{25}Ga_{25-x}Fe_x$ microwire

It is found that for as-prepared microwire, with the increase of Fe content, the degree of phase transformation hysteresis decreases. It is generally believed that the thermal hysteresis of martensite phase transition is mainly caused by the friction of the phase interface between martensite and the parent phase during the process of progression, which is irreversible heat dissipation. When the austenitic phase is transformed to martensite in the cooling process, part of the energy is converted to phase interface friction heat. After the transformation, the lattice strain energy is stored in the martensite phase. On the contrary, when the inverse phase transition of martensite occurs during the heating process, it also needs to overcome the phase interface friction energy, leading to the hysteresis of phase transformation temperature. In phase transformation process, the less frictional heat generated in the process of phase interface movement, the more elastic strain energy stored in the form of lattice distortion, and thus the smaller the temperature hysteresis in the process of the inverse phase transformation of martensite. The thermal hysteresis of $Ni_{50}Mn_{25}Ga_{25-x}Fe_x$ ($x = 1$–6) microwire varies little with Fe content. When Fe content increases from 3 to 6%, the thermal hysteresis decreases by 1.8 K. The values of thermal hysteresis are listed in Table 4.6, showing the characteristics of typical thermal elastic MT. The change of enthalpy for phase change is caused by the increase of MT temperature, that is, the higher the temperature of phase change of alloy, the greater the enthalpy change of phase change process, and conversely, the smaller the enthalpy change, as shown in Fig. 4.17. The enthalpy of phase transition of the microwire was calculated and listed in Table 4.7. The results show that the enthalpy of phase transition is less than 2 J/g when $x < 4$, and the MT temperature is much lower than room temperature, which should belong to martensite. When $x > 5$, the enthalpy of phase transition is about 4.5 J /g, and M_s is close to room temperature, so martensite should be the second type of martensite.

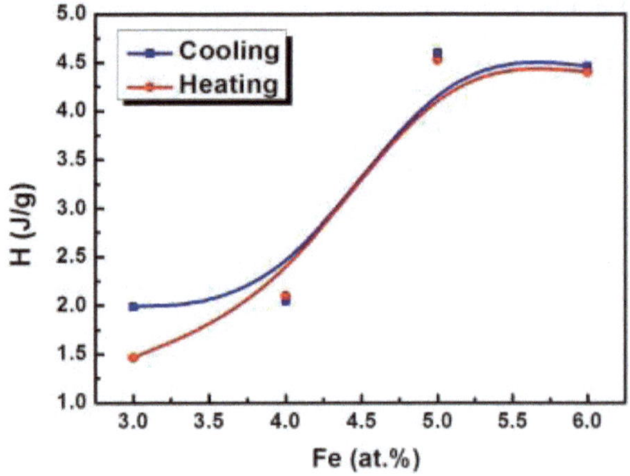

Fig. 4.17 Influence of Fe content on the phase transformation enthalpy of $Ni_{50}Mn_{25}Ga_{25-X}Fe_X$ microwire

4.3.3 Microstructure Evolution of Ni–Mn–Ga Microwire Prepared by Spinning Method

4.3.3.1 Diameter Distribution

In this section, NMG3 microwire is taken as an example to conduct a systematic study on the microstructure evolution in the spinning process [3, 4]. The continuous Ni–Mn–Ga microwire (length ~3–20 cm) with uniform diameter (diameter *D*, defined as the maximum size of the cross-section) was successfully prepared by spinning method using optimal process parameters. As shown in Fig. 4.4, the oxidation phenomenon didn't occur during the preparation process of the microwire, and the microwire presented silvery white metallic luster. In order to study the diameter distribution of the as-prepared microwires, 100 microwires were randomly selected and their diameter D was statistically analyzed, and the results are shown in Fig. 4.18.

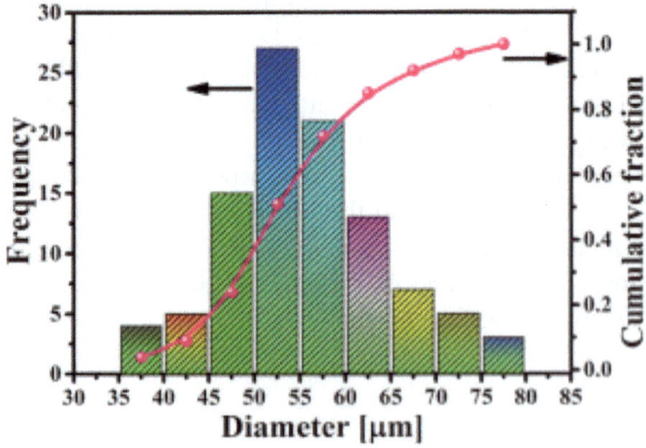

Fig. 4.18 Diameter distribution of NMG3 microwires

It can be seen that the diameter of the as-prepared microwires ranges from 35 to 80 m, and most of them are distributing in the range of 45–65 m.

4.3.3.2 Nucleation Area and Roundness

Based on the optimized preparation process parameters, the nucleation, growth and structure evolution characteristics in the preparation process of microwires are studied in this section. In order to study the microstructure evolution characteristics, the microwires were cut along the radius and axial direction. Figures 4.19 and 4.20 respectively show the SEM morphology and EBSD orientation image of the microwire in the cross-section and longitudinal section (or the outer surface of the plane part). The SEM images show the typical characteristics of polycrystalline Ni–Mn–Ga microwires prepared by spinning method. The transgranular fracture phenomenon can be observed in the cross-sectioned morphology, suggesting the

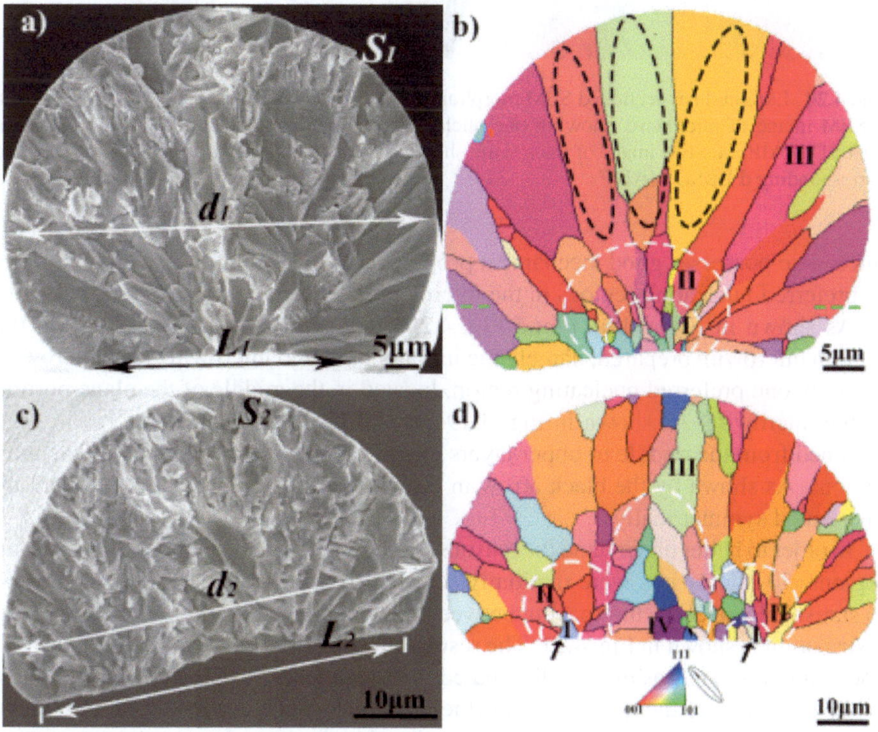

Fig. 4.19 Cross-sectioned SEM morphology and EBSD orientation of NMG3 microwires. **a** SEM image of microwire, showing one nucleation point in cross-section, and corresponding **b** EBSD MAIPF; **c** SEM image of microwire, showing two nucleation points in cross-section, and corresponding **d** EBSD MAIPF

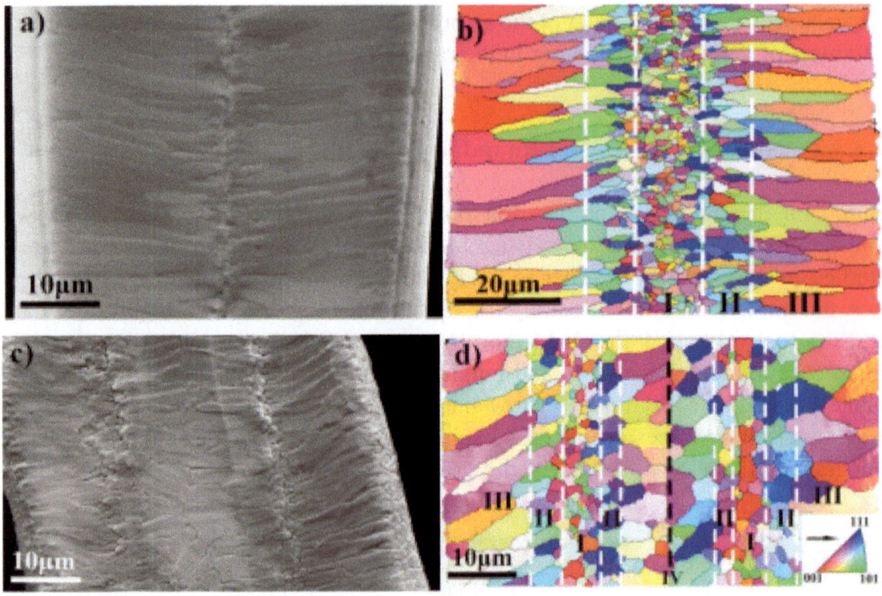

Fig. 4.20 Longitudinal-sectioned SEM morphology and EBSD orientation of NMG3 microwires. **a** SEM image of microwire, showing one nucleation point in cross-section, and corresponding **b** EBSD MAIPF; **c** SEM image of microwire, showing two nucleation points in cross-section, and corresponding **d** EBSD MAIPF

microwire possesses good mechanical properties. It is worth noting that the number of preferred nucleating regions in the microwire is different.

As shown in Fig. 4.19a, b and Fig. 4.20a, b, the microwire is taken from the bottom microwire prepared, namely the initially prepared microwire. The microwire has only one preferred nucleating region, located in the middle of the plane portion of the microwire. As shown in Fig. 4.19c, d and Fig. 4.20c, d, the microwires are obtained from the middle or upper layers, the microwire has two priority nucleating regions. As shown by the black arrow in Fig. 4.19d, the two nucleating regions are distributed symmetrically.

In order to better understand the reasons for the change in the number of nucleating regions, 50 microwires were randomly selected from the middle and upper layers, to statistically analyze the spacing between the two nucleating regions. The spacing distribution is shown in Fig. 4.21. The results show that the distance between the two nucleating regions was mainly distributed in the range of 21–24 μm, which is close to the tip radius of the copper wheel (the tip radius of the copper wheel is 20 μm, as shown in Fig. 4.3. The results indicate that the number of nucleating areas may depend on the heat exchange between the alloy melt and the copper wheel tip, which can be explained as follows: In the early stage of preparation, the copper wheel tip temperature is low. When the copper wheel tip contacts the molten alloy, the heat is transferred from the alloy to the copper wheel tip quickly, thus nucleating rapidly. In

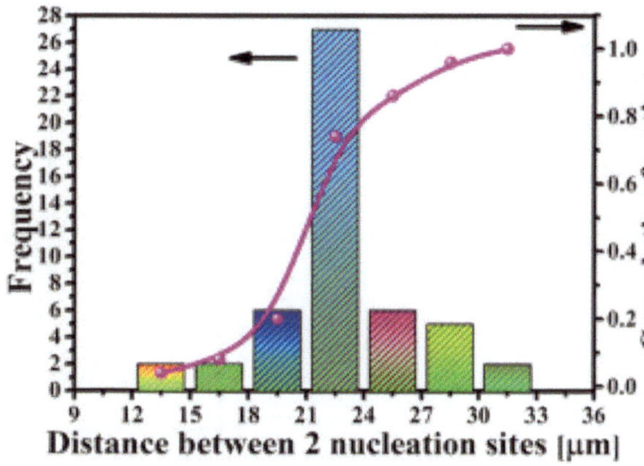

Fig. 4.21 Frequency distribution histogram and cumulative distribution curve of nucleation region spacing for NMG3 microwire

this case, only one nucleating region is formed at the center of the two contact regions. As the preparation process goes on, the temperature at the tip of the copper wheel starts to rise, and at the moment of contact with the alloy melt, sufficient subcooling degree cannot be provided to nucleate. At this time, both sides of the copper wheel tip contact with the gas as cold ends, so the alloy melt nucleates here, thus forming two nucleating regions. As the preparation progresses, the temperature of the copper wheel tip begins to rise. Therefore, the hotter copper wheel cannot provide enough supercooling for the alloy melt to nucleate. However, due to contacting with gas, the temperature of the two sides of the copper wheel tip is lower. Thus, the alloy liquid can nucleate on the side of the copper wheel, forming two nucleating regions.

In addition, it is found that the roundness of microwire cross-section is closely related to the number of nucleation areas. In this paper, the roundness of the microwire is defined as the ratio of the cross-section arc length (S) to the chord length (L), S/L, as shown in Fig. 4.19a. It is well known that when the cross-section is circular, $L = 0$, so S/L tends to infinity; and when it's a semicircle, $L = d$, $S = \pi d/2$, so $S/L = \pi/2$. Thus, if the cross-section is greater than half circle, $S/L > \pi/2$, and conversely, $S/L < \pi/2$. Therefore, the characteristic of S/L increasing with the roundness determines that S/L can be used as the characteristic parameter to measure the roundness of the cross-section. In order to better study the relationship between microwire roundness and nucleation region, 50 microwires were selected for statistical analysis of S/L, and the results were shown in Fig. 4.22.

It can be seen from the figure that S_1/L_1 is mainly distributed between 2.7 and 3.6 in the case of one nucleation region, while S_2/L_2 is mainly distributed between 1.8 and 2.7 in the case of two nucleation regions. The results indicate that the roundness of microwires with one nucleating region is higher than that of microwires with two nucleating regions. Moreover, there is an intersection point at $S/L = 2.7$, which also verifies the transition point from one nucleating region to two nucleating regions in the preparation process. In this paper, SEM observation was also carried out on more than 50 microwires, and it was found that the microwires with two nucleating regions had smaller Rayleigh waves, that is, smaller diameter fluctuation. This phenomenon

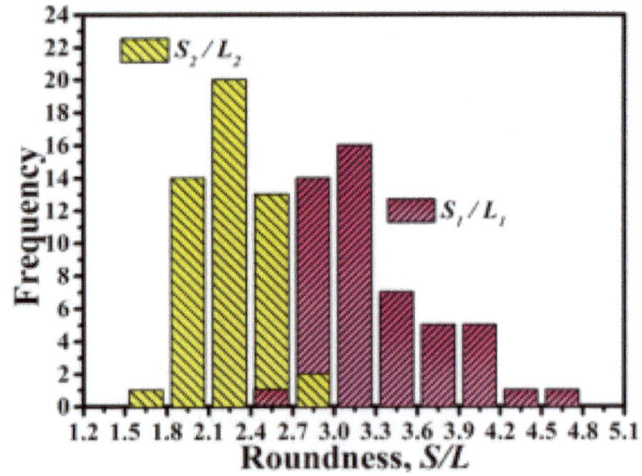

Fig. 4.22 Histogram of frequency distribution of chord arc ratios (S_1/L_1 and S_2/L_2) of two microwires with different nucleation regions

is mainly due to the fact that in the later stage of microwire preparation, the melt nucleates on both sides of the copper wheel tip, so the microwire plane part is wider. This feature makes the alloy melt more stable in the solidification process to form a free surface, reduces the influence from surface energy and gravity, and thus reduces the formation of Rayleigh waves.

To sum up, in the early stage of the spinning process, due to the tip of the copper wheel is cooler, a microwire with a preferred nucleation region was formed, which had better roundness but relatively uneven diameter. As the preparation process progresses, the alloy melt begins to nucleate on both sides of the copper wheel tip, forming a microwire with two nucleation regions, which has a more uniform diameter but a relatively low roundness. With the increase of wheel temperature, even on both sides of copper round tip cannot provide effective super-cooling degree to make metal liquid nucleate, the molten alloy tend to tip along the copper round face downward flow on both sides, into the location of the supercooling degree of nucleation, this kind of microwire SEM morphology and EBSD orientation as shown in Fig. 4.23. With the further development of the preparation process, both

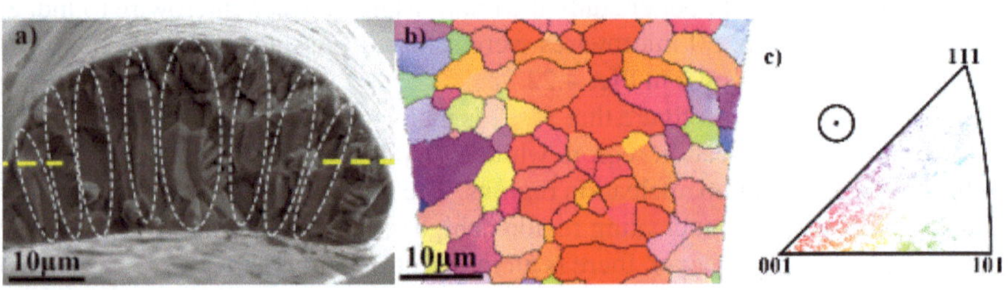

Fig. 4.23 SEM image of the cross-section and EBSD orientation of the longitudinal section of the NMG3 microwires at the later stage of preparation **a** SEM image; **b** EBSD orientation diagram; **c** discrete antipolar diagram of figure (**b**)

sides of the copper wheel tip cannot provide effective undercooling to nucleate the alloy molten liquid. At this time, the molten alloy tends to flow downward along the copper wheel plane on both sides of the copper wheel tip, and nucleate at the location where sufficient undercooling is achieved. The SEM morphology and EBSD orientation of the microwire taken from the top layer of the as-prepared microwire are shown in Fig. 4.23. At this point, the microwire still has two nucleating regions, but the plane part of the microwire is transformed from straight to slightly curved. More importantly, as shown in cross-section, the grains in the microwire tend to grow perpendicular to the plane part, resulting in a certain texture. This feature is discussed in detail in the next section.

4.3.3.3 Microstructure Evolution During Heat Transfer

As heat is transferred between the alloy, the copper wheel and the surrounding gas environment, the grains begin to nucleate and grow. Firstly, it should be emphasized that the direction of heat flow and the direction of grain growth are located in the cross-section of the microwire during spinning process of microwire. In this paper, EBSD technology was used to analyze the grain morphology of the cross-section and the vertical section of the prepared microwire, revealing the grain growth process of the as-prepared microwire [3, 4].

For the microwire with one nucleating region, the grains within the microwire can be divided into three regions, as shown in Fig. 4.19b and Fig. 4.20b: Region I is the nucleating region and contains fine equiaxed grains with a grain diameter of approximately 0.4–2 μm, resulting from rapid solidification. As heat is transferred from the molten alloy to the copper wheel tip, the temperature of the copper wheel tip increases so that no new nucleating particles can be formed. In this case, those nuclei with favorable orientation begin to grow along the heat flow direction to form columnar grains with a width between 2 and 10 μm, as shown in region II. These columnar grains combine and grow, forming columnar grains with a width of 10–35 μm, as shown in region III. Therefore, the grains in the Ni–Mn–Ga microwire prepared by spinning process grow towards the free surface from the nucleation region, and the special grain distribution is determined by the radial heat flow direction.

For microwires with two preferred nucleating regions, the grain distribution inside the microwire is slightly different. As shown in Fig. 4.19d and Fig. 4.20d, both the cross and longitudinal sections of the microwire consist of two perfectly symmetrical parts growing in opposite directions and touching each other. The two parts are separated by black dotted lines. This grain, first of all, also contains region I-III found in the microwire with one nucleating zone, but on the other hand, due to the grains grow from the surface nucleation area towards the microwire's centerline, and finally met in the intermediate position of the microwire's plane part, we called this region area as IV, as shown in Fig. 4.19d and 4.20d. This is different from the case of one nucleating region, and it is this growth mode of grain that leads to the two almost symmetrical components.

4.3.3.4 Texture Evolution of in the Cross-Section of Microwire

Ni–Mn–Ga polycrystalline alloy with a strong texture can reduce the motion resistance of the twin boundary by reducing the restriction between adjacent grains, thus facilitating the acquisition of MFIS [25]. As mentioned above, the direction of grain growth in the microwire is radial in the cross-section, which is determined by the direction of heat flow. This epitaxial growth of grains may result in a certain degree of texture in microwire's cross-section. This section describes the texture evolution of the Ni–Mn–Ga microwire in cross-section [3, 4].

In order to study the texture in microwire's cross-section, a new IP(F) color scheme is introduced. In general, the EBSD orientation diagram reflects the orientation of the grains in the diagram by IPF color matching. Taking [001] IPF color matching as an example, this EBSD orientation diagram reflects the crystallographic orientation of the grains in the scanning area and the orientation of the sample [001] (normal direction of the sample scanning surface). Different colors reflect the crystallographic orientation parallel to the orientation of the sample [001]. However, for the cross-section of microwire in this paper, due to the special growth mode of grains, the ordinary IPF diagram obviously cannot accurately reflect the orientation information. Therefore, a color matching method called "Major Axis Inverse Pole Figure" (MAIP(F) was adopted in this paper to reflect the crystal orientation of the microwire. As shown in Fig. 4.19b, the grains in the cross-section of microwire are abstract into elliptic shapes (black dotted ellipse). It is well known that the ellipse has a long axis direction, therefore, MAIPF color reflects the orientation relationship between crystal orientation and the major axis direction of the ellipse in this region, that is, the sample orientation used for color matching is not a single orientation, but a series of major axis directions of the ellipse. Since the direction of grain growth in the microwire's cross-section is consistent with the long axis direction of the ellipse, this method can perfectly express the grain orientation of the microwire cross-section. EBSD orientation diagram with MAIPF color can accurately reflect the crystallographic direction of grain growth. Figure 4.19b, d is the EBSD orientation diagram with MAIPF color. The grains in the diagram are mostly red. According to the color matching scheme, it can be seen that in the microwire cross-section, the grains grow along the crystallography <001> .

The general IPF color matching method can be used to observe grain orientation information in the longitudinal section of microwire. Figure 4.20b, d shows the EBSD orientation diagram with IPF color matching in the axial direction of the vertical microwire in the longitudinal section of the microwire. It can be seen from the figure that most of the grains also appear red, indicating that the same conclusion can be drawn from the angle of the longitudinal section, that is, the grains grow along the direction <001> in crystallography. From the perspective of interface dynamics, such growth along the direction of <001> in the crystal structure of body-centered cube tends to occur [33].

As mentioned earlier, in the late stage of microwire preparation, due to the change of microwire nucleation position, the flat part of the microwire changes to a slight curve, this change leads to the grain growth direction of the basic in the microwire cross-section perpendicular to the plane section, cross-section grain abstract into elliptic curve (white), as shown in Fig. 4.23, can be observed in all the direction of the axis of the ellipse basic vertical flat part of the trace. In order to further study the orientation of grains in the microwire, the longitudinal section of the microwire was analyzed by EBSD. Figure 4.23b is the EBSD orientation diagram with the normal IPF color matching of the longitudinal section of the microwire, and (c) is the corresponding discrete IPF antipole diagram. The large area of red in Fig. 4.23b shows that the crystallographic direction <001> , that is, the growth direction of the grain is parallel to the normal section, indicating the strong texture, which can be confirmed by Fig. 4.23c. In this case, if 5 M Martensite is assumed to be formed in the microwire, only the crystallography [100] and [001] variants parallel to the normal section can be formed, that is, the number of martensite variants is greatly reduced, thus reducing the incompatibility between variants and increasing the possibility of obtaining large MFIS [33]. More detailed and in-depth studies are needed in the future.

4.3.4 Magnetic Properties of Ni–Mn–GA Microwire Prepared by Melt Spinning

This section takes Ni–Mn–Ga prepared microwire as an example for the introduction [3, 4]. The magnetization curve (M–H) of a typically prepared microwire in a martensite state is shown in Fig. 4.24. According to the martensitic phase transition temperature of NMG3 microwires, the magnetization curve was tested at 223 K to ensure that the microwires were in martensitic state. It can be seen from the figure that, with the increase of external magnetic field, the magnetization intensity of the microwire increases sharply at first and slows down after reaching a certain critical value, which is a typical magnetization curve characteristic of ferromagnetic alloy. It

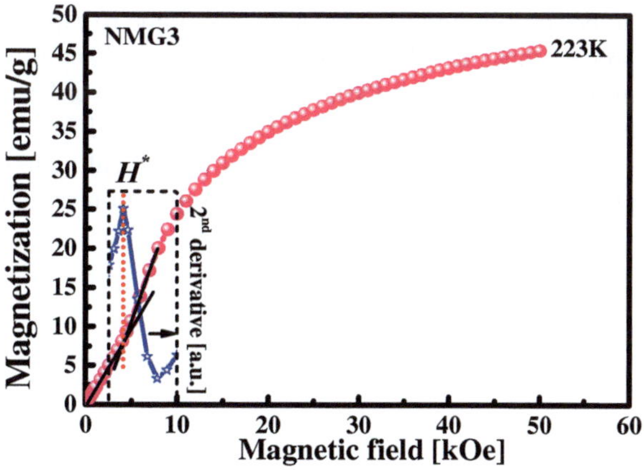

Fig. 4.24 Magnetization curve of NMG3 microwire at low temperature of 223 K

can be seen from the figure that there is an inflection point of the curve in the region of an abrupt increase in magnetization, and the corresponding magnetic field is the critical magnetic field H* of magnetic-field-induced mutation reorientation (MIR) [25]. After the second-order derivative of the curve, the solid blue line in the figure shows that the critical magnetic field H* of MIR-at 223 K of NMG3 microwire is 4.5 kOe. The value of H* decreases with the increase of temperature, which is related to the decrease of the critical motion stress at the twin boundary with the increase of temperature [34]. At the maximum tested magnetic field (50 kOe), the microwire had a maximum magnetization of 45.5 EMU/g, which was not saturated, due to the low degree of atomic order formed during rapid solidification.

Magnetic crystal anisotropy is another important parameter of Ni–Mn–Ga ferromagnetic shape memory alloy, which determines the force of magnetic field on the twin boundary [35].

First of all, under the magnetic field of 1 kOe strength, the sample was rotated by two different rotation modes above 180°, and the maximum and minimum positions of magnetization were found from the relation curve of magnetization and rotation angle, that is, the directions of easy and difficult magnetization. Then the magnetization curve is measured in these two directions. The area surrounded by the two magnetization curves is the anisotropic energy of the magnetic crystal. Due to the influence of sample shape, the measured magnetization curve needs to be corrected for the demagnetization factor, while the relationship between magnetization intensity and rotation angle is only to find the extremum point of magnetization, which does not need to be corrected.

4.4 Effects of Chemical Ordering on Microstructure and Properties of Ni–Mn–Ga Microwire

Chemical ordered heat treatment refers to the full heat preservation annealing near the disordered-ordered transition temperature of the alloy so that Mn and Ga atoms do not occupy the correct lattice point due to rapid solidification in the preparation process that fully diffuse at the temperature, and thus reoccupy the correct position in the solid solution lattice and improve the order degree of the alloy. In the heat treatment process, 998 K and 973 K are held for 2 h for sufficient atom diffusion, and 773 K is held for 20 h for de-stress annealing to release the internal stress in the process of rapid solidification and preparation of microwire and reduce defects.

4.4.1 The Microstructure and Phase of Microwire Treated by Ordering Heat Treatment

According to the research needs, this paper only conducts the chemical ordered heat treatment on NMG1 and NMG4. The composition pairs of microwires before and after the heat treatment are shown in Table 4.8, and the letter "C" represents the ordered state [1, 4]. As shown in Table 4.8, the composition of microwire refers to mean composition, and the mean composition and standard deviations were statistical results of 6 microwire surfaces and 4 microwire cross-sections. Due to the addition of Mn in the heat treatment process, the loss of Mn element is small, while the loss of Ga element is more obvious. When Mn content remains unchanged, the decrease of Ga content will lead to an increase in phase transition temperature, which can be verified in subsequent phase transition temperature tests. Within the allowable error range, the composition difference before and after ordering is very small while the composition standard deviation is reduced, indicating that the composition uniformity is improved after ordering treatment.

Figure 4.25 shows the SEM morphology of the microwires before and after ordering. (a) and (b) show the cellular grains on the free surface of NMG1 microwire. It can be seen that the grain size (1–5 μm) and morphology don't change before and after ordering. Figure 4.25c and d show the longitudinal-sectioned morphology of NMG1 microwire after polishing. Except that the ordered microwire is martensite at room temperature, the size and distribution of grains are basically the same before and after the ordering, except that the martensite twin variants can be observed. Figure 4.25e and f show the fracture morphology of NMG4 microwire before and after ordering treatment. It can be seen that the distribution and size of grains in the microwire do not change significantly, but the microwire tends to intergranular fracture after ordering treatment. To sum up, the ordered heat treatment under 998 K did not cause the growth of microwire grains due to the low temperature, but caused the grain boundary weakening to some certain extent, leading to the occurrence of intergranular fracture.

Figure 4.26 shows the XRD curves of NMG1C and NMG4C microwires at room temperature. The analysis results show that at room temperature, they are respectively the 5 M martensite with public structure [36] and the 7 M martensite with non-public structure [6]. The crystal structure information of ordered microwires obtained by

Table 4.8 The composition of microwires before and after ordered heat treatment

Microwires	Compos (at.%)					
	Ni	ΔNi	Mn	ΔMn	Ga	ΔGa
NMG1	49.9	0.3	28.6	0.5	21.5	0.5
NMG1C	50.1	0.2	28.5	0.3	21.4	0.3
NMG4	50.5	0.3	29.1	0.4	20.4	0.4
NMG4C	50.4	0.2	29.4	0.2	20.2	0.1

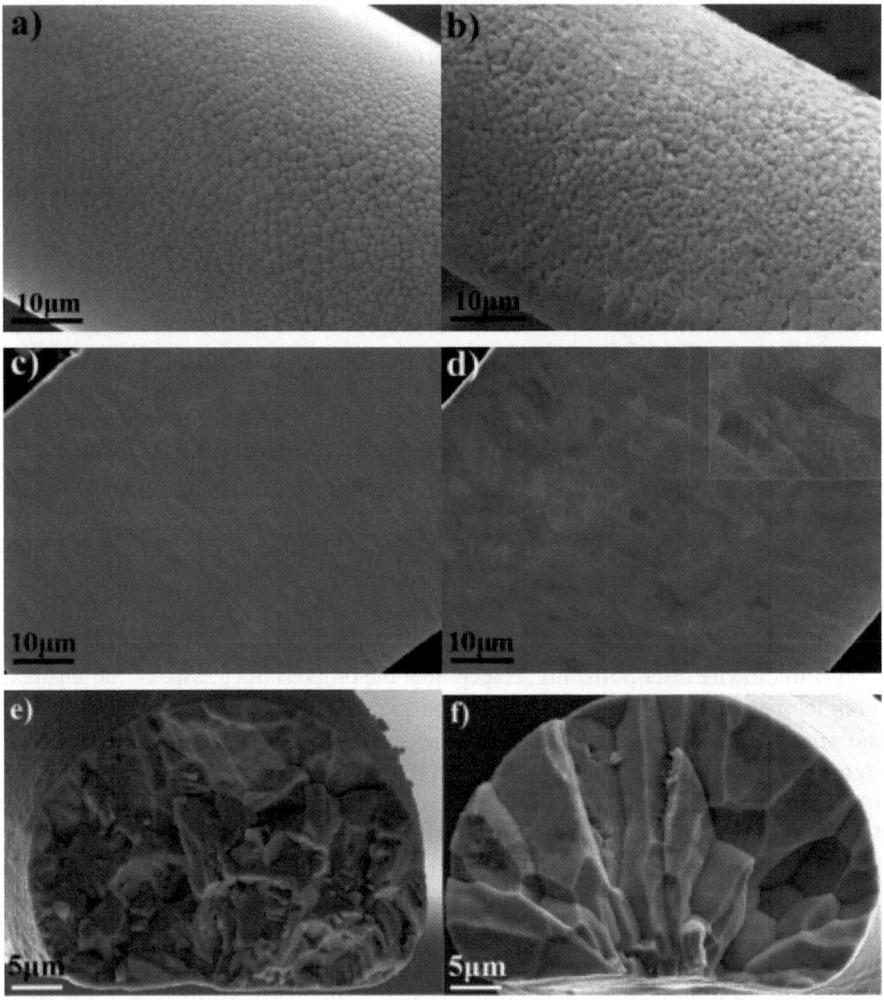

Fig. 4.25 Comparison of SEM morphology of microwires before and after ordered heat treatment. **a, b** Outer surface; **c, d** Longitudinal section; **a, c** and **e** before heat treatment, **b, d** and **f** after heat treatment

XRD analysis is summarized in Table 4.9. As indicated above, the as-prepared NMG1 microwire at room temperature possesses austenite structure with cubic L21 structure, and the lattice parameter is $a = 5.83$ Å, and thus $V_a = 12.38$ Å^3. When the temperature drops to 263 K, the martensitic transformation occurs in the microwire, generating the non-public 7 M martensite. The lattice parameters are $a = 4.25$ Å, $b = 5.53$ Å, $c = 43.34$ Å, $\beta = 93.6°$, and thus $V_a = 12.4$ Å^3. After ordering heat treatment, the NMG1 microwire presented a 5 M martensite structure at room temperature, with lattice parameters $a = 4.22$ Å, $b = 5.57$ Å and $c = 20.96$ Å, $V_a = 12.29$ Å^3.

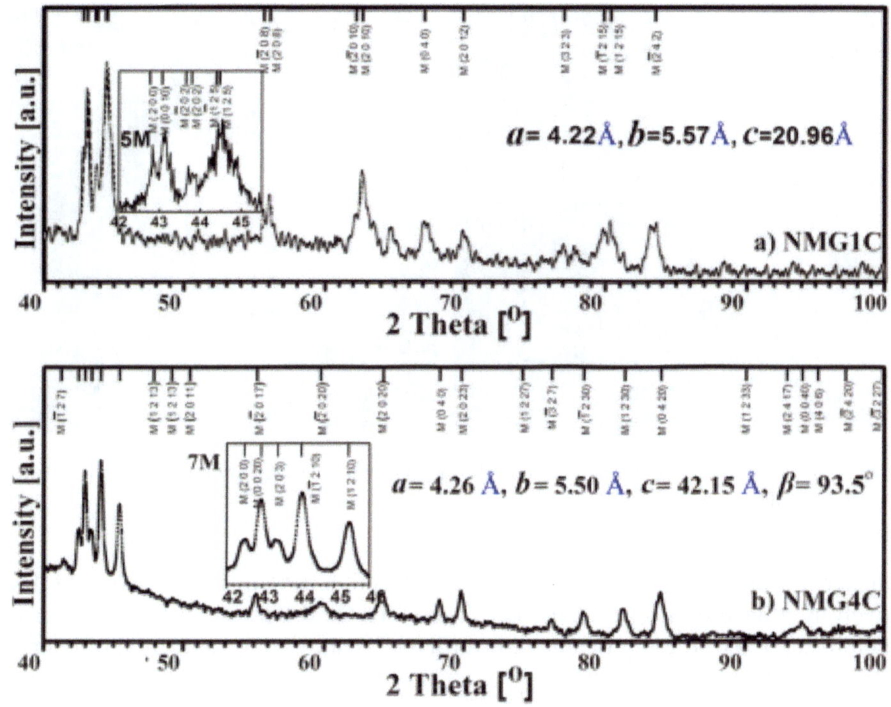

Fig. 4.26 XRD patterns of ordering treated microwires at room temperature. **a** NMG1C; **b** NMG4C

Table 4.9 Phase composition and lattice constants of the microwires after ordering treatment

Microwires	Phase composition	Lattice constants						
		a(Å)	b(Å)	c(Å)	α(°)	β(°)	γ(°)	V_a(Å^3)
NMG1C	Room temperature 5 M	4.22	5.57	20.96	90	–	90	12.29
NMG4C	Room temperature 7 M	4.26	5.50	42.15	90	93.5	90	12.32

Figure 4.27 shows the TEM morphologies and electron diffraction spots of NMG1 microwires before and after ordering treatment under ~288 K. As shown in (b), NMG1 is a mixed microstructure of austenite and martensite under ~288 K. In the figure, not only can the martensite twin structure with straight interface in austenite grains be observed, but also a large number of dislocation defects within the grains can be observed. (a) is the TEM morphology of NMG1C microwire. The martensite twin lath is clearly visible, and the electron diffraction spots confirm the microwire is composed of 5 M martensite, which is consistent with the XRD results. The yellow solid line in the figure shows the twin boundary, and the thin twin laminate may be related to the modulation structure. The pink dotted line shows the 90° domain boundary, and the martensite variants on both sides of the domain boundary have

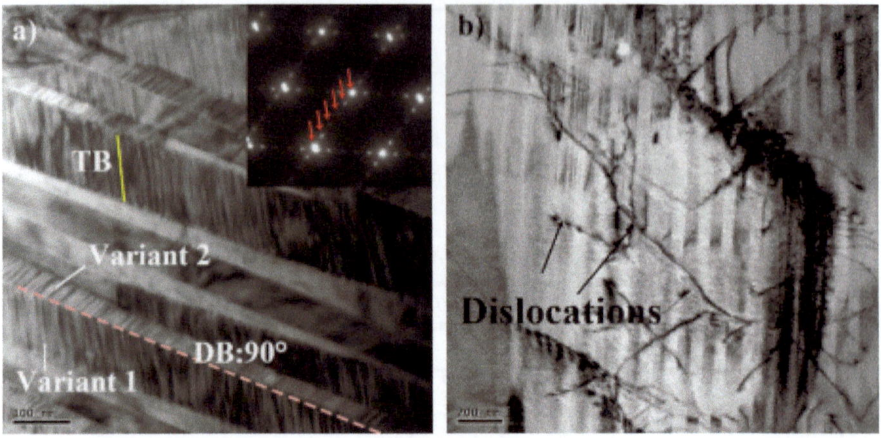

Fig. 4.27 TEM morphologies and electron diffraction spots of NMG1 microwire before and after ordering treatment. **a** Ordered state; **b** As-prepared state (~288 K)

different orientations. It is found that the dislocation density can be reduced effectively after the ordering heat treatment. Therefore, the difference in V_a of martensite before and after ordering treatment is attributed mainly to the reduction of internal stress and preparation defects.

In order to further study the internal microstructure of the microwire after ordered treatment, the structure and orientation of the twins in the cross-section and longitudinal section of the microwire were studied by EBSD, as shown in Fig. 4.28. NMG1C microwire is 5 M structure martensite at room temperature, so the microwire EBSD is calibrated with 5 M structure martensite in the cubic coordinate system. Figure 4.28a shows the EBSD orientation diagram with the normal IPF color matching of NMG1C microwire cross-section. The MAIPF is different from that of the prepared cross-section because the focus here is on the orientation relationship between the direction <001> in the microwire cross-section and the axial orientation of the microwire. Figure 4.28b shows the EBSD orientation diagram for the longitudinal section of NMG1C with microwire axial IPF color matching. It can be seen that due to the wide scanning range, the scanning step is larger than the width of the twin grain lath, thus it is difficult to find the twin structure.

Although the size of twin lamellae is small and the twin distribution and lamellae morphology cannot be observed, the misorientation between crystallographic orientation and sample orientation can be obtained from the EBSD orientation diagram. Figure 4.28c and e respectively shows the misorientation distribution of <001> and <100> in Fig. 4.28a. It can be seen that the misorientation distribution of < 001 > in the range of $[-45°, 45°]$ is less than 20%, while that of <100> (5 M structure <010> and <100> are equivalent) in the range of $[-45°, 45°]$ is more than 70%, indicating the c axis of grains (<001>) is more inclined to be arranged along the radial direction of the microwire than along the axial direction. Figure 4.28d and f respectively shows the misorientation distribution of <001> and <100> in Fig. 4.28b. It can be

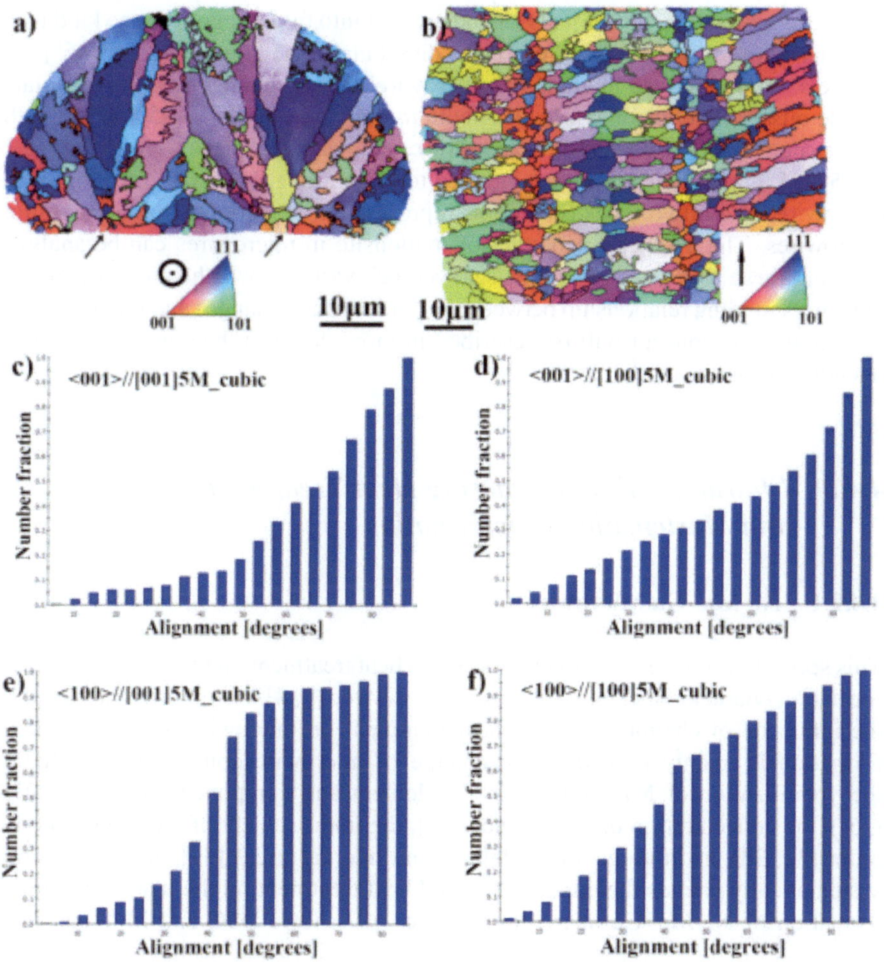

Fig. 4.28 Orientation distribution of grains at transverse/longitudinal sections of NMG1C microwire with 5 M structure. **a** EBSD orientation diagram of cross section; **b** EBSD orientation diagram of longitudinal section; **c** <001> //[001] ratio; **d** <001> //[100] ratio; **e** <100> //[001 ratio]; **f** <100> //[100] ratio

seen that the misorientation distribution of < 001 > in the range of [−45°, 45°] is less than 35%, while that of <100> in the range of [−45°, 45°] is about 70%, indicating the c axis of grains(<001 >) is more inclined to be arranged along the radial direction of the microwire than along the axial direction. However, the orientation information obtained by this method only reflects the orientation information in the plane, which cannot reflect the orientation of the twins inside the entire microwire. To obtain complete orientation information, the microwire should be sliced layer by layer, and the 3D EBSD analysis should be carried out in the future. However, the orientation information given in this paper inspires us that, although martensitic variants are very fine, the variants can be divided into two categories, one kind is the variants whose misorientations of <001> distribute in the range of [−45°, 45°], and their c axes are inclined to along the microwire's axial, another kind is the variants whose misorientations of <001> distribute in the range of [45°, 135°], namely, the misorientation of <100> (<010> and <100> equivalent) distribute in the range of [−45°, 45°], and their c axes are inclined to align along the radial direction of the microwire. These two types of variants represent all the variants in polycrystalline microwires. The average orientation of martensite in microwires can be analyzed by using these two types of "statistical average" variants, which can easily explain the corresponding relationship between the performance changes and microstructure evolution. This concept will be described in more detail in the construction of the phenomenological model later.

4.4.2 Characteristics of Ordering Heat Treatment and Martensitic Transformation

4.4.2.1 Ni–Mn–Ga Microwire

This section introduces the effect of ordering heat treatment on the martensitic transformation characteristics of Ni–Mn–Ga microwires [1, 4]. As reported in literature, the influence of chemical ordered heat treatment on microwire martensitic transformation is mainly related to the change of microwire composition (including the volatilization of Mn element or Ga element and component homogenization [25]), the concentration of Ga vacancy [34], the improvement of long-range atomic order [37, 38], and the release of defects and internal stress [39], etc. This section systematically studies the effect of ordered heat treatment on the martensitic transformation of Ni–Mn–Ga microwire. Figure 4.29 shows DSC and M-T curves of

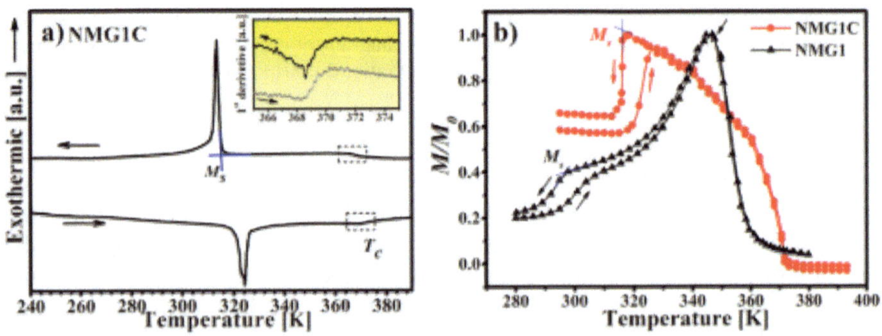

Fig. 4.29 DSC and VSM phase transition temperature curves of microwires. **a** DSC curve, the inset is the first derivative curve of DSC curve; **b** VSM M-T curve

Table 4.10 Phase transition information of NMG1C microwire acquired by DSC and VSM

Wire	Methods	A_s	A_f	M_s	M_f	T_c	Hysteresis	$Q_{cooling}$	$Q_{heating}$
		K						J/g	
NMG1C	DSC	321.8	325.9	316.0	311.7	367.5	10.6	−5.5	4.5
	VSM	316.9	325.5	317.9	315.8	371.8	1.1	–	–

NMG1C microwire after ordering heat treatment. For comparison, the magnetization is normalized, where M_0 is the maximum magnetization in the M-T curve. Differing from the DSC curves of as-prepared microwire, the enthalpy change of ferromagnetic–paramagnetic transition can be observed in the DSC curve of the microwire after ordered treatment. The value of T_c can be obtained by taking the first derivative of DSC near the magnetic phase transition of lifting and cooling curve. As shown in Fig. 4.29b, the sharp peaks near T_c do not appear after ordering, indicating that ordered heat treatment alleviates the interaction between internal stress formed during rapid solidification and strong magnetoelasticity. Moreover, there is a steep drop in magnetization intensity when martensitic transformation occurs during the cooling process. The reason is that martensite with low symmetry has a large magneto-crystalline anisotropy, and it is more difficult to magnetize than austenite under the same external field action, so the magnetization intensity is lower (Tables 4.10 and 4.11).

According to DSC and M-T curves of the microwire, the martensitic phase transition information of NMG1C microwire is summarized, and the results are shown in Table 4.12. Compared with Fig. 4.14, the martensitic phase transition temperature moved towards high temperature after ordered treatment, and M_s was ~317 K so that the microwire composed of the martensitic at room temperature, which also explained martensitic twin laminar structure of the NMG1C microwire shown in Fig. 4.26d. As listed in Table 4.12, the phase transition temperature of NMG1C microwire is in the range of the phase transition temperature of typical modulation structure martensite, and its phase transition enthalpy shows that it belongs to the second type of martensite range. There is a slight difference in enthalpy value in the process of rise/drop, which is mainly related to phonon emission in the process of martensitic transformation [40].

According to DSC results in Table 4.10 and Fig. 4.29, the temperature range of martensite phase transformation of NMG1C microwire decreases from 20.4 K to 4.3 K after ordered heat treatment while the temperature range of martensite inverse phase transition decreases from 21.0 K to 4.1 K. The decrease of phase transformation width is related not only to the homogenization of composition after heat treatment but also to the release of internal stress. In as-prepared microwire, the existence of internal stress is conducive to the formation of the preferentially oriented martensite variants, which will lead to the formation of more stored elastic strain energy, thus requiring more supercooling to complete the phase transformation, thus increasing the width of the phase transformation.

Table 4.11 Phase transformation temperature of heat-treated microwires

Fe	A_s(K)	A_f(K)	A_p(K)	M_s(K)	M_f(K)	M_p(K)	T_0	T_0'	A_p-M_p (K)	$Q_{Cooling}$ (J/g)	$Q_{Heating}$ (J/g)
$x = 1$	230	233.38	231.8	221	218.5	214.1	227.2	224.3	17.7	2.632	2.298
$x = 2$	257.1	268.5	262.2	254.7	246.4	248.2	261.6	251.8	14	1.8	1.518
$x = 3$	281.4	288.6	284	282.5	271.7	274.9	285.6	276.6	9.1	4.274	3.657
$x = 4$	304	311.6	309	311	298	300.7	311.3	301	8.3	6.266	5.697
$x = 5$	311.2	331.1	333	323.2	304	325	327.2	307.6	8	3.692	3.433
$x = 6$	354	363.4	361.1	361.4	350.4	353.6	362.4	352.2	7.5	10.76	9.638

Table 4.12 The composition for NMG2 microwire before and after heat treated at 1323 K for 3 h

Wires	Position	Composition (at.%)					
		Ni	ΔNi	Mn	ΔMn	Ga	ΔGa
NMG2	Average	50.6	0.4	28.8	0.3	20.6	0.4
NMG2A1323/3 h/0.5 atm	Inner parts	50.9	0.1	28.6	0.4	20.5	0.3
	Outer surface	51.0	0.4	28.8	0.5	20.2	0.4

The phenomenon that phase transformation hysterics (temperature difference of peak phase transformation $|A_p\text{-}M_p|$) decreases from 12.2 K for NMG1 to 10.6 K for NMG1C is related to energy loss: On the one hand, the elastic strain energy stored in the phase transformation process will be lost as the strain at the martensite-austenite phase interface is released or the interface passes through dislocations or precipitations. On the other hand, in order to overcome the resistance when the phase interface moves, it is necessary to overcome the friction to do work, resulting in friction loss. The energy dissipation via these two paths will lead to the increase of martensitic transformation hysteresis. Studies have shown that after ordered heat treatment, the energy loss will be reduced due to the reduction of internal stress and defects. Since the stored elastic strain energy in the phase transformation process increases with the increase of external stress or internal stress, the stored elastic strain energy is the driving force of the inverse phase transformation [41]. Therefore, the internal stress existing in as-prepared microwire will increase the elastic strain energy stored in the phase transformation process and thus become the driving force for the reverse phase transformation, leading to the reduction of phase transformation hysteresis. However, the existence of internal stress and defects, on the other hand, increases the loss of elastic strain energy stored during the phase interface movement, which leads to the increase of phase transformation hysteresis. For the microwire treated by ordered heat treatment, the elimination of internal stress and defects reduces the loss of stored elastic strain energy in the phase transformation process. Therefore, even though the stored elastic strain energy is relatively less, the stored elastic strain energy is sufficient to drive inverse phase transformation, thus reducing the phase transformation hysteresis. On the other hand, the reduction of internal stress and defects also reduces the friction loss, thus reducing the phase transformation hysteresis of the ordering-treated microwire.

In addition, the T_c obtained by the two methods after the ordered treatment were 367.5 K and 371.8 K respectively, which are higher than that of the T_c of as-prepared microwire (360.4 K), indicating that the ordered heat treatment effectively improved the magnetic properties of the microwire.

The martensite formed during the cooling process changed before and after ordering. Based on the parameters of microwire phase transformation process, the change can be explained from the perspective of thermodynamics. Firstly, according to the literature [42], the change of volume free energy of the microwire during the phase transformation from austenite to martensite can be expressed as:

$$\Delta G_v = \frac{Q \Delta T}{T_0} \quad (4.5)$$

where Q is the enthalpy change in the phase transformation process, T_0 is the theoretical phase transformation temperature, and ΔT is the supercooling degree of the phase transformation process, which can be expressed by (T_0-M_P) and approximately equal to (A_P-M_P)/2, where M_P and A_p are the peak temperature during phase transformation process. Therefore, Eq. (4.6) can be expressed as:

$$\Delta G_v = Q * (A_p - M_p)/(A_p + M_p) \quad (4.6)$$

Combined with Tables 4.6 and 4.10, it can be seen that during the phase transformation from austenite to 7 M or 5 M martensite, they have the equivalent phase transformation enthalpy Q, and the value of (A_P-M_P) is very small in comparison with that of ($A_p + M_p$). Therefore, the value of ΔG_v mainly depends on the numerical value of ($A_p + M_p$). Substitute the data into Eq. (4.7), and the calculation results are as follows:

$$\Delta G_v^{P \to 7M} < \Delta G_v^{P \to 5M} < 0 \quad (4.7)$$

where P represents parent phase austenite. Formula 4.7 shows that 7 M has lower free energy than 5 M in the phase transformation process, that is, 7 M is more stable than 5 M. Therefore, no matter the martensite phase transformation or its reverse phase transformation occurs, the microwire with 7 M martensite should have higher phase transformation temperature. However, in this paper, we find a completely opposite phenomenon, which is also related to the higher internal stress in the prepared microwire. Since 7 M is stable compared with 5 M, 7 M can be obtained by the phase transformation of 5 M through applying external stress. Therefore, the higher internal stress of the as-prepared microwire provides a driving force to form 7 M martensite during the phase transformation. After the ordered heat treatment, the internal stress was released and only 5 M martensite was formed during the phase transition.

4.4.2.2 Ni–Mn–Ga-Fe Microwire

(1) Influence of Fe on M_T temperature of microwire treated by ordered heat treatment

Figure 4.30 shows the DSC curves of the microwire after heat treatment, and the phase transformation temperature of microwires with different Fe content ($x = 1 \sim 6$) is measured, indicating that the phase transformation temperature of microwire after heat treatment is higher than that of the as-prepared microwire [17]. It can be seen from the figure that, similar to the as-prepared microwire, there is only one endothermic and one exothermic peak in the heating and cooling process, indicating that the M_T of this group of microwires is one-step thermoelastic M_T. In addition, the

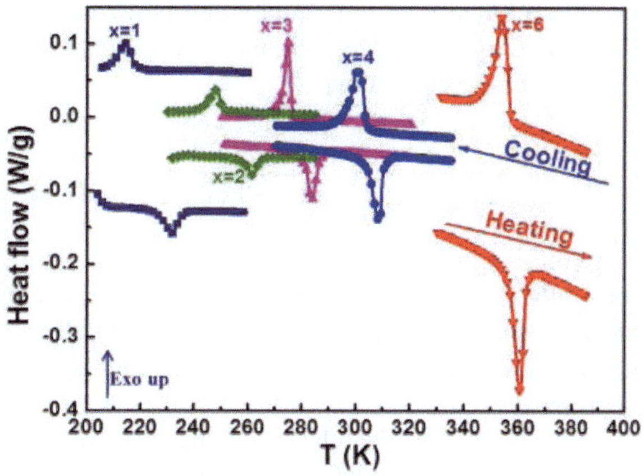

Fig. 4.30 DSC curves of the $Ni_{50}Mn_{25}Ga_{25\text{-}x}Fe_x$ ($x = 1 \sim 6$) microwire after heat treatment

large span of phase transformation temperature also indicates that the M_T temperature of this kind of microwire is very sensitive to the composition. The phase transformation temperature, phase transformation hysteresis and phase transformation enthalpy are listed in Table 4.11.

Figure 4.31 shows the effect of Fe content on the M_T temperature of $Ni_{50}Mn_{25}Ga_{25\text{-}x}Fe_x$ microwire. According to the degree of decrease of critical temperature and the classification described by Wayman, when $x < 3$, $A_f > A_s > T_O > T_O^{'} > M_s > M_f$, so the MT of the alloy should belong to the first kind of M_T. With the increase of Fe content, when $x > 3$, $A_f > T_O > M_s > A_s > T_O^{'} > M_f$, that is, Fe doping causes the type of MT to change. Similarly, for a fixed atomic percentage ratio between Ni and Ga, when Fe was used to replace Mn, and the martensite positive and negative phase transition temperature (M_s, M_f, A_s, A_f) of the alloy increased linearly with the increase of Fe content. When Fe content is increased from 1 to 6%,

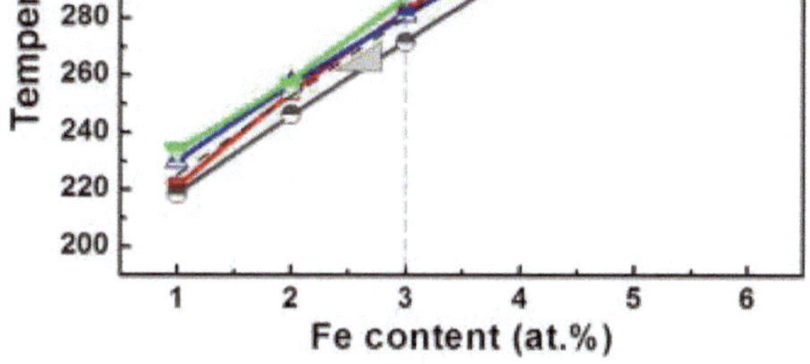

Fig. 4.31 Influence of Fe content on M_T temperature of heat-treated $Ni_{50}Mn_{25}Ga_{25\text{-}x}Fe_x$ microwire

the phase transformation temperature is increased by 140.4 K. In the heat-treated $Ni_{50}Mn_{25}Ga_{25\text{-}x}Fe_x$ ($x = 1$–6) microwire, the increased rate of phase transformation temperature was as follows:

$$\frac{\Delta T}{\mathrm{Fe}(at.\%)} = \left|\frac{T_6 - T_1}{\mathrm{Fe}_6 - \mathrm{Fe}_1}\right| = \left|\frac{361.4 - 221}{6 - 1}\right| = 28.1K/at.\% \quad (4.8)$$

In other words, after heat treatment, when Ga atoms are replaced by Fe atoms, for 1 at%. increase of Fe atoms, the M_T temperature is increased by 28.1 K, the M_T temperature of the alloy changes dramatically with the increase of Fe atom, and the increasing rate of M_T temperature is slightly slower than that of the as-prepared microwire.

At the same time, the relationship between M_T temperature and the composition of Ni–Mn–Ga-Fe microwire was analyzed, and the empirical expression of M_T temperature and composition was also divided into two regions considering the change of M_T type with the increase of Fe content:

for $x < 3$

$$T_{M_s} = -49.87X_{N_i} + 57.75X_{M_n} + 47.75X_{G_a} + 76.8X_{F_e} \quad (4.9)$$

for $x > 3$

$$T_{M_s} = -0.944X_{N_i} + 11.85X_{M_n} - 1.32X_{G_a} + 20.2X_{F_e} \quad (4.10)$$

In the formula, X_{Mn}, X_{Ga} and X_{Fe} are the atomic percentage of Mn, Ga and Fe in the alloy, respectively. It is found that the atomic percentage coefficient of Fe is relatively large, so the Fe content can be substantially adjusted the phase transition temperature. Moreover, the M_s of different M_T types of quarternized alloys after heat treatment can be roughly estimated by this formula.

Figure 4.32 shows the influence of electron concentration on the phase transformation temperature of $Ni_{50}Mn_{25}Ga_{25\text{-}x}Fe_x$ alloy. It can be seen from the figure that with the increase of electron concentration, the M_T temperature of the alloy increases monotonically. Similar to the case of as-prepared microwire, the empirical formula of phase transformation temperature and valence electron concentration of microwire after Fe doping ordered heat treatment is:

$$M_s(K) = 7.5(e/a) - 5113 \quad (4.11)$$

Compared with the empirical formula of as-prepared microwire, it is found that the changing trend of ternary alloy [32] is closer to that of phase transition temperature with the increased coefficient of electron concentration after ordering.

(2) Phase transformation hysteresis and enthalpy of ordered heat-treated microwire

Figure 4.33 shows the variation of the phase transformation hysteresis of the heat-treated microwire. It is found that the higher the Fe content is, the higher the phase

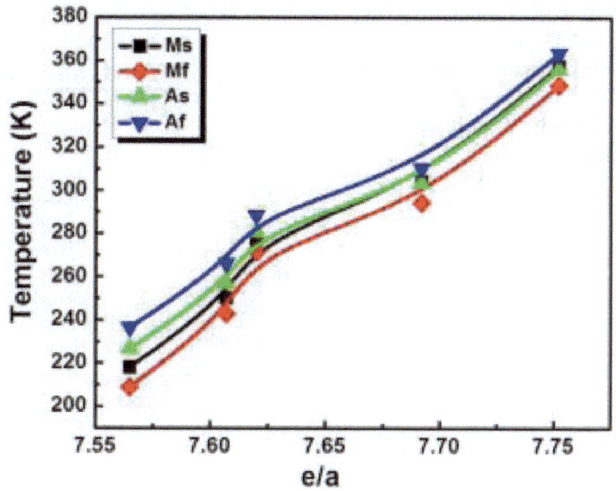

Fig. 4.32 Effect of electron concentration (*E*/*A*) on M_T temperature of $Ni_{50}Mn_{25}Ga_{25-x}Fe_x$ microwire after heat treatment

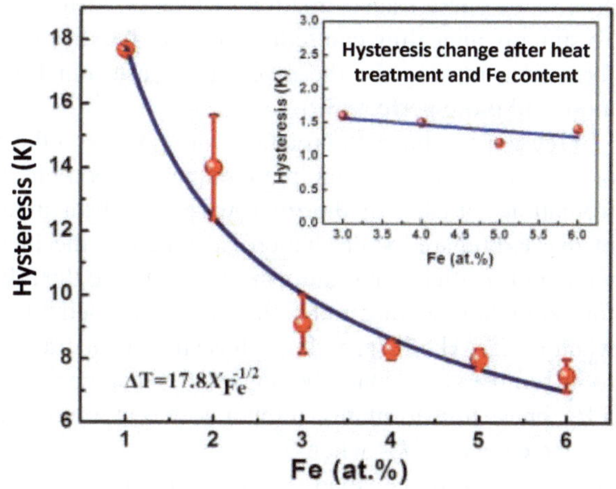

Fig. 4.33 Influence of Fe content on phase transformation hysteresis of $Ni_{50}Mn_{25}Ga_{25-x}Fe_x$ microwire after heat treatment

transition temperature is, and the lower the phase transformation hysteresis. The fitting results show that the phase transformation hysteresis temperature is inversely proportional to the 1/2 power of the Fe atomic percentage content, i.e.

$$\Delta T = 17.8X_{Fe}^{-1/2} \tag{4.12}$$

The inset in Fig. 4.33 shows the variation of phase transformation hysteresis before and after heat treatment with Fe content. It can be seen that the average phase transformation hysteresis is decreased by 1.5 K after heat treatment. The reason can be explained as follows: thermal hysteresis is due to friction in the process of phase interface moving. With the progress of annealing, the number of defects within grains is decreased while the internal stress is decreased as well, thus the phase interface

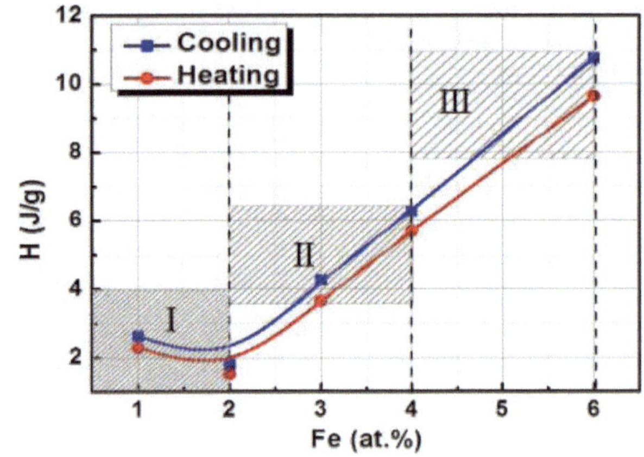

Fig. 4.34 Influence of Fe content on the phase transformation enthalpy of $Ni_{50}Mn_{25}Ga_{25-x}Fe_x$ microwire after heat treatment

friction decreases. Therefore, the energy of interface friction to be overcome in the process of phase transformation decreases, resulting in smaller thermal hysteresis. To sum up, annealing can reduce the thermal driving force required for phase transformation, making the thermoelastic phase transformation properties of materials more and more perfect [39].

Figure 4.34 shows the relation curve between the phase transformation enthalpy of the microwire after heat treatment and Fe content. It is found that the phase transformation enthalpy of the microwire after heat treatment is also higher than that of the as-prepared state, which is related to the increase of phase transformation temperature after heat treatment. According to the relationship between phase transformation temperature and enthalpy, it is divided into the following three categories: when $x < 2$, $\Delta H \approx 2$ J/g, and the phase transformation temperature is much lower than room temperature, which belongs to first category M_T; when $3 < x < 4$, ΔH at about 5 J/g, phase transformation temperature is at room temperature, which belongs to the second category M_T; when $5 < x < 6$, $\Delta H > 8$ J/g, the phase transformation temperature is much higher than room temperature, which belongs to the third category M_T.

(3) Microscopic mechanisms for the variation of phase transformation temperature

Three factors are generally considered to influence the phase transformation temperature of Heusler alloy martensite: (1) The change of electron concentration in the matrix; (2) the size of primitive cell; (3) precipitates. The above analysis shows that the stability of $L2_1$ crystal structure for Ni–Mn–Ga alloy doped with Fe is closely related to the conductive electron concentration in the crystal.

According to the free electron model of Fermi-Sommerfeld electronic theory referred in energy band theory [43], the movement of each atom can be approximately regarded as independent (near free electron approximation) with a series of determinate eigenstate, thus a single electron approximation system macroscopic state can be described by the statistical distribution of electrons. For the equilibrium state of the system, the quantum free electron theory believes that the state of the free electron obeys the quantum statistical law of Fermi-Driac. The energy of free electrons in metals is quantized and constitutes a quasi-continuum spectrum.

The probability of a state with energy E being occupied by an electron is f(E), which can be described by a definite Fermi distribution function:

$$f(E) = \frac{1}{\exp\left[\frac{E-E_F}{k_B T}\right] + 1} \tag{4.13}$$

where E_F is Fermi energy, k_B is Boltzmann constant, T is kelvin's temperature.

When $E = E_F, f(E) = 1/2$; $E \gg E_F + k_B T, f(E) \approx 0$; $E \ll E_F + k_B T, f(E) \approx 1$.

The above equation suggests that when T → 0 K, the Fermi surface is the interface of electron-occupying region and electron-non-occupying region in k space; when T ≠ 0 K, most electrons are distributed in Fermi surface and the energy is less than E_F, only a few electrons are located on the Fermi surface. If the Fermi surface is reduced, according to the quantum statistical law of Fermi-Driac, the energy of each electron will be reduced, which will inevitably lead to the energy reduction of the whole system, and then the system will be in a relatively stable state. According to Hume-Rothery et al. [44], at a certain electron/atom ratio, when the Fermi surface is in contact with the (110) Brillouin region, the periodic structure of the crystal can exist stably. From the perspective of band theory, when the electron concentration reaches a critical value and the Fermi surface contacts the boundary of Brillouin area, the electrons above the Fermi surface are likely to jump to the corner of the Brillouin region, making the energy of the system increase sharply. In order to reduce the free energy, the lattice will be distorted, so that a new Brillouin region will be formed to contain those high-energy electrons. When the percentage of Ni and Mn is constant, the valence electron concentration of the alloy increases with the increase of Fe content, which leads to the increase of the number of conductive electrons in the crystal, resulting in the overlap between Fermi surface and Brillouin region boundary [44]. According to the different ratio of electron/atom, the crystal lattice parameters will change in the periodic direction, and the corresponding crystal structure will be unstable, thus the M_T temperature rises. In addition, size is also one of the factors that affect the temperature of SMAS MT. As aforementioned, the atomic radius and valence electron number of Ga is 1.81 Å and 3, respectively, and that of Fe is 1.72 Å and 8, respectively. Therefore, replacing Ga atoms by Fe atoms with small radius and high valence state will cause the shrinkage of cell volume and the increase of electron concentration E/A, which will increase the MT temperature of the alloy. In addition, no precipitates were found in XRD, SEM and TEM, so it is considered that the increase of M_T temperature is the result of the combined effect of the increase of electron concentration and the shrinkage of cell volume.

By comparing the phase transition temperature of the as-prepared and as treated microwires, it is found that the ordering of atom increases the phase transition temperature, which may be attributed to the followings: (1) the change of alloy composition, such as the loss of Mn during annealing, the aggregation of Ga vacancies [25]; (2) the change of microstructure, such as the elimination of defects and the homogenization of element distribution during annealing; (3) the improvement of atomic ordering degree and the release of internal stress make the shear deformation of the lattice easier [45].

4.4.3 Magnetic Properties of Microwires After Ordered Heat Treatment

4.4.3.1 Ni–Mn–Ga Microwire

In order to compare the magnetic properties of NMG1 microwires before and after ordering, the magnetization curves of microwires in martensite state were tested at the same temperature (M_s-T_{test}), and the results were shown in Fig. 4.35 [1, 4].

At the same temperature, the saturation magnetization of the ordered microwires is significantly increased from 40 to 64 emu/g. In addition, the saturation magnetic field intensity of the ordered microwire is ~7 kOe, while the prepared microwire does not reach saturation until 50 kOe, which is consistent with the magnetization curve of the as-prepared NMG3 microwire. Both the increase of saturation magnetization and the decrease of saturation magnetic field intensity prove that the magnetic properties of the microwire are improved.

The improvement degree of magnetic property of Ni–Mn–Ga alloy depends on the ordering degree of the alloy, or rather, whether the Mn atom occupation is proper or not in the solid solution. Proper occupation can make the electrons in the outer layer of Mn–Mn atoms interact with each other, so as to improve the ferromagnetism of the alloy [46]. According to the experimental results, the ordered heat treatment does cause the ordering of atoms in the microwire, which is the reason for the improvement of magnetic properties. In addition, differing from the as-prepared microwire whose H* is about 4.5 kO(e), the magnetization curve of as-treated microwire rises rapidly

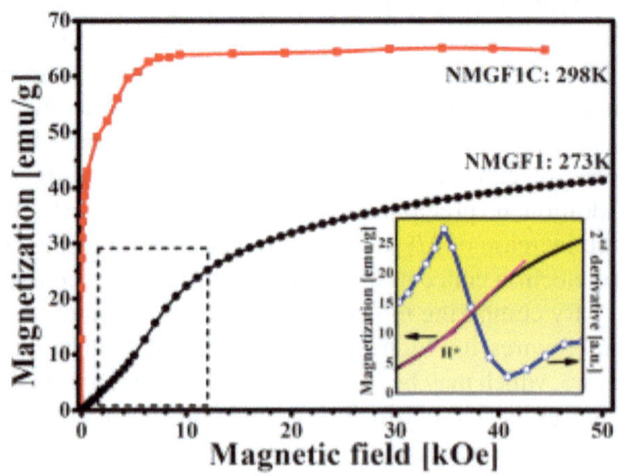

Fig. 4.35 Magnetization curve of NMG1 microwire before and after ordered heat treatment

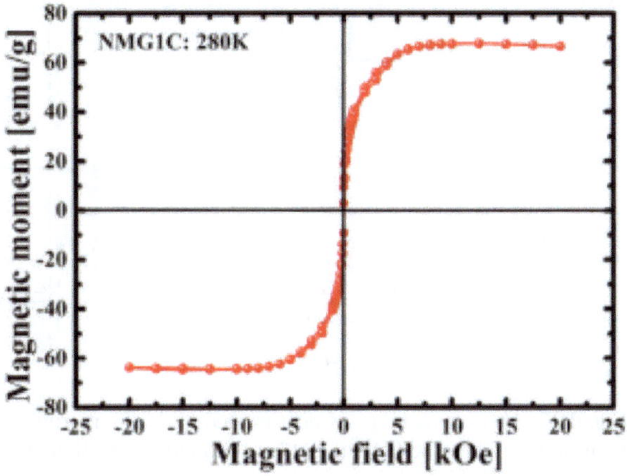

Fig. 4.36 The hysteresis loop of NMG1C microwire with martensite microstructure

before 0.5 kOe and reaches magnetization saturation at ~7 kOe. This indicates that the magnetocrystalline anisotropy of the microwires is improved after ordered treatment, and therefore the MIR process may occur at very low magnetic fields, i.e., the twin boundary can be activated at even lower magnetic fields. Figure 4.36 shows the hysteresis loop of martensite phase of NMG1C microwire at 280 K. It can be seen that the microwire almost has no hysteresis phenomenon. Reference [25] showed that the low saturation magnetic field intensity and high saturation magnetization after heat treatment suggest that the internal stress is released and the defects are eliminated. Therefore, the increase of saturation magnetization and the increase of saturation magnetic field intensity of the ordered NMG1C microwires indicate the internal stress in microwire is decreased.

4.4.3.2 Ni–Mn–Ga-Fe Microwire

The magnetic properties of NMGF6 single microwire and multiple microwire were tested in longitudinal field (the magnetic field direction is parallel to microwire axis, $H_{\|}$) and perpendicular field (the magnetic field direction is parallel to microwire axis, $H_{\perp}$). As shown in Fig. 4.37a [17, 47], the hysteresis loops obtained in $H_{\|}$ field have high initial magnetic sensitivities, and rapidly reach saturation at relatively lower magnetic field intensity (9 kOe). Moreover, in the high field zone, the saturation magnetization is slightly lower than that in $H_{\perp}$ direction (~3.7%). As presented in Fig. 4.37b, the changing trend of hysteresis loop for multiple microwire is consistent with that of single microwire. At high field zone, the saturation magnetization in $H_{\|}$ direction is slightly lower than that in $H_{\perp}$ direction (~6.3%. The inset shows that coercivity in $H_{\|}$ and $H_{\perp}$ direction is ~ 9.6 and 17.5 Oe, respectively. To sum up, compared with the $H_{\perp}$ direction, the $H_{\|}$ direction has a low magnetic crystal anisotropy, showing the characteristics of easy magnetization. Hereinafter, all the magnetic properties tests were carried out along the easy magnetization direction.

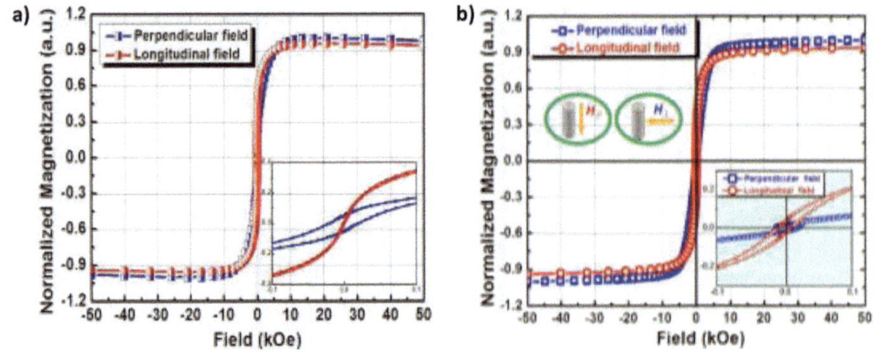

Fig. 4.37 Hysteresis loops of heat-treated NMGF6 microwires obtained under perpendicular and longitudinal magnetic field. **a** single microwire; **b** multiple microwires

4.5 Effect of Grain Growth on Microstructure and Properties of Ni–Mn–Ga–X Microwires

4.5.1 Heat Treatment Process for Grain Growth

In this section, Ni–Mn–Ga microwires with alternate or slub grain morphology were expected to be obtained through heat treatment of grain growth, in order to reduce the obstacle of grain boundaries to the movement of twin boundaries as well as to obtain microwires with larger MFIS through training. Therefore, we have mainly systematically studied the influencing factors of the heat treatment process of grain growth for Ni–Mn–Ga microwires and obtained the optimal process parameters [2, 4].

Grain growth process occurred in the Ni–Mn–Ga microwires during the course of heat preservation at 1273 K for 1 h when putting microwires into quart tubes as well as passing Ar gas as the protective atmosphere. The related results for the crystal grain morphology of the microwire under this heat treatment condition were shown in Fig. 4.38, showing that the crystal grains in the microwire have been effectively grown, and there were 1–2 crystal grains within the microwire diameter. Obvious martensite twins can be seen on the microwire surface, indicating that the microwires have already been in a martensite state at room temperature. However, it can be seen from the figure that the microwire cracks have appeared at the grain boundary, which was related to the loss of composition at the grain boundary and the uncoordinated deformation of the phase transformation process. In the process of the cooling process from the grain growth temperature to the martensitic transformation temperature, the martensitic transformation has occurred. However, because the martensite formed after phase transformation in different grains has different orientations, leading to the inconsistent deformation at the grain boundaries. On the other hand, due to the volatilization of the elements at the grain boundaries, the bonding strength at the grain boundaries has been reduced, and different grains will crack at the grain boundaries

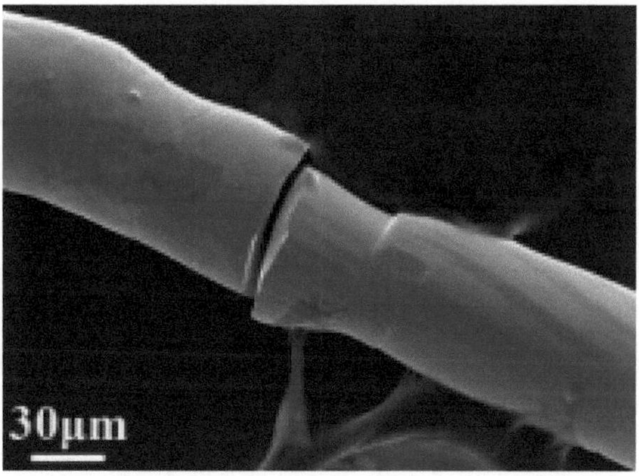

Fig. 4.38 The SEM morphology of microwire (NMG1A1273/1 h)

due to the different orientations of martensite deformation. In the case of severe fracture along the grain boundary, the microwire will break, as shown in the shape of the middle crystal grain. Although this situation was not conducive to the acquisition of polycrystalline MFIS, the single crystal particles were aligned in the resin under a magnetic field. In the collective, it was also expected to obtain larger MFIS in resin matrix composites. Overall, ensuring the bonding strength for grain boundary is essential for obtaining MFIS in polycrystalline Ni–Mn–Ga microwires.

In the process of optimizing the parameters for the heat treatment of grain growth, the main parameters considered included factors such as oxidation and element volatilization. High vacuum or inert gas protective atmosphere can be used to ensure that the microwires did not oxidize at high temperatures. Considering the problem of different martensite orientations in each grain during cooling, heat treatment assisted under stress can be designed to make the martensite variants preferentially oriented and reduce deformation incompatibility. In addition, the volatilization of elements can be achieved by increasing the air pressure in the quartz tube at high temperatures. In summary, the design of increasing the amount of Ar gas filled in the vacuum quartz tube, on the one hand, can prevent oxidation and increase the pressure to prevent element volatilization. On the other hand, the pressure of Ar atmosphere increases at high temperatures also induced pressure around the microwires.

Therefore, the design experiment was designed to judge the influence of air pressure on the heat treatment result, air pressures of 0, 0.1, 0.3 and 0.5 atm were selected to heat the microwire for 3 h at 1273 K. According to the universal gas law PV = nRT, the pressure will increase by about 4 times with the temperature increasing from 297 to 1273 K. Therefore, the observation of the influence of air pressure on the grain growth of microwires was carried at high temperatures, the pressure of 0, 0.4, 1.2 and 2 atm, respectively. In order to guarantee the consistency of the results, 6 microwires were selected and each microwire was divided into 4 sections, which were used for researching the heat treatment process under different air pressures. After the heat treatment, the EDS component analysis of the different microwires

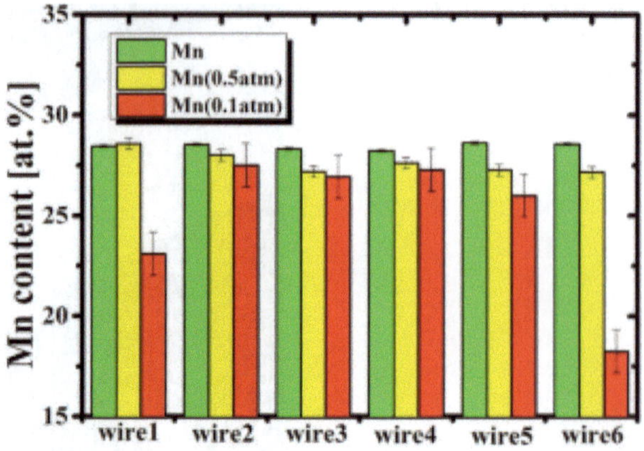

Fig. 4.39 The Mn content of NMG1 microwires after heat treatment under different Ar gas pressure states

was performed, and the results were compared with the prepared microwires. The microwires undergone serious oxidation and volatilization under 0 air pressure. The composition of the Mn element after heat treatment at 0.1 and 0.5 atm have been compared to that for the prepared state, and the results have been shown in Fig. 4.39.

The results have shown that under the pressure of 0.1 atm, the Mn content in microwire 1 and microwire 6 decreased significantly, and the Mn element in microwires 2–5 also decreased in different ranges. The large loss of the element of Mn element would seriously affect the performance of the microwires, so the Ar gas protective atmosphere of 0.1 atm cannot meet the requirements of the research. In the case of 0.5 atm, the Mn element component in microwire 2–6 has exhibited different degrees of loss, but the loss is less, and the Mn element content in microwire 1 is slightly higher than the prepared value, which may due to the volatilization of Ga element. Based on the test results, the Ar gas pressure was continued to be increased to 1 atm, but the results showed that the Ar gas atmosphere was too high to inhibit the growth of crystal grains. Therefore, an Ar atmosphere of 0.5 atm was selected for subsequent experiments.

Taking into account that the addition of Ar gas will inhibit the growth of grains to a certain extent, different temperatures of heat treatment and holding times have been selected to optimize the parameters for heat treatment. And the typical morphologies of microwires obtained under different heat treatment parameters have been shown in Fig. 4.40. Figure 4.40a and b showed the results of NMG1 microwire treated at 1273 K for 3 h and 12 h, respectively. It has been found that the prolonging time has not led to a bamboo-crystal state. Figure 4.40c and d exhibited the SEM morphologies of NMG1 microwires treated at 1323 K for 3 h. Such results indicated that a bamboo-crystal morphology has been achieved, which proved that the increasing heat treatment temperature can be more effective than extending the heat treatment time. A similar heat treatment at 1323 K for 3 h has been applied to the NMG2 microwire. As shown in Fig. 4.40e and f, the morphology of bamboo-crystal was also obtained. However, due to the difference in composition and size between different microwires, the activation energy for grain growth was also different. In many cases, it was difficult

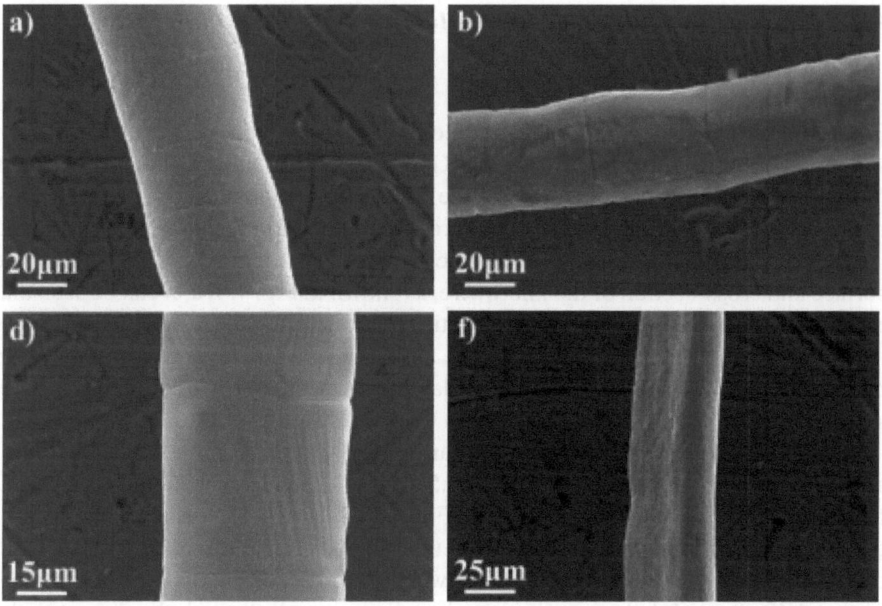

Fig. 4.40 The crystal morphology under different heat treatment temperature and time with Ar atmosphere of 0.5 atm. **a** 1273 K/3 h, **b** 1273 K/12 h and **c**, **d** 1323 K/3 h NMG1; **e**, **f** 1323 K/3 h NMG2

to grow continued after trigeminal grain boundaries have appeared in some positions, therefore, the completely bamboo-crystal state can hardly be achieved. The result for the optimization of hear treatment can only increase the proportion of bamboo-crystal structure in the microwire, so as to reduce the critical stress of the movement for twin boundaries.

The composition of the NMG2 microwire after heat treatment at 1323 K for 3 h was analyzed by EDS, and the results were shown in Table 4.12, which exhibited that the composition changed little after heat treatment, i.e., the content for Ga element decreased a loss of 0.4 at.% on the microwire surface comparing to the prepared state. Although the composition uniformity of Mn element has been slightly reduced (the standard deviation for Mn content rose from 0.3 to 0.4 and 0.5), the average composition has basically not changed. The result of both raised a slight increase for the content of Ni element, and the transformation temperature will increase after heat treatment.

4.5.2 Grain Morphology and Interface of Oligocrystalline Microwires

Figure 4.41 showed the SEM of the morphology for cross-section and outer surface in a typical oligocrystalline microwire (NMG1A1273/3 h/0.5 atm). Compared with the morphology of the prepared microwire, the microwire grains have grown significantly accompanying 1–3 grains within the microwire diameter range after heat treatment. After heat treatment, a coarse martensite twin layer structure can be seen on the surface and cross-section of the microwire, as shown in Fig. 4.41a, it can be seen that the martensite twin layer spanned the entire microwire diameter, and the thickness of a single twin layer can reach 6 μm. The middle twin wafer layer ended at the grain boundary, and the thickness of the lamellae grew with the growth of the grain size, shown in Fig. 4.41b.

The morphology and orientation of bamboo crystal (NMG2A1323/3 h/0.5 atm) and oligocrystalline (NMG1A1273/3 h/0.5 atm) microwires were studied by EBSD technology. The results are shown in Figs. 4.42 and 4.43, respectively. All of them were EBSD orientation maps from the sample normal direction to match IPF colors. The color of the grains in figures was colored according to their orientation. Figure 4.42 showed that grains in the microwire were all through the diameter, the interface among the grains was straight and clear, and the size of the crystal grains along the microwire axis was about 30–60 μm. According to the color of each grain, it can be seen that a large difference in orientation existed between adjacent crystal

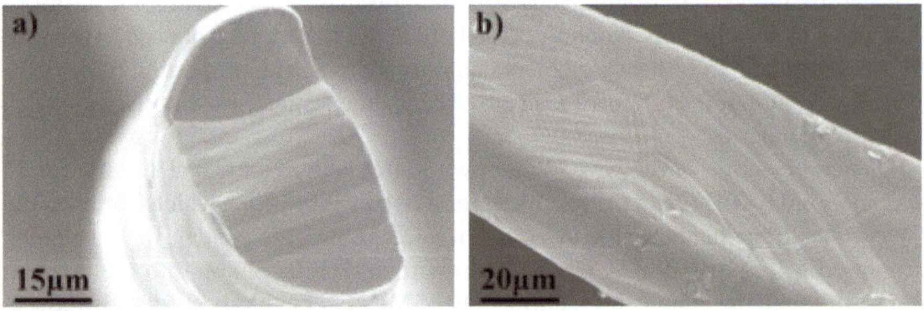

Fig. 4.41 SEM morphology of the oligocrystalline microwire

Fig. 4.42 EBSD map of the bamboo crystal microwire (NMG2A1323/3 h/0.5 atm)

Fig. 4.43 EBSD map of the oligocrystalline microwire (NMG1A1273/3 h/0.5 atm)

grains, indicating that the texture arose from the preparation process for microwires by the spinning method has not been inherited after heat treatment, so training was required in the subsequent research to reduce the misorientation of martensite variants in adjacent grains as well as the number of variants, which laid the foundation for obtaining MFIS.

Figure 4.43 showed that a large number of trigeminal grain boundaries were distributed on the right side of the microwire. There were 1–2 crystal grains in the radial direction of the microwire in the cross-section range with straight and clear grain boundaries, which has not approached bamboo crystal morphology. The size of the crystal grains that did not penetrate the microwire radial direction is small, about a dozen to 30 μm along the axial direction, while the size of the crystal grains that penetrate the microwire radial direction was larger, and the largest crystal grain size in the scanning range reached more than 70 μm. The orientations for the grains at the trigeminal grain boundaries were relatively consistent, but those for the grains on the left were quite different. Therefore, to obtain a larger MFIS in this kind of microwires, effective training was also required.

The characteristics of grain boundaries have a crucial influence on the mechanical properties for oligocrystalline microwires. Figure 4.44 showed the surface scanning result of EDS composition within four grain range of the oligocrystalline microwire (NMG2A1323/3 h/0.5 atm) for the outer surface. It can be seen from the figure that the content of the Mn element is slightly decreased at the grain boundaries, indicating that there was a small amount of volatilization loss of Mn element at grain boundaries. In contrast, the distribution for Ni and Ga elements were uniform throughout the scan range which proved that the weakening of the interface should be related to the volatilization of the Mn element. The volatilization of the Mn element induced a certain defect on the outer surface of the interface, i.e., stress concentration was generated at the grain boundary, which resulted in prioritized fracture at the grain boundary during the loading run as well as weaker mechanical properties of the microwire.

The weakening of the grain boundaries caused by the volatilization of Mn element mainly appeared on the surface. In order to study the bonding among the grain boundaries inside the microwire, FIB technology has been used to cut the grain boundary between the grain A and the grain B of the microwire (NMG2A1323/3 h/0.5 atm). The TEM microscopic morphology and electron diffraction results of the sample have

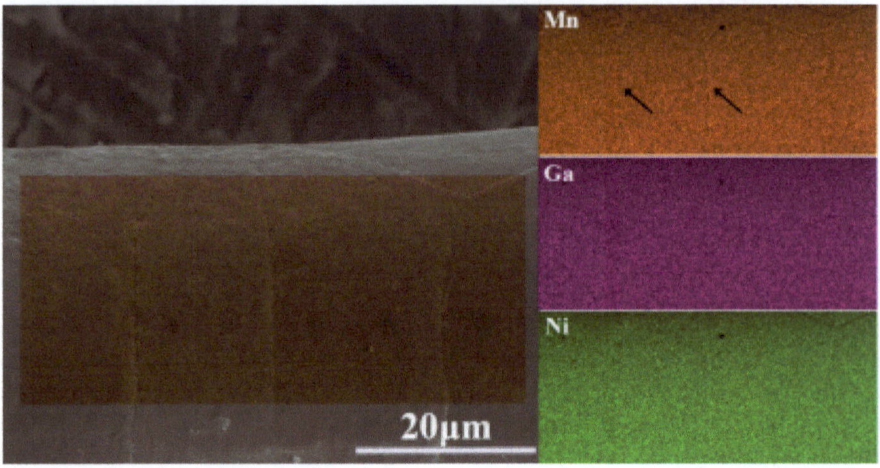

Fig. 4.44 Surface scanning spectrum of EDS analysis for the oligocrystalline microwire crossing the grain boundary (NMG2A1323/3 h/0.5 atm)

been shown in Fig. 4.45a. The interface in the figure was the grain boundary between grain A and B, which was clear and well bonded. The parallel stripes shown in grain A were secondary twin structures in the internal of the matensitic variant, which exhibited the characteristics of obvious modulation structure of martensite phase. The twin structure in the grain B appeared to be fuzzy, but its electron diffraction showed that it possessed a 5 M martensite structure.

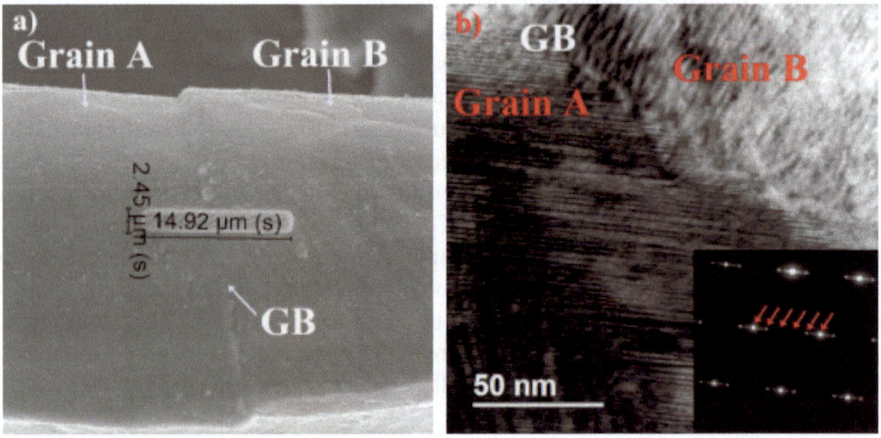

Fig. 4.45 Research on interface characteristics of the oligocrystalline microwire (NMG2A1323/3 h/0.5 atm). **a** the SEM image for grain position; **b** TEM bright field image at the grain boundary

4.5.3 *Crystal Microstructure and Phase Component of the Oligocrystalline Microwires*

With the aim of studying the microstructural characteristics of martensite phase, TEM observation technology has been applied on the thinning sample which was produced by FIB technology from a single grain of the microwire (NMG2A1323/3 h/0.5 atm), Fig. 4.46a showed the overall morphology of the grain under TEM, in which twin lamellas with straight interfaces and relatively uniform thickness have been observed. When enlarged the area, the obtained morphology was shown in Fig. 4.46b. Compared with the ordered microwire (as shown in Fig. 4.46a), the lamellar structure was more uniform, clear and regular. In which 90° magnetic domain boundaries were clearly visible, and the martensitic variants on both sides of the domain boundary behaved obviously with different orientations. It can be seen from the figure that there are only two oriented variants in a given grain, indicating that the heat treatment of grain growth reduced the number of martensite variants in a single grain. However, the prepared sample was too thin here, the electron diffraction spots were elongated, and clear structural information cannot be obtained.

In order to obtain the martensite structural information of the microwire (NMG2A1323/3 h/0.5 atm) after grain growth, XRD analysis was applied on the microwire at room temperature, and the related results were shown in Fig. 4.47. The standard positions for XRD peak of 5 M and 7 M martensite respectively have been marked on the upper frame in the figure. Comparing these peaks with the XRD spectrum of the microwire, it has been found that it was a mixed structure of 5 M and 7 M at room temperature.

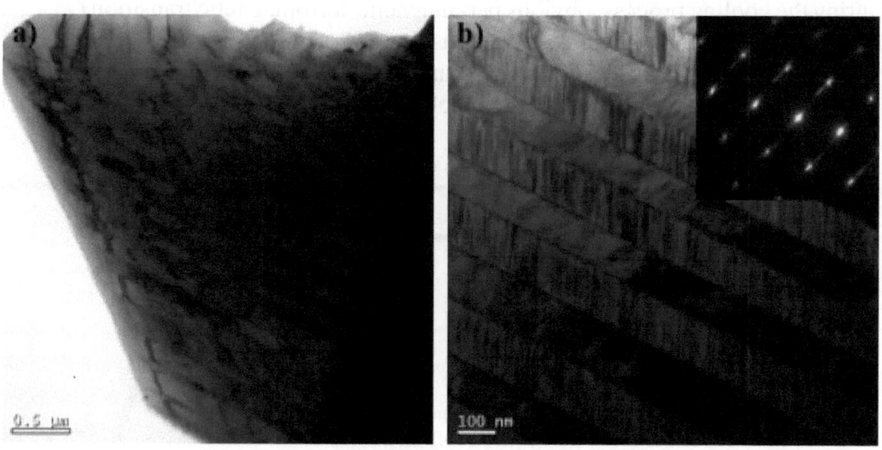

Fig. 4.46 The TEM morphology with twin structure and corresponding electron diffraction information in a single grain (NMG2A1323/3 h/0.5 atm)

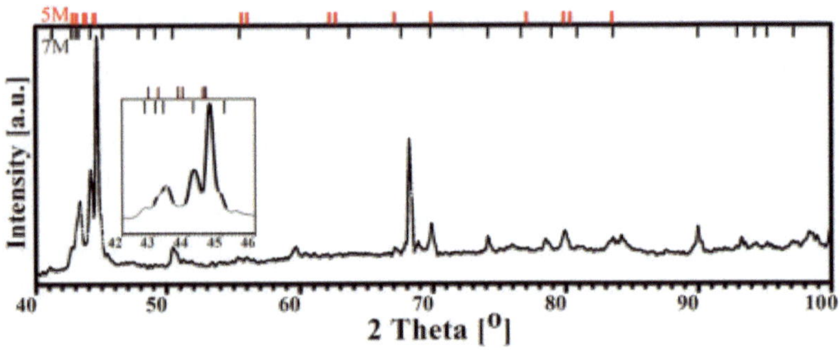

Fig. 4.47 The XRD spectrum of the oligocrystalline microwire (NMG2A1323/3 h/0.5 atm) at room temperature

4.5.4 *Characteristics of Martensitic Transformation in the Oligocrystalline Microwire*

As mentioned above, after the grains grow up, all the microwires were under martensite state at room temperature, therefore the heat treatment of grain growth increased the martensitic transformation temperature. In order to study the martensitic transformation characteristics of the microwires after the grain growth, we have taken the oligocrystalline microwire of NMG1A1273/3 h/0.5 atm as the research object. Figure 4.48 exhibited the M-T curve of the oligocrystalline microwire during the cooling process under the magnetic strength of 0.05 and 0.28 kOe as the previous one. For a better comparison, the magnetization was normalized. During the cooling process, obvious paramagnetic-ferromagnetic transition ($T_{(c)}$ and austenite–martensite structural phase transition (M_s) can be observed in the curves.

It can be seen from the figure that under the two size of magnetic fields, the results of T_c and M_s were identical, which were 353 and 370 K respectively. Such

Fig. 4.48 The M-T curves under the magnetization of 0.05 kOe and 0.28 kOe of the oligocrystalline microwire (NMG1A1273/3 h/0.5 atm)

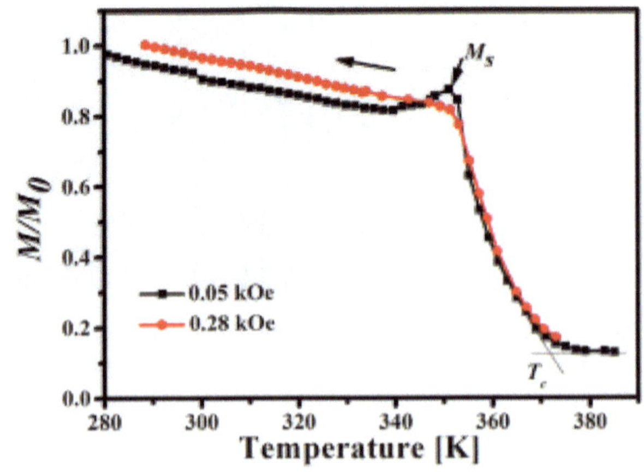

result indicated that the martensitic transformation temperature of the microwire after heat treatment of grain growth was significantly higher than that of the microwire after order heat treatment (M_s: 317.9 K), while the T_c was nearly equivalent to that of the microwire after order heat treatment (T_c: 371.8 K). As mentioned above, the composition change after heat treatment of grain growth was not very obvious, therefore the increase of martensitic transformation temperature was not only related to the composition change, but was also related to the further elimination of internal stress and defects after heat treatment. The microwire T_c is equivalent to that after the ordering heat treatment, which reflects that the ordering heat treatment after the grain growth heat treatment has achieved the effect of improving the magnetic properties of the microwire.

The equal T_c values of the microwires after two kinds of heat treatment reflected that the ordering heat treatment accompanying with the heat treatment of grain growth has achieved the effect of improving the magnetic properties. It can also be seen from the figure that under the magnetization of 0.05 kOe, the phenomenon of the decrease in the decrease of magnetization due to the higher magnetocrystalline anisotropy in martensite phase can be observed on the *M-T* curve during cooling process. However, such phenomenon disappeared at 0.28 kOe showing that martensite can reach the state of magnetization saturation under a very small magnetic field, which further proved the improvement of magnetic properties.

In this paper, EBSD technology under high temperature was used to study the morphology and orientation characteristics for austenite phase as well as the corresponding martensite phase at the same trigeminal grain boundary position of the oligocrystalline microwire. The EBSD orientation map with microwire axial IPF color matching was shown in Fig. 4.49. The color of each grain in the picture was colored according to its orientation.

Figure 4.49a showed the grain morphology characteristics for the austenite of the microwire at 373 K, reflecting the oriented relationship between the upper three grains and the microwire axis. The red grains represented the grains that were easy to magnetize. The phenomenon that the crystal axis of <001> was arranged along the direction of [100] axis of the microwire, i.e., <001> //[100]. It can be seen that the orientation difference between the three grains at the trigeminal grain boundaries was relatively large. EBSD analysis was performed on the martensite structure in the same area at room temperature, and the 5 M martensite structure was used for calibration, which was described in Fig. 4.49b corresponding to the area in the black dashed line box in Fig. 4.49a. It can be seen from the color of the martensite variants that the interior of each original austenite grain was mainly composed of two different oriented martensite variants which were alternately distributed. Cong et al. [48] reported that the color of the variant in the EBSD oriented map will change in depth if excessively local internal stress existed. Therefore, no excessively local internal stress has been induced here considering the uniform color of the variants, as shown in Fig. 4.49b. In addition, it can also be seen from Fig. 4.49b that the thickness of the twin lamellas was on the order of micrometers, which was significantly larger than that of the ordered microwires.

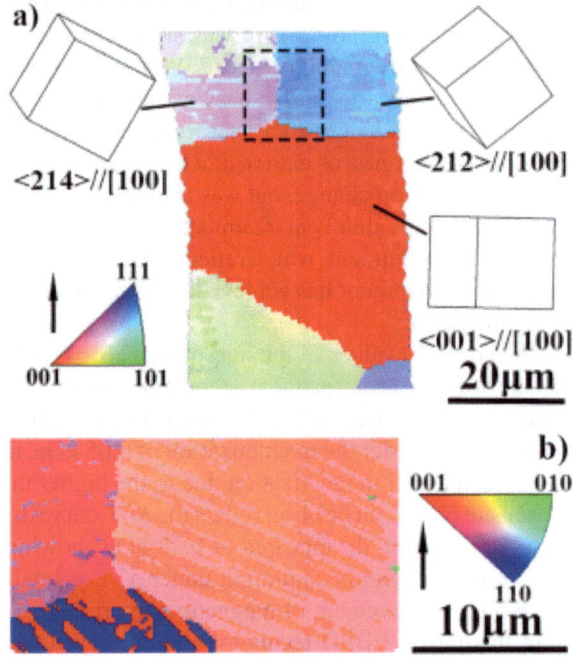

Fig. 4.49 The EBSD spectrums at the trigeminal grain boundaries of the crystalline microwire under high (low) temperature. **a** austenite phase under high temperature; **b** martensite phase under low temperature

4.6 Conclusions

Microwires can be prepared by melt-extraction. At high copper wheel speed and high heating power, particles can be produced. While at lower heating power, broad wires can be created. The uniformity of the wire diameter is related to the formation of the Rayleigh wave; the homogeneity of the wire diameter becomes worse with decreasing copper wheel speed. The optimum technological parameter for the production of wire is as follows: heating power 20 kW, rotate speed 24 m/s, feed rate 60–90 μm/s. The wire diameter is mainly distributed in 45–65 μm. The preferential nucleation region changes with the preparation process. Cellular grains are formed on the surface of the wire, with an intergranular orientation angle less than 1°. The internal austenite is characterized by tweed strain contrast and a large number of dislocations. The martensite has a multistage twin structure with a straight interface.

After the chemical ordering heat treatment, the composition loss of the wire is small and the compositional uniformity is increased. In addition, the grain size and morphology are not changed and the martensitic transformation temperature and T_c are increased with decreasing transformation width and phase transition hysteresis. Ordering treatment can change the martensite type and the volume of the cell, improve the magnetic properties of the wire and the magnetic properties of the magnetic crystal. Under the Ar gas protective atmosphere of 0.5 atm, the wire has bamboo-like grains. The orientation difference between the adjacent grains of less crystalline wire is large, and the martensite microchip layer is thick and clear.

Chemical ordering reduces the critical stress and hysteresis of stress-induced martensite, increases the superelastic recovery rate and stability, and decreases the slope of the transformation platform and the slope of the Clausius-Clapeyron relationship. The maximum theoretical strain of fiber before and after ordering is 0.017 and 0.018, respectively. Heat treatment changes the initial martensite orientation in the wire, resulting in different recovery paths of shape memory curves. *c* axis in martensite tends to be more obvious aligned along the axial direction.

References

1. M.F. Qian, X.X. Zhang, L.S. Wei et al., Effect of chemical ordering annealing on martensitic transformation and superelasticity in polycrystalline Ni–Mn–Ga microwires. J. Alloy. Compd. **645**, 335–343 (2015)
2. M.F. Qian, X.X. Zhang, L.S. Wei et al., Structural, magnetic and mechanical properties of oligocrystalline Ni–Mn–Ga shape memory microwires. Mater. Today Proc. **2S**, S577–S581 (2015)
3. M.F. Qian, X.X. Zhang, L.S. Wei et al., Microstructural evolution of Ni–Mn–Ga microwires during the melt-extraction process. J. Alloy. Compd. **660**, 244–251 (2016)
4. M.F. Qian, *Ni–Mn–Ga Investigation of the Microstructure, Thermal-induced and Magnetocaloric Characteristics in Ni–Mn–Ga Microwires* (Harbin Institute of Technology, Harbin, 2016).
5. V.A. Chernenko, E. Cesari, V.V. Kokorin et al., The development of new ferromagnetic shape-memory alloys in Ni–Mn–Ga system. Scripta Mater. **33**(8), 1239–1244 (1995)
6. L. Righi, F. Albertini, E. Villae et al., Crystal structure of 7M modulated Ni–Mn–Ga martensitic phase. Acta Mater. **56**(16), 4529–4535 (2008)
7. M.A. Taha, N.A. Elmahallawy, M.F. Abdelgaffar, Geometry of melt-spun ribbons. Mater. Sci. Eng. A **134**, 1162–1165 (1991)
8. H. Wang, D. Xing, X. Wang et al., Fabrication and characterization of melt-extracted Co-based amorphous wires. Mater. Sci. Eng., A **42A**(4), 1103–1108 (2011)
9. T. Nagase, K. Kinoshita, Y. Umakoshi, Preparation of Zr-based metallic glass wires for biomaterials by arc-melting type melt-extraction method. Mater. Trans. **49**(6), 1385–1394 (2008)
10. N.I. Baik, Y. Choi, K.Y. Kim, Fabrication of stainless steel and aluminum microwires by PDME method. J. Mater. Process. Technol. **168**(1), 62–67 (2005)
11. J.C. Bennett, C.V. Hyatt, M.A. Gharghouri et al., In situ transmission electron microscopy studies of directionally solidified Ni–Mn–Ga ferromagnetic shape memory alloys. Mater. Sci. Eng., A **378**, 409–414 (2004)
12. X.X. Zhang, M.F. Qian, Z. Zhang et al., Magnetostructural coupling and magnetocaloric effect in Ni–Mn–Ga–Cu microwires. Appl. Phys. Lett. **108**(5), 52401 (2016)
13. C.B. Jiang, G. Feng, S.K. Gong et al., Effect of Ni excess on phase transformation temperatures of NiMnGa alloys. Mater. Sci. Eng. A **342**, 231–235 (2003)
14. J. Marcos, L. Manosa, A. Planes, et al., Multiscale origin of the magnetocaloric effect in Ni–Mn–Ga shape-memory alloys. Phys. Rev. B. **68**, 094401 (2003)
15. J. Wang, H. Bai, C. Jiang et al., A Highly plastic Ni50Mn25Cu18Ga7 high-temperature shape memory alloy. Mater. Sci. Eng. A **527**(7–8), 1975–1978 (2010)

16. P. Li, J. Wang, C. Jiang, Magnetic field-induced inverse martensitic transformation in NiMnGaCu alloys. J. Phys. D (44), 285002 (2011)
17. Y.F. Liu, *Transformation Behavior and Properties of $Ni_{50}Mn_{25}Ga_{25-x}Fe_x$ Shape Memory Alloy micromicrowires* (Harbin Institute of Technology, Harbin, 2015)
18. J.Q. Li, Z.H. Liu, H.C. Yu et al., Martensitic transition and structural modulations in the heusler alloy Ni2FeGa. Solid State Commun. **126**(6), 323–327 (2003)
19. J.C. Bennett, C.V. Hyatt, M.A. Gharghouri et al., In situ transmission electron microscopy studies of directionally solidified Ni–Mn–Ga ferromagnetic shape memory alloys. Mater. Sci. Eng., A **378**(1–2), 409–414 (2004)
20. W. Wang, X.B. Ren, G. Wu, Martensitic microstructure and its damping behavior in $Ni_{52}Mn_{16}Fe_8Ga_{24}$ single crystals. Phys. Rev. B **73**(9), 092101–092104 (2006)
21. S. Sarkar, X. B. Ren, K. Otsuka, Evidence for strain glass in the ferroelastic-martensitic system $Ti_{50-x}Ni_{50+x}$. Phys. Rev. Lett. **95**(20), 205702 (2005)
22. N.V.R. Rao, R. Gopalan, V. Chandrasekaran et al., Microstructure, magnetic properties and magnetocaloric effect in melt-spun Ni–Mn–Ga ribbons. J. Alloy. Compd. **478**(1–2), 59–62 (2009)
23. H.B. Xu, Y. Li, C.B. Jiang, Ni–Mn–Ga high-temperature shape memory alloys. Mater. Sci. Eng. A **438–440**, 1065–1070 (2006)
24. T. Kanomata, S. Murakamia, D. Kikuchia et al., Magnetic properties of Ni–Mn–Fe–Ga ferromagnetic shape memory alloys. Int. J. Appl. Electromagnet Mech **2008**(27), 215–224 (2008)
25. O. Heczko, P. Svec, D. Janickovic et al., Magnetic properties of Ni–Mn–Ga ribbon prepared by rapid solidification. IEEE Trans. Magn. **38**(5), 2841–2843 (2002)
26. X. Zhou, H. Kunkei, G. Williams et al., Phase transitions and the magnetocaloric effect in Mn rich Ni–Mn–Ga Heusler alloys. J. Magn. Magn. Mater. **305**(2), 372–376 (2006)
27. Y.F. Liu, X.X. Zhang, D.W. Xing et al., Martensite transformation and magnetic properties of $Ni_{50}Mn_{25}Ga_{25-x}Fe_x$ ferromagnetic microwires for application in microdevices. Phys. Status Solidi A **212**(4), 855–861 (2015)
28. Y.F. Liu, X.X. Zhang, D.W. Xing et al., Martensite transformation and superelasticity in polycrystalline Ni–Mn–Ga–Fe microwires prepared by melt-extraction technique. Mater. Sci. Eng. A **636**, 157–163 (2015)
29. C.M. Wayman, Introduction to the Crystallography of Martensitic Transformations (Macmillan, 1964)
30. D. Soto, F. Hernandez, H. Flores-Zuniga, et al., Phase diagram of Fe-doped Ni–Mn–Ga ferromagnetic shape-memory alloys. Phys. Rev. B **77**(18), 184103 (2008)
31. I. Glavatskyy, N. Glavatska, O. Soderberg et al., Transformation temperatures and magnetoplasticity of Ni–Mn–Ga alloyed with Si, in, Co or Fe. Scripta Mater. **54**(11), 1891–1895 (2006)
32. V.A. Chernenko, Compositional instability of β-Phase in Ni–Mn–Ga alloys. Scripta Mater. **40**(5), 523–527 (1999)
33. U. Gaitzsch, R. Chulist, L. Weisheit et al., Processing routes toward textured polycrystals in ferromagnetic shape memory alloys. Adv. Eng. Mater. **14**(8), 636–652 (2012)
34. J. Gutierrez, J.M. Barandiaran, P. Lazpita et al., Magnetic properties of a rapidly quenched Ni–Mn–Ga shape memory alloy. Sensors Actuat. A Phys. **129**, 163–166 (2006)
35. A.A. Likhachev, A. Sozinov, K. Ullakko, Influence of external stress on the reversibility of magnetic-field-controlled shape memory effect in Ni-Mn-Ga, in *Proceedings of the Society of Photo-optical Instrumentation Engineers*, (Newport Beach, 2001), 4333: 197–206
36. L. Righi, F. Albertini, L. Pareti et al., Commensurate and incommensurate "5M" modulated crystal structures in Ni–Mn–Ga martensitic phases. Acta Mater. **55**(15), 5237–5245 (2007)
37. Y.F. Liu, X.H. Liu, L. Ma et al., Effect of chemical ordering annealing on Ni–Mn–Ga–Fe shape memory microwires. Chinese J. Nonferrous Metals **28**(04), 749–757 (2018)
38. V. Sanchez-Alarcos, V. Recarte, J.I. Perez-Landazabal et al., Correlation between atomic order and the characteristics of the structural and magnetic transformations in Ni–Mn–Ga shape memory alloys. Acta Mater. **55**(11), 3883–3889 (2007)

39. F. Albertini, S. Besseghini, A. Paoluzi et al., Structural, magnetic and anisotropic properties of Ni_2MnGa melt-spun ribbons. J. Magn. Magn. Mater. **242**(2), 1421–1424 (2002)
40. E.P. Da, Calorimetric observations on an NiTi alloy exhibiting two-way memory effect. Mater. Lett. **38**(5), 341–343 (1999)
41. R.F. Hamilton, H. Sehitoglu, Y. Chumlyakov et al., Stress dependence of the hysteresis in single crystal NiTi alloys. Acta Mater. **52**(11), 3383–3402 (2004)
42. C. Jiang, Y. Muhammd, L. Deng et al., Composition dependence on the martensitic structures of the Mn-rich NiMnGa alloys. Acta Mater. **52**(9), 2779–2785 (2004)
43. S. Tian, *Physical Properties of Materials* (Beijing University of Aeronautics and Astronautics Press, Beijing, 2004)
44. J. Smit, Magnetism in Hume-Rothery alloys. J Phys F Metal Phys **8**, 2139 (1978)
45. Y. Feng, J.H. Sui, L. Chen et al., Martensitic transformation behaviors and magnetic properties of Ni–Mn–Ga rapidly quenched ribbons. Mater. Lett. **63**(12), 965–968 (2009)
46. A. Ayuela, J. Enkovaara, K. Ullakko et al., Structural properties of magnetic Heusler alloys. J. Phys.-Condensed Matter **11**(8), 2017–2026 (1999)
47. Y.F. Liu, X.X. Zhang, D.W. Xing, Y.F. Liu, et al., Magnetocaloric effect (MCE) in melt-extracted Ni–Mn–Ga–Fe Heusler microwires. J. Alloys Compounds **616**, 184–188 (2004)
48. D.Y. Cong, *Study on Crystal Structure and Microstructure of Novel Ni–Mn–Ga Magnetic Shape Memory Alloys* (Northeastern University, Shenyang, 2008)

References

Chapter 5
Properties of Magnetic Shape Memory Alloy Microwires

5.1 Introduction

Ni–Mn–Ga–X ferromagnetic shape memory alloys have attracted much attention due to their thermal-driven shape memory effect (SME), superelasticity (SE), magnetic-induced strain (MFIS), damping and magnetocaloric effects (MCE). However, the intrinsic brittleness of Ni–Mn–Ga bulk alloys often induces serious intergranular fracture. Because of the fine grains and internal stress formed in the process of rapid solidification, the intergranular fracture tendency of Ni–Mn–Ga–X microwire is weakened. It is possible to produce MFIS in the microwire under complex stress conditions.

It can be seen from the previous paper that heat treatment has a significant impact on the microstructure and phase transformation of the microwire. Therefore, it is useful to study the effect of heat treatment on the properties of the microwire. In addition, one-dimensional microwire has a high specific surface area and low martensite transition hysteresis. So, the microwire may act as promising magnetic refrigerants. This chapter mainly introduces the martensitic transformation, superelasticity and magnetocaloric effect of Ni–Mn–Ga–X microwires [1–12].

5.2 Superelasticity (SE)

This section takes NMG1 microwire before and after heat treatment as an example to illustrate the effect of chemical ordering treatment on SE properties [2, 9]. The phase transition temperature of the two microwires is determined by the temperature-dependent damping spectrum (heating and cooling rates 1 K/min). The result is shown in Fig. 5.1. The phase transition temperatures of as-prepared microwire are $M_s = 287.7$ K, $M_f = 277.2$ K, $A_s = 294.4$ K, $A_f = 305.8$ K; the phase transition temperatures of ordered state are $M_s = 309.9$ K, $M_f = 300.8$ K, $As = 316.6$ K, $A_f = 334.4$ K. According to the phase transition temperature of the microwire, the SE test

X. Zhang and M. Qian, *Magnetic Shape Memory Alloys*,
https://doi.org/10.1007/978-981-16-6336-9_5

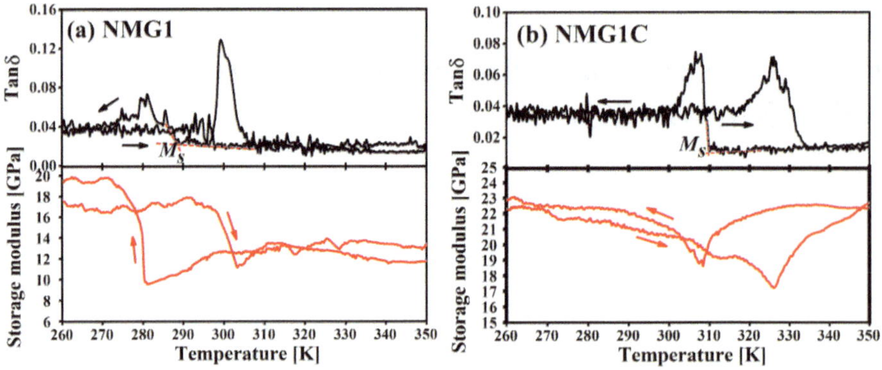

Fig. 5.1 Determination of phase transition temperature of single NMG1 microwire before and after ordering treatment. **a** NMG1; **b** NMG1C

temperature of as-prepared microwire is determined to be 288, 291, 293, 296, 298, 301, 303 and 306 K, and that of ordered microwire is 310, 313, 315, 318, 320 and 323 K. Here, the difference between test temperature T_{test} and M_s is used to express the test temperature, that is, T_{test}-$M_s \approx 0, 3, 5, 8, 10, 13$ and 15 K.

The SE curves of microwires before and after ordering at different temperatures are shown in Fig. 5.2. The brittleness of the microwire is significantly lower than that of the bulk polycrystal. At 293 K, the maximum strain of NMG1 microwire reaches 3.63% at 330 MPa, while the strain of NMG1C microwire is slightly smaller due to the overall low maximum stress value. Therefore, at 310 K, the maximum strain reaches 2.06% at 165 MPa.

As shown in Fig. 5.2b, a SE curve can be divided into four stages: Stage I is the elastic deformation stage of austenite, and then, with the increase of stress, a stress plateau appears, or the curve slope is smaller, that is stage II. Due to the non-uniformity of microwire diameter, martensitic transformation will take place at the position with a smaller diameter due to greater stress under the same load, thus forming a slope with gradually increased stress. In stage II, NMG1 and NMG1C microwires transform from austenite to 7 M and 5 M martensite, respectively. With the test, after austenite in the small diameter area has completely transformed into martensite, if the rest of the position cannot drive martensitic transformation due to insufficient stress level or the transformation position is very few, then the trend of the next curve is a smooth curve with increased slope. This process should include the elastic deformation of retained austenite and new martensite, and the ratio of residual austenite and new martensite to martensite is also discussed. Less stress induces the formation of martensite, which is stage III. If the stress continues to increase without fracture, there should be martensitic transformation. However, if the stress in some areas is enough to drive the generation of martensite in the process of stress increase, the step like morphology will appear when T_{test}-M_s is 0 K in Fig. 5.2. The reason is that the critical stress of stress-induced martensite is lower, which can drive more martensite. The unloading process includes elastic strain recovery and strain recovery

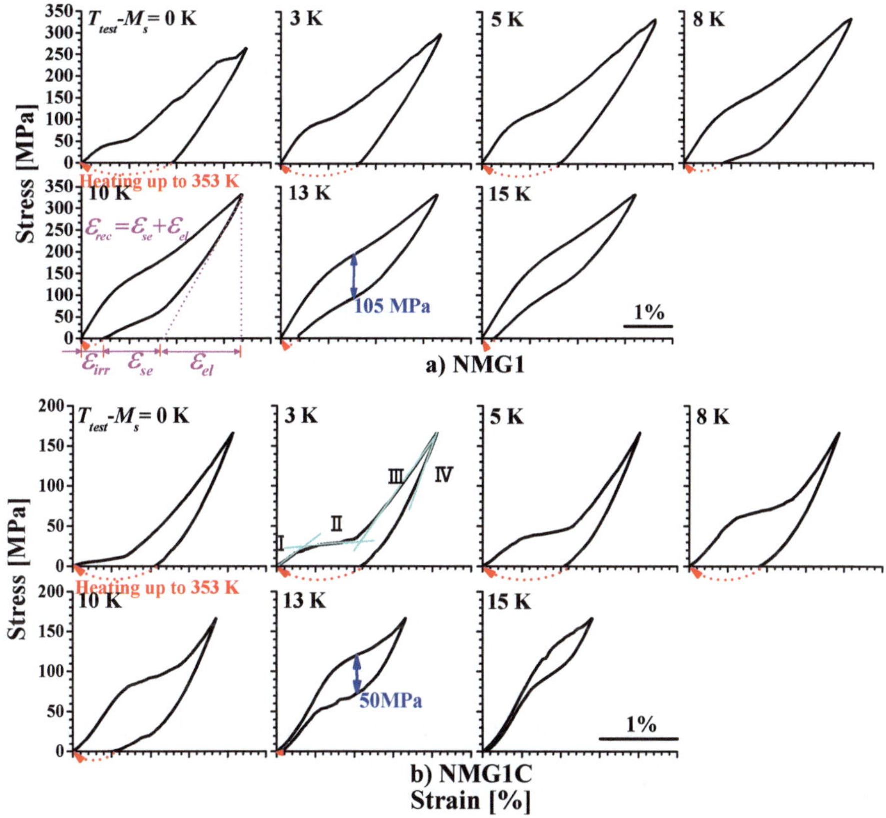

Fig. 5.2 SE comparison of NMG1 microwires at the same temperature level (T_{test}-M_s) before and after ordering treatment. **a** NMG1; **b** NMG1C

due to reverse phase transformation. We define the elastic strain recovery process as stage IV. The strain which cannot be recovered in the annealing process completely recovered, indicating that there is no plastic strain.

Compared with Fig. 5.2a, b, it is found that the critical stress of martensite induced by ordered microwires is much smaller than that of as-prepared microwires at the same temperature level (T_{test}-M_s). Firstly, the types of martensite produced by the transformation are different. Because 7 M martensite is more stable than 5 M martensite, more energy is needed for the transformation of as-prepared microwire austenite to 7 M martensite. In addition, the increase of atomic ordering degree will reduce the shear resistance of martensite, and the reduction of defects and internal stress after ordering treatment will reduce the resistance to the formation of new phase growth core so that SIM process can be produced at a lower-energy level, that is, σ_{Ms} reduction.

Figure 5.3 summarizes the slopes of the SE phases. The slope of stage I represents the elastic modulus of austenite in the microwire. The austenite modulus of the

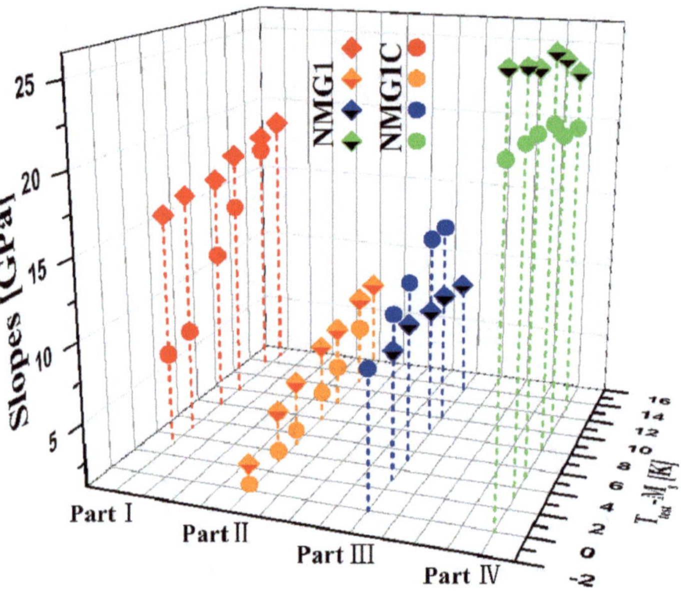

Fig. 5.3 Variation of characteristic slope of SE curve with temperature before and after ordering treatment

two microwires increases with the increase of temperature, which is related to the condensation of soft phonon modes caused by the preexisting modulation structure in austenite. At stage II, i.e., SIM process, the stress increases gradually, showing the characteristics of polycrystalline and non-uniform diameter. As can be seen from Fig. 5.3, NMG1C microwire has a smaller slope value than NMG1 in this process. This phenomenon is related to the decrease of internal stress after ordering [13]. In NMG1 microwire, the existence of internal stress hinders the formation of adaptive martensite, so that higher elastic strain energy is stored, which is the resistance of martensitic transformation. Therefore, SE needs more energy to complete. Furthermore, with the increase of ordering degree, the shear process of martensitic transformation becomes easier [14], which leads to the SE process of NMG1C microwire under low stress. Finally, the homogenization of composition and the reduction of defects make the resistance of martensitic twins in NMG1C microwire decrease, which is beneficial to the SE process.

After the end of the SE phase, the slope of the curve rises again, as shown in part III in Fig. 5.3. Although this part is generally considered as the elastic deformation stage of martensite, it also includes the unfinished SE process, the elastic deformation of retained austenite and the transformation between martensite and martensite. However, further research is needed to confirm this aspect.

Stage IV is the stage of elastic recovery, including the elastic recovery of martensite and austenite. Since the modulus of austenite increases with the increase of temperature, and the slope as shown in Fig. 5.2 decreases with the increase of temperature, it is considered that the decrease of modulus is caused by the increase of twinning mobility of martensite with the increase of temperature. This phenomenon is related to Ni–Mn–Ga single crystal, and other ferromagnetic shape memory alloys such as Ni–Fe–Ga and Ni–Fe–Co–Ga are similar [15, 16].

The hysteresis of SE cycle is a sign of energy loss. In order to compare the difference of SE hysteresis of microwires before and after ordering, the same temperature level (T_{test}-M_s = 13 K) was selected for comparison. The hysteresis is measured by the stress difference between the loading and unloading platforms of the SE curve [17]. As shown in the "13 K" SE curve in Fig. 5.2, the value of NMG1 microwire is 105 MPa and that of NMG1C microwire is 50 MPa. This result is consistent with the phase transition hysteresis in the DSC result of the microwire and is also related to the reduction of energy loss in the microwire phase transformation process after heat treatment.

To sum up, the difference of SE curve of ordered microwire compared with that before treatment is mainly concentrated in the lower critical stress σ_{Ms}, lower phase transition platform and small hysteresis.

In shape memory alloys, the recovery rate of SE is important for practical application. Figure 5.4 shows the change curve of SE recovery rate of microwires before and after ordering treatment at different temperature levels. The recovery curve can be divided into three parts: A, B and C. The recovery rate increased slowly with temperature. SE recovery of NMG1 microwire does not occur until T_{test}-M_s = 8 K. Therefore, the strain recovered in part A is only the elastic strain. In addition, due to the large external load of NMG1 microwire, the elastic strain produced is more recovered during unloading, which leads to the recovery rate of in part A higher than that of NMG1C.

After T_{test}-M_s = 8 K, the SE recovery rate of both microwires increased rapidly, as shown in part B. With the increase of temperature, more energy is needed, that is, more external stress is needed to induce the SE process. As a result, the elastic

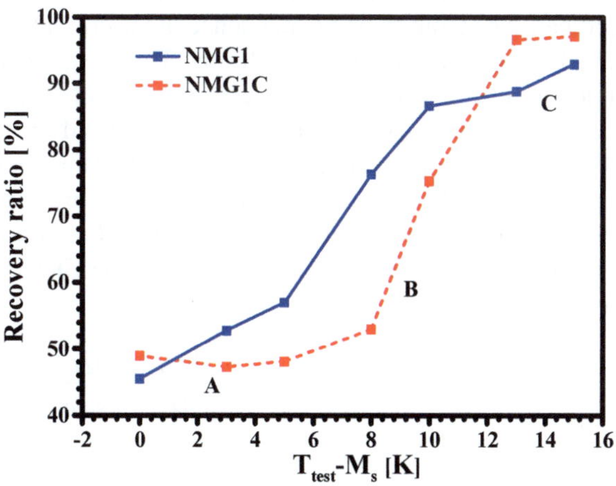

Fig. 5.4 SE recovery rate with temperature of NMG1 microwire before and after ordering treatment

strain energy stored increases, and this elastic strain becomes the driving force of the reverse phase transition during unloading, thus increasing the proportion of SE strain ε_{se} in the recovery strain. However, in this process, the recovery rate of NMG1 microwire is still higher than that of NMG1C.

With the increase of temperature to T_{test}-M_s = 13 K, the rising trend of recovery rate becomes slower when entering part C. At this time, the external stress required to drive the SE process continues to increase, so the SE process cannot be fully driven under the test stress. However, with the increase of the applied stress, the stored elastic strain energy increases, which promotes the occurrence of inverse phase transition, and thus ε_{se} continues to increase. As shown in Fig. 5.2b, the SE curve of NMG1C microwire almost completely recovered at the time of unloading when T_{test}-M_s = 13 K. However, NMG1 microwire only partially recovered at the same temperature level. As shown in Fig. 5.4, the recovery rate of NMG 1C microwire reaches 97% at T_{test}-M_s = 15 K. As mentioned before, the driving force of reverse phase transformation is the elastic strain energy stored in the loading process and the difference of chemical free energy between two phases, and the same temperature level determines the same chemical free energy difference. Therefore, the reason why the microwires are more likely to recover after ordering treatment is that the stored elastic strain energy is less consumed in the phase transformation process, which may be related to the different diameters, but more. The reason is that the decrease of the internal stress in the microwire defects after the ordering treatment reduces the energy loss of the phase interface movement, and the yield strength increases after the ordering treatment, which reduces the probability of irreversible plastic deformation and reduces the loss of elastic strain energy. In addition, the improvement of the recovery of NMG1C microwire is also related to the improvement of atomic order after ordering heat treatment.

As shown in Fig. 5.2b T_{test}-M_s = 3 K, the values of SE critical stress (σ_{Ms}), martensitic transformation end stress (σ_{Mf}), reverse transformation start stress (σ_{As}) and reverse transformation end stress (σ_{Af}) in SE curve were selected by tangent method. The results show that the onset critical stress of SE satisfies the Clausius–Clapeyron relation with temperature, and the slope of the relationship ($d\sigma_{Ms}/dT$) indicates the temperature dependence of the critical stress of SE, that is, the difficulty of SE at different temperature levels. The $d\sigma_{Ms}/dT$ values of NMG1 and NMG1C microwires are 7.6 and 7.1 MPa/K, respectively. Figure 5.5 shows the temperature dependence of σ_{Ms} of Ni–Mn–Ga microwire and a variety of materials including Ni–Mn–Ga single crystal [14, 18] and other ferromagnetic shape memory alloy single crystals [19, 20]. It is obvious that the temperature dependence of SE critical stress of Ni–Mn–Ga single crystal is much less than that of polycrystalline alloy. The reason for this phenomenon can be explained by the Clausius–Clapeyron. The equation is as follows:

$$\frac{\mathrm{d}\sigma_{M_s}}{\mathrm{d}T} = \frac{Q\rho}{\varepsilon T_0} \tag{5.1}$$

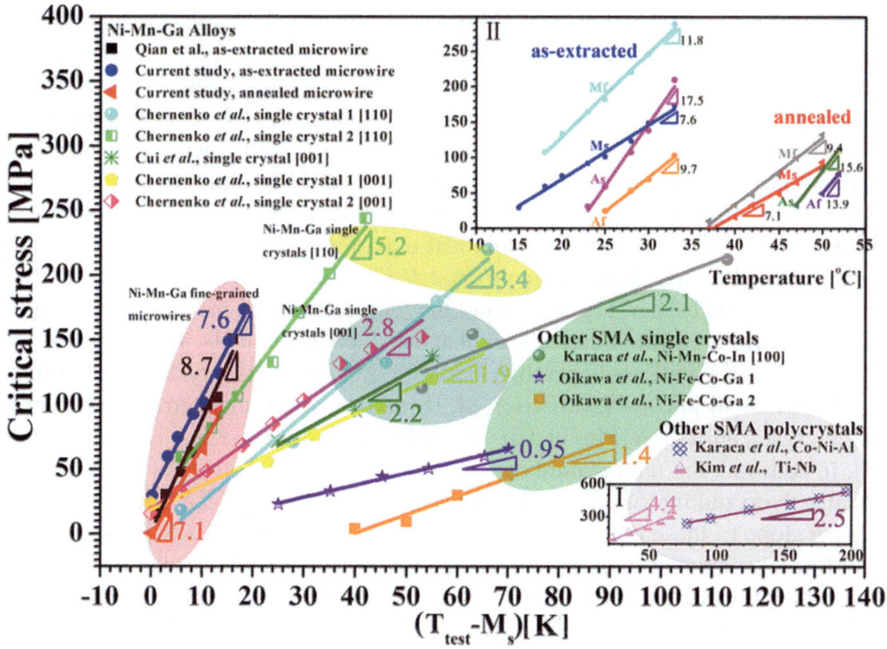

Fig. 5.5 Temperature dependence of SE critical stress of NMG1 microwire before and after ordering treatment and its comparison with other kinds of SMAs

where Q and T_0 are the enthalpy change of transformation process and the theoretical transformation temperature between austenite and martensite, respectively, and T_0 is approximately equal to $(A_p + M_p)/2$, ρ is the density of the material, and ε is the theoretical maximum strain of the twin boundary movement in the loading direction.

It can be seen from Table 5.1 that the average enthalpies Q of NMG1 and NMG1C are 4.7 and 5.0 J/g, respectively. The density of the material is 8.2 g/cm^3. The T_0 values of the two microwires are 294.2 K and 317.5 K, respectively. The theoretical maximum strain in single crystal can be obtained by the lattice parameter of martensite. However, the ε value in polycrystalline cannot be obtained by this method. However, due to the limitation of grain boundary and the existence of various variants in martensite, the ε value in polycrystal must be lower than that in single crystal, which leads to the increase of slope ($d\sigma_{Ms}/dT$). Moreover, due to the different values of Q, T_0 and ε, the $d\sigma_{Ms}/dT$ of the microwires before and after ordering are also different. Conversely, according to formula (5.1), the theoretical maximum strain

Table 5.1 Theoretical maximum strain of microwire before and after ordering treatment

Microwire	T_0 (K)	Q	ρ(g/cm^3)	Slope (MPa/K)	ε
NMC1	294.2	4.7	8.2	7.6	0.017
NMG1C	317.5	5.0		7.1	0.018

values of the two microwires in the tensile direction can be obtained by substituting the known $d\sigma_{Ms}/dT$, Q and T_0 values. The results are shown in Table 5.1. The theoretical strain value here refers to the strain induced by stress or magnetic field due to the reorientation of martensitic twins, excluding the strain caused by elastic deformation or plastic deformation. It can be seen from Table 5.1 that the theoretical maximum strain values of NMG1 and NMG1C are 0.017 and 0.018, respectively. Since the grains do not grow up during ordering heat treatment, the theoretical strain values are not much different, and the theoretical maximum strain values of polycrystals are much smaller than those of single crystals [21, 22].

As mentioned earlier, Fig. 5.5 shows the temperature dependence of SE critical stress for various alloys. It can be seen that the slope value of Ni–Mn–Ga microwire ($d\sigma_{Ms}/dT$ ~ 7.6 MPa/K) is larger than that of Ni–Mn–Ga single crystal (along [001] ~ 2.8 MPa/K; along [110] ~ 5.2 MPa/K) [14, 18] and other ferromagnetic shape memory single-crystal alloys (Ni–Mn–In–Co ~ 2.1 MPa/K; Ni–Fe–Co–Ga ~ 1.4 MPa/K) [19, 20]. In addition, the temperature dependence of SE critical stress of other conventional polycrystalline SE alloys is shown in embedded Fig. 5.5 . It can be seen that the slope of these polycrystalline SE alloys (Co–Ni–Al ~ 2.5 MPa/K; Ti–Nb ~ 4.4 MPa/K) is smaller than that of Ni–Mn–Ga microwire, which may be related to the coarse grain size of these alloys [21, 22]. In conclusion, the critical stress of SE in Ni–Mn–Ga microwire is more temperature-dependent, which should be closely related to the smaller grain size in the microwire.

Embedded Fig. 5.5 II shows the temperature dependence of the four SE critical stresses (σ_{Ms}, σ_{Mf}, σ_{As} and σ_{Af}) of NMG1 and NMG1C microwires with temperature. Among the four kinds of temperature dependence of critical stress, the temperature dependence of critical stress at the beginning of reverse phase transformation, i.e., $d\sigma_{As}/dT$, is the highest, which indicates that the temperature dependence of the reverse phase transformation process is the largest, and it decreases after ordering heat treatment, indicating that ordering makes reverse phase transformation easier.

It can also be found from Fig. 5.5 that there is an intersection point between the relationship between σ_{Ms} and σ_{As} and temperature at different temperatures. In fact, due to the fact that the fitting lines are not parallel, if the stress is increased or the temperature is increased, there will be an intersection point between them. However, the microwire will be damaged when the stress is increased or the temperature is increased, so there is only one intersection point at present. This intersection point shows that when the phase transition occurs in the process of temperature rise and fall under a large stress, the elastic strain energy formed by the external stress increases, and the elastic strain energy is the driving force of the inverse phase transformation so that the reverse phase transition can occur at a lower temperature.

5.3 Magnetic-Field-Induced Strain of Ni–Mn–Ga Microwires

5.3.1 Twin Boundary Motion Characteristics of Oligocrystalline Microwires

It is necessary to obtain MFIS in Ni–Mn–Ga alloy with low critical stress σ_{tw} of twinning boundary movement and high magnetocrystalline anisotropy Ku. The magnetocrystalline anisotropy of the alloy can be effectively improved by reducing the sample size and conducting ordering heat treatment [23], which has been described in the previous work. There are two ways to reduce the critical stress σ_{tw} of twin boundary movement, one is to make the microwire grain grow up by heat treatment, and the other is to reduce the number of variants and reduce the *tw* through stress thermal cycle training. In this section, the influence of two methods on σ_{tw} is explored by studying the twin boundary motion of oligocrystalline microwire under stress.

5.3.1.1 Analysis of Twin Boundary Motion Characteristics of Oligocrystalline Microwire

NMG1A 1273/3 h/0.5 atm oligocrystalline microwire was selected to study the characteristics of twin boundary motion under stress [9, 24]. In order to study the maximum strain value of oligocrystalline microwire, the method of breaking the microwire was used. The results are shown in Fig. 5.6. As MFIS is measured at room temperature, tensile test is also selected at room temperature. The results show that the maximum fracture strain is 4.55%, which is slightly less than the theoretical maximum strain of 5 M martensite (6.2%). In the process of drawing, it can be seen that there are three plateaus of twin boundary movement caused by different twin

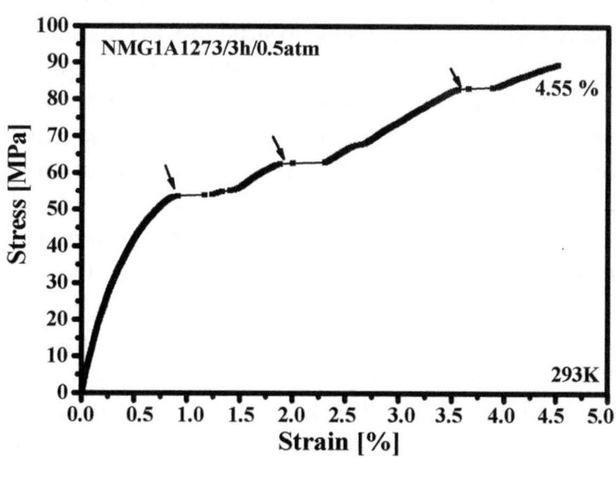

Fig. 5.6 Stress–strain curve of oligocrystalline microwire under tensile condition (tensile fracture)

Fig. 5.7 Stress–strain curve of oligocrystalline microwire under tensile condition (298 K, tensile unloading)

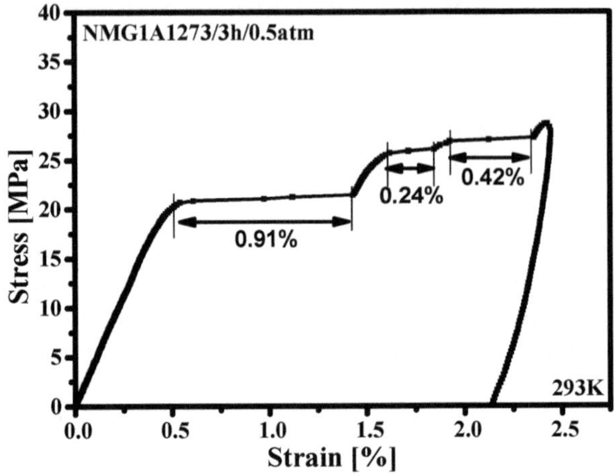

opening stress and uneven microwire diameter. The critical stress of twin boundary opening is 55 MPa, 62 MPa and 82 MPa, respectively. In this case, an effective training process is necessary to obtain MFIS in microwire.

Due to the large initial value of the oligocrystalline microwire, it is necessary to train the microwire in stress thermal cycle. Figure 5.7 is the stress–strain curve of the microwire before training. It is found that the microwire has the same phenomenon of multiple stress platforms as that of Fig. 5.6, which is also a feature of most of the oligocrystalline microwires. This paper makes a detailed analysis here. In order to prevent the microwire from fracture, the maximum strain is 2.5%. There are three moving stress platforms in the twin boundary, which together lead to ~1.57% strain. The critical stress of the first platform is ~20 MPa, and 0.91% strain is produced after starting. After that, with the increase of stress, the second and third platforms start at 25 MPa and 26 MPa, respectively, and produce 0.24% and 0.42% strains, respectively. The difference in critical stress may be related to the different orientation of twins in different grains, which leads to different critical stress. In addition, since the microwire is not a perfect slub structure, the resistance of twin boundary movement in each grain varies with the grain boundary and grain size, resulting in different strain in each plateau. For the grains with more free surfaces, the phase transformation discordance between adjacent grains decreases, which is conducive to the movement of twin boundaries. Furthermore, the twin variants with different grain sizes and orientations and the twin variants with the most favorable orientation and with the minimum critical stress of twin boundary movement tend to start first, corresponding to the situation shown in the first platform in the figure. With the increase of stress, the twins in disadvantageous orientation also begin to move, thus showing another plateau in the stress–strain curve. It should be noted that the strain calculation in the figure is based on the gage length of the whole microwire. However, in the actual deformation process, the deformation parts must be concentrated in several specific parts, so the actual strain must be greater than the strain shown in the figure.

5.3.1.2 Effect of Thermomechanical Training on Twin Boundary Motion Characteristics

It can be seen from Fig. 5.7 that the critical stress of the initial twinning boundary movement of the microwire reaches ~20 MPa, which is far greater than the force σ_{mag} that the magnetic field can act on the twin. It is ~2 to 3 MPa for 5 M martensite and ~1.5 MPa for 7 M martensite. Therefore, the critical stress of twin boundary movement is reduced by stress thermal cycle training. The microwire is trained by stress thermal cycle five times across the phase transition region at 25 MPa. The stress–strain plot of the microwire after training is shown in Fig. 5.8.

Firstly, the critical stress of initial martensitic deformation of the microwire decreases to ~9 MPa, and then the critical stress of the second and third steps is ~16 MPa and ~20 MPa respectively, which indicates that the training effectively reduces the critical stress of some twin boundary movement, but the strain value of each step decreases, which is related to the tensile test. The stress thermal cycle training makes the martensite twin variants in the microwire preferred orientation. Therefore, when the temperature is reduced to the martensite state test temperature (298 K), the orientation of martensite variants in the microwire is the state that most of the long axes are arranged along the axis. At this time, even though the critical stress of twin boundary movement decreases, the axial stretching cannot make the microwire further increase, so the stretching after training appears the curve that shows the decrease of step strain. Therefore, in general, in bulk polycrystalline alloys, the relationship between the training direction and the test direction is orthogonal [25]. However, the effect of training does not affect the measurement of MFIS, because when the magnetic field is applied along the microwire axis, the c axis tends to be arranged along the axial direction, so that the microwire size shrinks along the axial direction. If the initial martensite in the microwire is oriented along the axial direction, the absolute value of the microwire MFIS increases under the magnetic field. Therefore, this training method is conducive to the acquisition of MFIS.

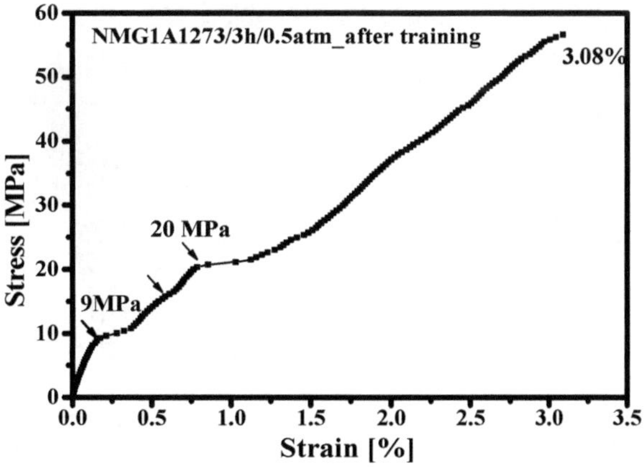

Fig. 5.8 Stress–strain curve of oligocrystalline microwire after thermal cyclic training (298 K)

It is found that the surface twinning state of oligocrystalline microwire before and after drawing can be observed by SEM, as shown in Fig. 5.9, which shows different parts of the same microwire where (a) and (b) correspond to the twin morphology on the free surface of the microwire before stretching and after unloading in Fig. 5.9c, d and are the twin morphology of the microwire plane before stretching and after fracture shown in Fig. 5.8, and the white solid lines in the figure are used to mark the boundary contour of twin. The movement of twin boundary can be observed by SEM.

Figure 5.9a, b shows that the grain is monomorphic, and the twin boundary is almost perpendicular to the microwire axis before stretching, while it is parallel to the microwire axis after tension unloading, which proves the movement of the twin boundary during stretching. As mentioned above, the strain of the microwire is smaller when calculated by the gage length of the whole microwire. It can also be calculated from the figure that if each plateau corresponds to the movement of twin boundary in a grain, for example, 0.91% of the plateau is caused by the movement of twin boundary in the microwire in (a) and (b), and the real strain of the microwire is calculated to be 45.5%, which is obviously larger than any martensite. It is proved that a step on the stress–strain curve not only corresponds to the movement of twin boundaries in a grain. In addition, if all the grains in the oligomeric microwire move in twin boundaries and contribute to the final strain, the strain obtained will be at the same level as that of the single crystal [26].

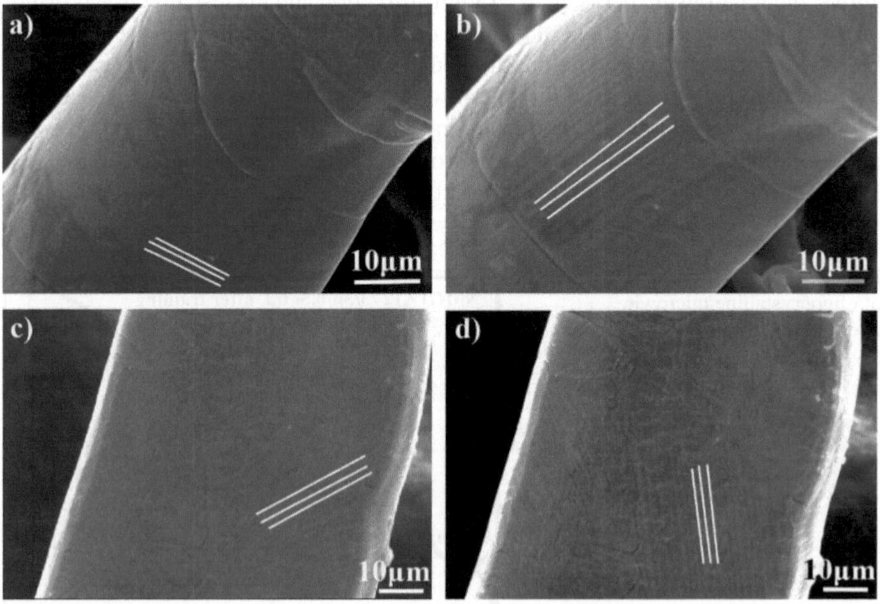

Fig. 5.9 SEM twin morphology of oligocrystalline microwire before and after stretching, **a**, **b** are free solidification surface before and after stretching; **c**, **d** are copper wheel contact surface before and after drawing

5.3.2 *Magnetic-Induced Strain of Ni–Mn–Ga Microwire*

At present, direct method is still used to measure the magnetic-induced strain in one-dimensional Ni–Mn–Ga microwire prepared by spinning method. However, Scheerbaum et al. [27] measured ~ 1% of magnetic-induced strain in $Ni_{50.9}Mn_{27.1}Ga_{22.0}$ by an indirect method in 2008. The diameter of the microwire is about 60–100 μm, and the grain shape is slub crystal, as shown in Fig. 5.10.

Scheerbaum et al. measured ~1% of the magnetic-induced strain by using the changes of grain orientation and local strain of the microwire in EBSD under the action of parallel and vertical magnetic fields and believed that ~1% of the magnetic-induced strain in the microwire only came from three local regions (as shown in Fig. 5.11). If the microwire is effectively trained and the magnetic field is applied to the whole microwire, the theoretical maximum strain can reach ~6%.

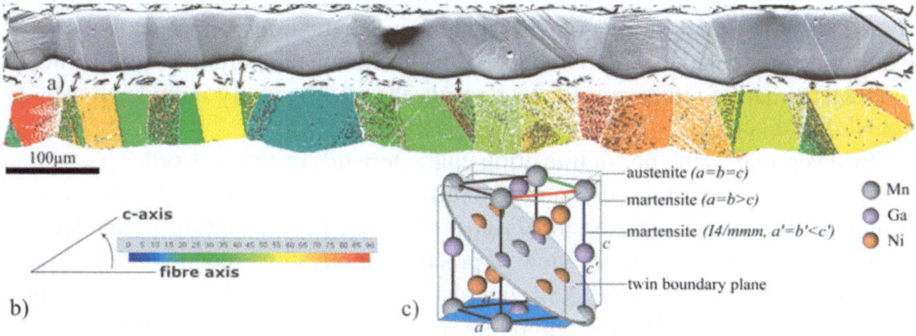

Fig. 5.10 Grain orientation and morphology of Ni–Mn–Ga microwire. **a** BSE and EBSD orientation diagram; **b** EBSD orientation criterion; **c** austenite and martensite crystal structure model

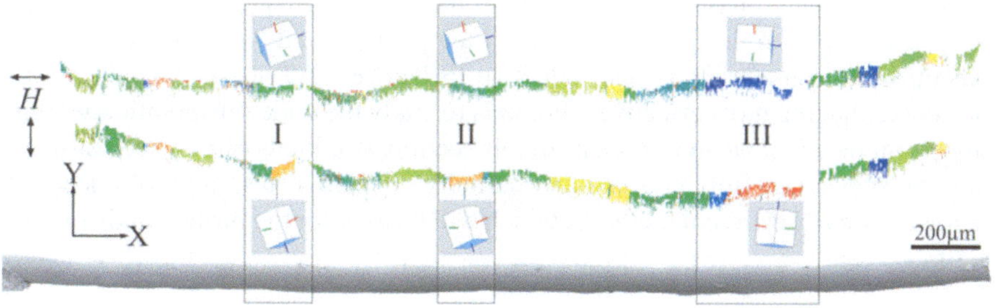

Fig. 5.11 EBSD pattern of Ni–Mn–Ga microwire under the action of 20 kOe magnetic field parallel to and perpendicular to Ni–Mn–Ga microwire, in which I, II and III are considered to be the regions with the largest deformation

5.4 Magnetocaloric Properties of Ni–Mn–Ga–X Alloy Microwires

5.4.1 Magnetic/Structural Phase Transition Design of Ni–Mn–Ga Microwires

When the first-order and second-order phase transitions of magnetic refrigeration materials are coupled, that is, the first-order magnetic phase transition occurs, the change of magnetic order state will produce a larger difference of magnetization ΔM with the discontinuous change of lattice structure, which leads to the generation of giant magnetic entropy change ΔS_m. There are two kinds of first-order magnetic transformation; one is that martensite and austenite have different magnetic states, so the first-order structure transformation process is accompanied by magnetic transformation at the same time. This phenomenon mostly exists in Ni–Mn–In, Ni–Mn–In–Co and Ni–Mn–Sn alloys; the other is that the magnetic transformation temperature of austenite is equal to that of martensitic transformation temperature by adjusting the two transformation temperatures. As a result, the alloys mainly include Ni–Mn–Ga and Ni–Mn–Ga Cu.

However, due to the phase transition characteristics of the first-order phase transition, the phase transition region is generally narrow. In the case of Ni–Mn–Ga, for example, ΔS_m is concentrated in δT_{FWHM} of 1–3 K [28], resulting in the decrease of RC value. From the point of view of practical application, magnetic refrigeration materials need not only large ΔS_m, but also wide δT_{FWHM}. The researchers obtain the stable ΔS_m in the whole cycle range by making materials with different phase transition temperatures into composite materials. The implementation of this method needs to accurately adjust the mass ratio of each component. However, due to the limitation of the material and the temperature range, the ΔS_m value decreases, which limits its application [29]. In addition, some scholars applied pressure [30] and introduced intermediate phase transition [31] to increase ΔT_{FWHM} [32]. Adjustment of two-phase transition distances in two high temperature and two kinds of magnetic phase transition materials is also effective to increase ΔT_{FWHM}. At present, it is considered that the most effective method is to study the wide refrigeration working range from the single material itself. In this section, two kinds of magnetic structure coupling states, i.e., fully coupled and partially coupled states, are introduced into the microwires by different methods, which will be used for the further study of MCE [3, 9].

5.4.1.1 Design of Magnetic Structure Coupling State

In Ni–Mn–Ga alloy, T_C and martensitic transformation temperature can be changed in a wide temperature range by adjusting the composition [33]. The change of composition can be expressed by the outer electron concentration *e*/*a*. Generally speaking, when *e/a* < 7.7, martensitic transformation occurs below T_c, conversely, when *e/a* > 7.7, martensitic transformation occurs above T_c. When *e/a* approaches 7.7, the two kinds of phase transitions tend to occur at the same time, resulting in the coupling of magnetic and structural transformation. Therefore, the state of magnetic structure coupling can be obtained by adjusting the composition of Ni–Mn–Ga alloy until *e/a* approaches 7.7. In addition, Mn is the main element affecting the magnetic properties of the alloy, so Mn-rich state has better magnetic properties. As mentioned above, the ordered heat treatment can effectively improve the magnetic properties of the microwires but does not change the composition of the alloy. To sum up, in this paper, Mn-rich alloy NMG4 is selected to design Ni, Mn and Ga alloys. The *e/a* of microwires before and after ordering treatment is 7.7.

As shown in Fig. 5.12, under the condition of $e/a = 7.7$, the martensitic transformation temperature ($M_p = 354.1$ K) is lower than T_c (366.2 K) due to the influence of defects and internal stress. The reason is that the existence of defects and internal stress hinders the martensitic transformation. After ordering treatment, the DSC curve and M-T curve of NMG4C microwire during the rising/cooling process and the M-T curve at low field 0.2 kOe are shown in Fig. 5.12. The corresponding martensitic transformation temperature and T_c information are shown in Table 5.2. Due to the decrease of defects and internal stress after heat treatment, the martensitic transformation temperature and T_c of NMG4 microwire have been improved.

In fact, DSC is sensitive to the change of heat flux in the process of heating/cooling and accompanied by the appearance of endothermic and exothermic peaks, and it is

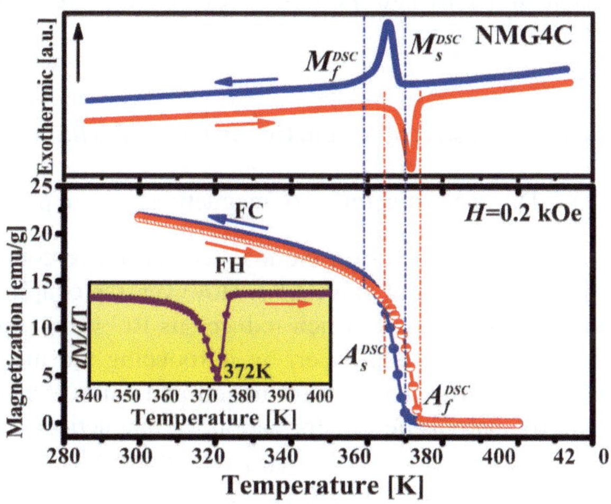

Fig. 5.12 DSC and M-T curves of NMG4C microwire

Table 5.2 Phase transition information of NMG4C microwire

Wire	A_s	A_p	A_f	M_s	M_p	M_f	T_c	A_f-A_s	$Q_{cooling}$	$Q_{heating}$
	K								J/g	
NMG4C	366.6	371.5	373.7	368.6	365.4	361.1	372.0	7.1	6.8	7.1

convenient to obtain the information of structural phase transition. On the contrary, the intensity of the magnetic moment signal in the process of structural transformation depends on the relationship between the magnetic interaction in martensite or austenite and the internal Mn–Mn atomic spacing, so it is difficult to distinguish the structural transformation from the magnetic transformation in the coupled state. Therefore, two kinds of testing methods are selected to compare and judge the relationship between structural phase transition and magnetic phase transition. From Fig. 5.12, it can be seen that the start temperature of reverse martensitic transformation (A_s^{DSC}) and the final temperature (A_f^{DSC}) of reverse martensitic transformation are located in the ferromagnetic region (high magnetization) and paramagnetic region (low magnetization), respectively, in the heating process of DSC curve. On the contrary, in the cooling process, the start temperature of martensitic transformation M_s^{DSC} and the final temperature M_f^{DSC} are located in the paramagnetic region and the ferromagnetic region, respectively. It is the mutual transformation between the ferromagnetic martensite and the paramagnetic austenite, that is, the coupling of the first-order structural transformation and the second-order magnetic transformation results in the first-order magnetic transformation.

From previous Chap. 4, XRD analysis of NMG4C microwire at room temperature shows that the microwire is a single 7 M martensite phase with lattice parameters $a = 4.25$ Å, $b = 5.53$ Å, $c = 43.34$ Å and $\beta = 93.6°$ at room temperature. Therefore, only the transformation between 7 M martensite and parent austenite occurs during the first-order magnetic transformation. Because martensite and austenite are ferromagnetic and paramagnetic, respectively, a large magnetization difference ΔM occurs in the transformation process. It can be predicted that giant ΔS_m will appear in the microwire.

5.4.1.2 Design of Magnetic Structure Partial Coupling State

(1) Distance adjustment of magnetic structure phase transition

As mentioned above, the magnetic structure coupling phase transition can lead to the generation of giant ΔS_m. However, the coupling is generally concentrated in the narrow δT_{FWHM}, which reduces its RC value, thus limiting its practical application value. In this paper, by introducing the method of adjusting the distance between martensitic transformation and Curie transformation and broadening the width of martensitic transformation, the magnetic structure transformation is partially coupled, so as to widen δT_{FWHM} and improve RC value. Therefore, it can be divided into two steps: (1) adjusting the transformation distance and (2) broadening the martensitic transformation width. This section mainly introduces the adjustment of phase transition spacing.

In addition to the composition adjustment, heat treatment is also a means to fine-tune the phase transition temperature of Ni–Mn–Ga microwire. It can be seen that the peak temperature A_p of NMG1 microwire increases from 297.8 K to 324.3 K, with an increase of 27 K and T_c value by ~7 K. For NMG1C microwire, the distance between

T_c and A_P is ~43 K. According to the DSC curve, (a) it can be seen that the difference between the peak of martensitic transformation and Curie transformation is far, which cannot reach the state of partial coupling. Therefore, it is necessary to select the microwires with slightly higher phase transition temperature in the preparation state. In this paper, a variety of microwires with different phase transition temperatures are selected for the experiment. Finally, NMG2 microwire is the most ideal, which is illustrated by this example. The A_p of NMG2 microwire is 303.1 K. According to the results of NMG1 microwire, it is predicted that A_p of NMG2 microwire after ordered treatment is about 330 K. If $T_c = 360$ K, the distance between them is about 30 K. The results show that this kind of transformation spacing can obtain better effect of magnetic structure partial coupling state through subsequent treatment. The final results will be described in the next section combined with the broadening treatment of martensitic transformation width.

(2) Broadening treatment of structural transformation

To obtain high magnetic properties for MCE, the microwires need ordering heat treatment. However, under normal heat treatment process, 0.2–0.3 atm Ar gas atmosphere and Mn particles should be added to the quartz tube to prevent the volatilization of Mn. This ordering heat treatment method will lead to the homogenization of the composition and narrow the phase transition width. However, to improve martensitic transformation the width of microwires, it is necessary to completely change the idea of heat treatment, that is, to deliberately create the loss of Mn element to achieve the desired effect. Therefore, a new ordered heat treatment method is adopted, that is, to cancel the protective atmosphere and Mn particles, and directly conduct the ordered heat treatment on the microwires under vacuum conditions. The surface area of one-dimensional microwire is large, and the saturated vapor pressure of Mn element is high. Therefore, the loss of Mn element on the surface of one-dimensional microwire is inevitable under high temperature and vacuum environment.

The results show that the distribution of Mn in NMG2 microwire is gradiently distributed from surface to interior after vacuum Mn free ordering heat treatment. The average composition of microwire is shown in Table 5.3. It is found that the uniformity of microwire composition is poor, which meets the design requirements. Then, the cross section of NMG2C microwire was polished, and the EDS energy spectrum analysis was carried out twice at the center of the microwire and on the contour lines 4, 6, 10, 14 μm and 4, 6, 9, 10, 12, 14, 15, 17 μm away from the microwire surface. The results were analyzed point to point, and the results are shown in Fig. 5.13a, b. The chromatogram on the right side of the figure shows that different components correspond to different colors, which shows that the gradient

Table 5.3 Composition of NMG2C microwire

Microwire	Composition (at.%)					
	Ni	ΔNi	Mn	ΔMn	Ga	ΔGa
NMG2C	50.6	0.2	28.1	0.6	21.3	0.5

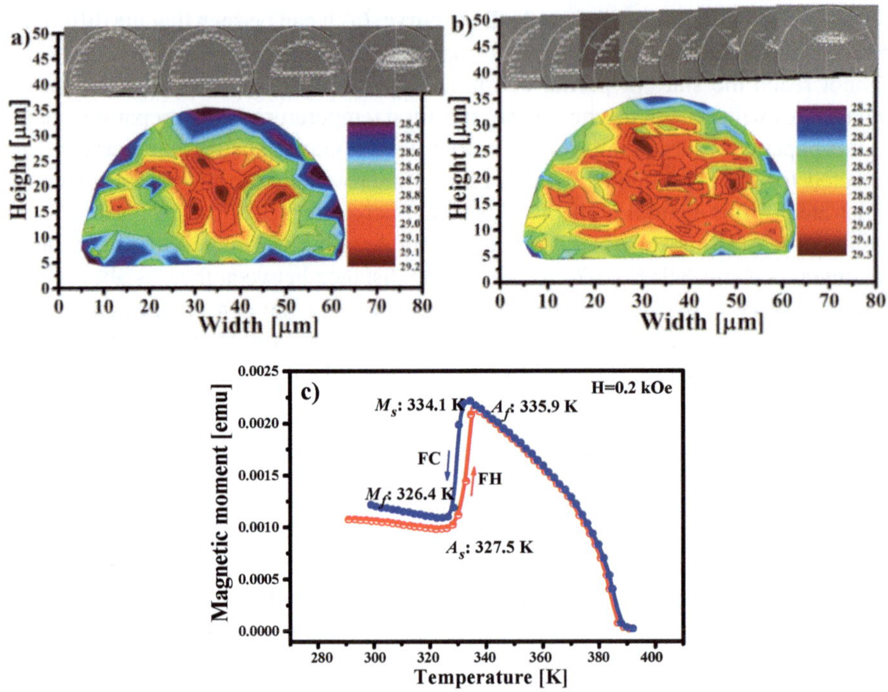

Fig. 5.13 Mn composition distribution of cross section of NMG2 microwire and M-T curve of single microwire after vacuum Mn free ordering heat treatment. **a**, **b** The distribution of NMG 2 cross street show off; **c** M-T curve of single microwire

distribution state of Mn element composition gradually increases from surface to interior, and the difference between surface and internal components is about 1.1 at.%. Here, the relative accuracy of EDS has been calibrated by ICP-OES method, and the relative value of components is considered to be accurate under the same test conditions.

The results of the microwire transition curve are shown in Fig. 5.13c. The results show that the martensitic transformation temperatures M_s, M_f, A_s and A_f are 334.1, 326.4, 327.5 and 335.9 K, respectively, and the transformation widths $|A_f - A_s|$ and $|M_f - M_s|$ are 8.4 and 12.3 K, respectively, which are much larger than the transformation width after normal ordering, which verifies the results of composition gradient.

In order to further determine the overall phase transformation characteristics of NMG 2 microwire after ordering heat treatment, the DSC curves of NMG 2 alloy ingot, as-prepared microwire and two kinds of ordered heat treatment methods were tested. The results are shown in Fig. 5.14. The martensitic transformation temperatures obtained by the tangent method are shown in Table 5.4. Firstly, it should be pointed out that NMG2 alloy ingot has a more heterogeneous composition distribution than microwire, so it has the widest transformation width.

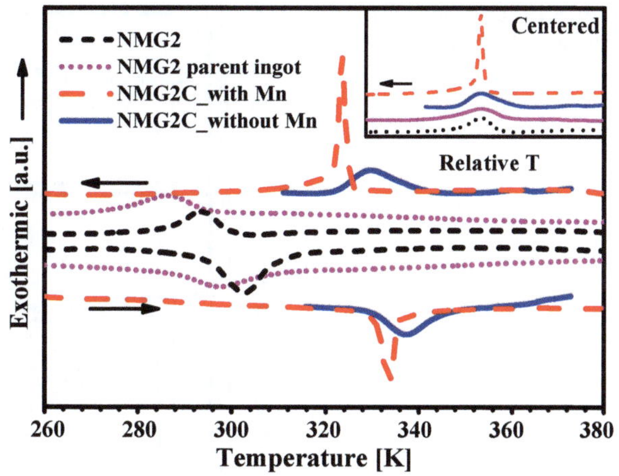

Fig. 5.14 DSC curves of NMG2 alloy ingot, as-prepared and different ordered microwires

Table 5.4 Transformation temperature of NMG2 alloy ingot, as-prepared and different ordered microwires

Object	A_s	A_p	A_f	M_s	M_p	M_f	T_c	A_f-A_s	M_s-M_f
	K								
Parent ingot	287.3	296.9	309.1	294.8	286.5	273.8	–	21.8	21.0
NMG2	294.7	303.1	309.7	299.8	293.8	285.6	360.5	15.0	14.2
NMG2C(Mn)	332.0	334.1	336.1	325.9	323.5	321.8	–	4.1	4.1
NMG2C	329.8	337.2	345.0	342.4	330.1	321.8	365.0	15.2	20.6

As shown in Table 5.4, the phase transition width of the heating/cooling process after normal ordering heat treatment is 4.1 K, while that after vacuum Mn free ordering heat treatment is 15.2 K and 20.6 K, respectively, which is increased by ~4 to 5 times. These two values are slightly larger than the VSM results of a single microwire, which may be due to different test methods. The phase transition width of VSM test results is much narrower than that of DSC, but the difference of composition between different microwires is not excluded. In conclusion, the phase transition width of NMG2C microwire is significantly increased after vacuum Mn free ordering heat treatment.

The DSC curve and M-T curve of NMG2C microwire (several microwires) during the rising/cooling process and the M-T curve under low field 0.2 kOe are shown in Fig. 5.15. The embedded diagram is the first derivative of the heating process curve, indicating that T_c is 365 K. Therefore, the temperature distance between martensitic transformation and magnetic transformation, namely T_c-A_p, is ~28 K. At this temperature interval, the increase of comprehensive martensitic transformation width makes the microwire to form a state of magnetic structure partial coupling. As shown in VSM curve in Fig. 5.15, there is a cross point between the two kinds of transformation, that is, the magnetic transformation occurs immediately after the end

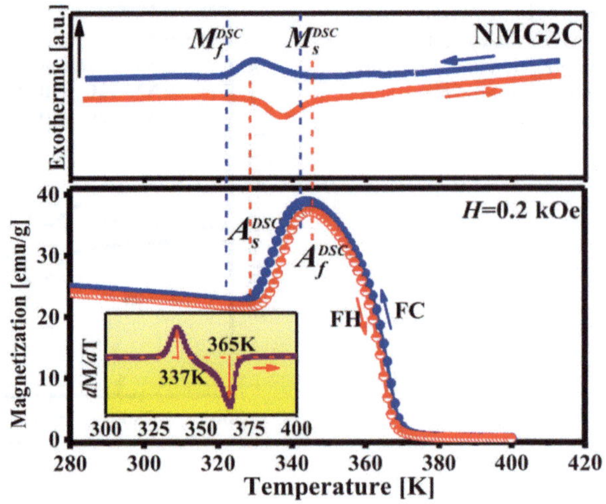

Fig. 5.15 DSC and M-T curves of NMG2C microwire

of martensitic transformation in the heating process, and the magnetic transformation and martensitic transformation have almost the same transformation width under 0.2 kOe magnetic field. In addition, the general second-order phase transition is not accompanied by hysteresis. Therefore, the hysteresis of magnetic phase transition during heating/cooling implies the existence of partial magnetic structure coupling phase transition.

5.4.2 Magnetocaloric Effect of Ni–Mn–Ga Microwire in Magnetostructural Coupled State

5.4.2.1 Thermal Hysteresis and Post Hysteresis Characteristics

The difference between microwires and bulk polycrystalline materials is mainly due to the high specific surface area of one-dimensional microwire materials. For a simple calculation, assuming that the diameter of the microwire is 50 μm and the length is 50 mm, the comparative area of the microwire is 6.2 times that of the same volume cubic block material [3, 6, 9]. Figure 5.16 shows the M-T curve of the rising/cooling process of NMG4C microwire at 0.2 and 50 kOe. It can be seen from the figure that the thermal hysteresis of phase transformation of NMG4C microwire is ~2.8 K at 50 kOe, which is far less than that of bulk polycrystalline material, ~10 K [34]. The thermal hysteresis of phase change process is an important problem affecting the performance of magnetic refrigeration materials. It is shown that the change of cell volume occurs during the phase transformation process. If the volume change cannot be effectively released, the phase transformation process will be delayed. However,

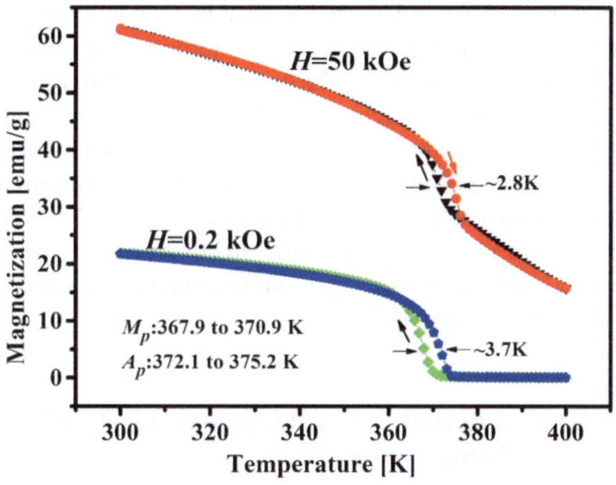

Fig. 5.16 M-T curves of NMG4C microwire at 0.2 and 50 kOe

in this paper, high specific area and high proportion of grain free surface of one-dimensional microwire can reduce the restriction between adjacent grains, which can effectively release the volume change in the process of phase transformation through the free surface, thus reducing the thermal hysteresis during phase transformation. The thermal hysteresis of the first-order phase transformation is also related to the defects and internal stress in the microwire. Therefore, the ordered heat treatment is an effective means to reduce the thermal hysteresis. In addition, the decrease of atomic diffusion distance in small size materials can also reduce the phase transition thermal hysteresis.

In addition to the phase transition thermal hysteresis, the hysteresis produced by magnetic refrigeration materials under the action of magnetic field is also an important factor affecting the magnetocaloric performance [9]. Figure 5.17 shows the magnetization curves of NMG4C microwire at different temperatures near the

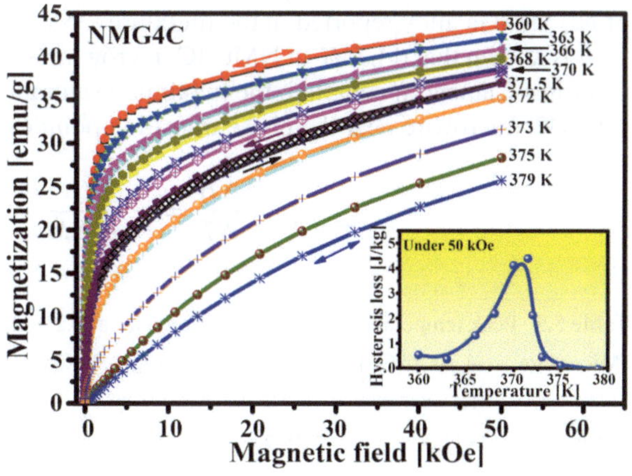

Fig. 5.17 Magnetization curves of NMG4C microwire at different temperatures and temperature dependence of hysteresis

phase transition when the maximum magnetic field is 50 kOe. The area enclosed by the curve is the hysteresis. The temperature curve after hysteresis is obtained by calculation, as shown in the embedded diagram in Fig. 5.17. The average hysteresis value of NMG4C microwire at 50 kOe is ~2 J/kg, which can be ignored and far better than the reported value [30]. Similar to thermal hysteresis, the decrease of hysteresis depends on the high specific surface area of the microwire and the reduction of defects and internal stress after heat treatment, which leads to the decrease of domain wall resistance under magnetic field. In conclusion, the magnetic structure coupling state of NMG4C microwire has a small thermal hysteresis and almost negligible hysteresis, which promotes MCE.

5.4.2.2 Effect of External Magnetic Field on Phase Transition Temperature

It can be seen from Fig. 5.16 that the transformation temperature of NMG4C microwire in the magnetic structure coupling state moves toward high temperature with the increase of magnetic field, which indicates that there is a magnetic-driven transformation from paramagnetic austenite to ferromagnetic martensite in the microwire. The maximum values of the first derivative of the heating/cooling process curve are taken as the peak temperature A_p and M_p to evaluate the change range of the phase transition temperature with the magnetic field. The peak temperatures of phase transition at 0.2 and 50 kOe of NMG4C microwire obtained from the derivation of Fig. 5.16 are shown in Table 5.5 [9].

When the magnetic field increases from 0.2 to 50 kOe, the changes of ΔT_{0h} and ΔT_{0c} are 3.9 and 2.8 K, respectively. Since only the heating process is considered in MCE, the rate of temperature variation with magnetic field $\mathrm{d}A_p/\mathrm{d}H$ is obtained to be ~0.08 K/kOe. Marcos et al. [35] reported that the MCE near the first-order magnetic phase transition mainly depends on the value of $\mathrm{d}A_p/\mathrm{d}H$, which reflects the strength of phonon spin coupling in the microstate system and increases with the increase of e/a value. The $\mathrm{d}A_p/\mathrm{d}H$ value of NMG4C microwire is equivalent to that of Ni–Mn–Ga alloy reported in the literature [35].

It is known that $\mathrm{d}A_p/\mathrm{d}H$ of NMG4C microwire is ~0.08 K/kOe. According to the Clausius–Clapeyron equation, the maximum saturation magnetization difference of NMG4C microwire near the phase transition point can be calculated as follows:

$$\frac{\Delta H}{\Delta T} = -\frac{Q}{T_0 \Delta M} \tag{5.2}$$

Table 5.5 Peak temperature of NMG4C microwire at 0.2 and 50 kOe

Microwire	A_p(0.2 kOe)	A_p(50 kOe)	M_p(0.2 kOe)	M_p(50 kOe)	ΔT_{0h}	ΔT_{0c}
	K					
NMG4C	371.3	375.2	368.1	370.9	3.9	2.8

In the formula, the left term of the equal sign is replaced by dA_p/dH

$$\frac{dA_p}{dH} = -\frac{T_0 \Delta M}{Q} \tag{5.3}$$

where Q and T_0 are the enthalpy change of transformation process and the theoretical transformation temperature between austenite and martensite, respectively, and T_0 is approximately equal to $(A_p + M_p)/2$, and ΔM is the theoretical maximum saturation magnetization difference between austenite and martensite near the transformation point.

According to the DSC curve in Fig. 5.12, the phase transition enthalpy Q of the heating process is 7.1 J/g. From Table 5.2, it can be seen that T_0 of the microwire is 368.5 K. In addition, $dA_p/dH = \sim 0.08$ K/kOe. The parameters are brought into Eq. (5.2), and $\Delta M = 15.4$ emu/g is obtained. As shown in Fig. 5.16, it is considered that the NMG4C microwire has reached the magnetization saturation state at 50 kOe, and the saturation magnetization difference between austenite and martensite near the first-order magnetic transformation is ~16 emu/g, which is in good agreement.

5.4.2.3 Magnetocaloric Properties

In this paper, MCE is characterized by ΔS_m, which can be obtained by measuring the isothermal magnetization curves at different temperatures and calculating by Maxwell relation. In recent years, the applicability of this method in the vicinity of the first-order magnetic phase transition has always been controversial, because the virtual high value will be generated when the ΔS_m near the magnetic phase transition is calculated by using the isothermal magnetization curve [36]. However, for the first-order phase transition with weak magnetoelastic coupling ($\lambda_1/\lambda_0 < 2$), Maxwell's relationship is fully applicable [36]. For Ni–Mn–Ga alloys, $\lambda_1 \sim 10^7$ and $\lambda_0 \sim 10^{10}$ erg cm^{-3} are fully met [37].

According to Maxwell's relation, the size of alloy ΔS_m depends on the value of $\partial M/\partial T$, so the temperature rise M-T of NMG4C is first-order differential. There is an extreme value of $\partial M/\partial T$ at 372 K, so a series of isothermal magnetization curves have been tested near 372 K. The temperature intervals of isothermal magnetization curves are selected as 0.5, 1, 2, 3 and 5 K. The results are shown in Fig. 5.18a. It can be seen from the figure that the saturation magnetization of the microwire decreases with the increase of temperature, and due to the coupling of martensitic transformation and magnetic transformation, the saturation magnetization of martensite at the same temperature level decreases compared with NMG1C.

Although the coupling of martensitic transformation and magnetic transformation of NMG4C microwire can be distinguished, the exact interaction mode of martensitic transformation and magnetic transformation near the transformation temperature cannot be known. In this paper, Landau's second-order transformation theory [38] is used for analysis. The magnetic properties of materials are mainly composed of

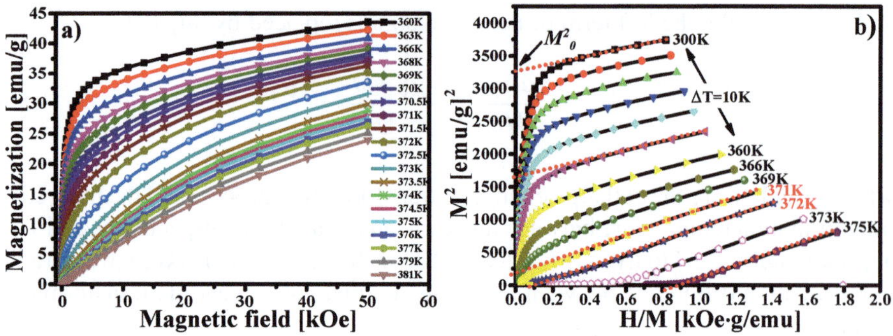

Fig. 5.18 Magnetization curves and Arrot curves of NMG4C microwire at different temperatures. **a** M-H curve; **b** Arrot curve

local magnetic moment and itinerant electron magnetic moment.

$$G = G_d + G_f - \mathrm{n} M_d M_f \tag{5.4}$$

where n is the molecular field coefficient and G_d is the free energy of the itinerant d-electron subsystem in the external field H

$$G_d = \frac{1}{2} a_1 M_d^2 + \frac{1}{4} a_3 M_d^4 + \frac{1}{6} a_5 M_d^6 - M_d H \tag{5.5}$$

G_f is the free energy of the local spin $4f$ electron subsystem

$$G_f = \frac{1}{2} b_1 M_f^2 + \frac{1}{4} b_3 M_f^4 + \frac{1}{6} b_5 M_f^6 - M_f H \tag{5.6}$$

With $M = M_f + M_d$ as a constant, the free energy can be reduced to:

$$G = \frac{1}{2} c_1 M^2 + \frac{1}{4} c_3 M^4 + \frac{1}{6} c_5 M^6 - M H \tag{5.7}$$

When the phase transition occurs, the energy is very small, i.e., $\partial G / \partial M = 0$, and the high-order term is ignored, the relationship between magnetization and magnetic field described by Landau phase transition theory near Curie point can be obtained as follows:

$$A + B M^2 = H/M \tag{5.8}$$

It can be seen from Fig. 5.18 that the martensitic transformation and magnetic transformation in the M-T curve of the magnetic structure coupling state coincide, so it is difficult to obtain T_c. According to formula (5.7), the method called Arrott curve [39], which is used to distinguish the relationship curve between M^2 and H/M, is drawn in Fig. 5.18b. Make the tangent of the curve under the high field intersect M^2, and mark the intersection point as $M_0{}^2$. For ferromagnetic memory alloy, when $M_0{}^2$ is positive, the alloy is in ferromagnetic state, i.e., $T < T_c$; when $M_0{}^2$ is positive, that is, when there is an intersection with *H/M*, the alloy is in paramagnetic state, i.e., $T > T_c$. When $T = T_c$, $M_0{}^2$ is 0, M^2 is linear with H/M, passing through the origin [39]. As shown in Fig. 5.18b, in the heating process of NMG4C, $M_0{}^2$ is positive before 371 K and negative after 372 K. Therefore, T_c is considered to be between 371 and 372 K, which is very consistent with the 372 K extreme value obtained by the first derivative method of *M-T* curve. According to Table 5.2, A_s and A_f are 366.6 K and 373.7 K, respectively, while T_c is ~371 to 372 K. Therefore, it can be determined that the martensitic structural transformation includes magnetic transformation, which further confirms the existence of magnetic structure coupling.

Based on the isothermal magnetization curves at different temperatures, ΔS_m was calculated by Maxwell equation:

$$\Delta S_m = S_m(H,T) - S_m(0,T) = \int_0^H \left(\frac{\partial M}{\partial T}\right)_H \mathrm{d}H \tag{5.9}$$

Based on the discontinuous temperature interval, the equation can be discretized

$$\Delta S_m \approx \frac{1}{\Delta T}\left[\int_0^H M(T+\Delta T)\mathrm{d}H - \int_0^H M(T)\mathrm{d}H\right] \tag{5.10}$$

The curves of the relationship between ΔS_m and temperature under different magnetic fields are obtained as shown in Fig. 5.19 [3]. It can be seen that the peak value of ΔS_m appears near the phase transition point ~372 K, and δT_{FWHM} is only ~3 K at 50 kOe. The values of ΔS_m of the microwires at 20 and 50 kOe are –

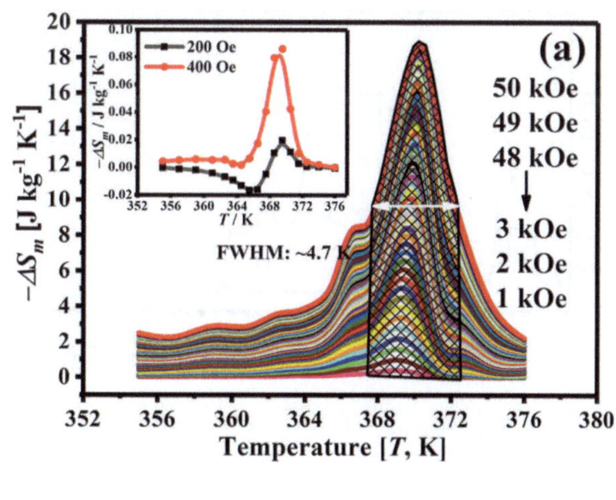

Fig. 5.19 Relationship between ΔS_m and temperature in different magnetic fields of NMG4C microwire

11.3 and −18.5 J/kg K, which are slightly lower than those of $Ni_{52}Mn_{26}Ga_{22}$ strip (−16.4 and −30.0 J/kg K) [28], but much higher than that of as-prepared tapes (−5.3 and −11.4 J/kg K) [40]. In addition, it is higher than that of $Ni_{55}Mn_{20.6}Ga_{24.4}$ and $Ni_{55}Mn_{19.6}Ga_{25.4}$ at 20 kOe (−9.5 and −10.4 J/kg K) [41]. The significant increase of ΔS_m in NMG4C microwire is mainly due to the improvement of atomic order and the effect of magnetic structure coupling. It can be seen from the figure that the maximum value of ΔS_m changes with the magnetic field. It can be seen from the diagram that the maximum value of ΔS_m changes with the magnetic field. Therefore, if the magnetic field continues to increase, the value of ΔS_m will continue to increase.

In order to evaluate the magnetocaloric properties of magnetic refrigeration materials, it is necessary to consider not only the value of ΔS_m, but also its temperature dependence. MCE is comprehensively evaluated by the RC value of the material, which is expressed by subtracting the hysteresis value from the integral within $\delta T_{FWHM}(T_1 - T_2)$ under the ΔS_m-T curve. The formula is shown in the relation (5.16). For NMG4C microwire, although the ΔS_m value is high and can be ignored after hysteresis, the net RC_{net} values at 20 and 50 kOe are only 25.1 and 63.6 J/kg, because the δT_{FWHM} value is only ~3 K. The RC value of NMG2C microwire in a partially coupled state will be discussed in detail.

5.4.3 *Magnetocaloric Effect of Ni–Mn–Ga Microwire in Partial Magnetic Coupling State*

5.4.3.1 Crystal Structure

As mentioned above, NMG2C has reached the state of magnetic structure partial coupling after vacuum Mn free ordering heat treatment. This section mainly studies the characteristics of microwires in this state [3, 9]. Figure 5.20 shows the XRD anal-

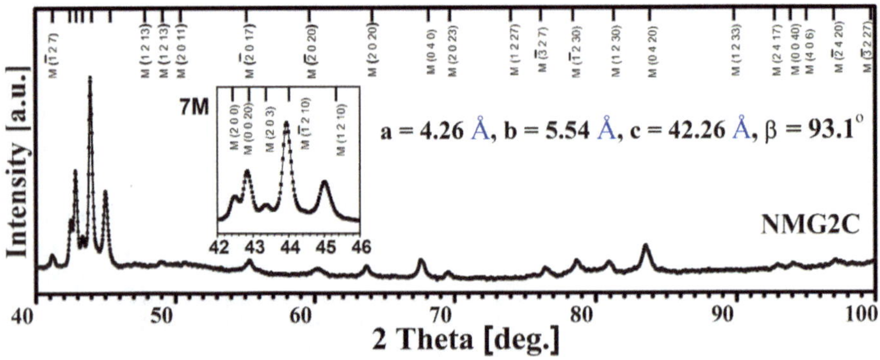

Fig. 5.20 Room-temperature XRD of 20 NMG2C microwire

Table 5.6 Structure information of NMG2C microwire

Microwire	Phase structure	Lattice constant						
		a (Å)	b (Å)	c (Å)	α (°)	β (°)	γ (°)	V_a (Å^3)
NMG2C	7 M under room temperature	4.26	5.54	42.26	90	93.1	90	12.45

ysis of NMG2C microwire at room temperature. The as-prepared NMG2 microwire exhibits austenite at room temperature and martensite at low temperature.

Through comparison, it is found that the martensite type of NMG2 microwire does not change after ordering treatment, and it is still 7 M monoclinic martensite with an incommensurate structure. Therefore, there is only a single transformation between 7 M martensite and austenite in the microwire. The lattice parameters of NMG2C microwire are also obtained by full-spectrum fitting of XRD results, as shown in Table 5.6. The lattice parameters of NMG2C are $a = 4.26$ Å, $b = 5.54$ Å, $c = 42.26$ Å and $\beta = 93.1,°$ respectively, and the calculated $V_a = 12.45$ Å^3. The V_a value fluctuates slightly at low temperature. As mentioned above, the martensitic structure of NMG1 microwire is changed after ordering, but the same situation does not appear in NMG2 and NMG2C. Through calculation, the *e/a* values of NMG1, NMG1C, NMG2 and NMG2C are 7.63, 7.69, 7.65 and 7.67, respectively. It can be seen that both 5 M and 7 M may appear in the *e*/*a* range of the four microwires. The range of the two martensites is similar, so it is reasonable to have 5 M or 7 M martensite.

5.4.3.2 Thermal Hysteresis and Post Hysteresis Characteristics

The hysteresis characteristics of partially coupled NMG2C microwires are also studied. Figure 5.21 shows the M-T curves of the rising/cooling process of NMG2C microwire at 0.2 and 50 kOe. It can be seen that the phase transition thermal hysteresis of the microwire is only 1.1 K at 50 kOe, which is smaller than that of NMG4C (~2.8 K) and NMGC1C (~4 K), which may be related to the coupling interaction between magnetic and structure. It can be seen from Figs. 5.16 and 5.21 that the thermal hysteresis decreases with the increase of magnetic field.

In addition to the phase transition thermal hysteresis, the hysteresis of magnetic refrigeration materials under the action of magnetic field is also an important factor affecting the magnetocaloric properties. Figure 5.22 shows the magnetization curves of NMG2C microwire at different temperatures near the phase transition when the maximum magnetic field is 50 kOe. The area surrounded by the curve is the hysteresis. The temperature curve after hysteresis is obtained by calculation, as shown in the embedded figure. It can be seen from the figure that the magnetization curves in the process of magnetization/demagnetization are basically coincident. The maximum value of the microwire after hysteresis at 50 kOe is only ~0.12 J/kg, which is an

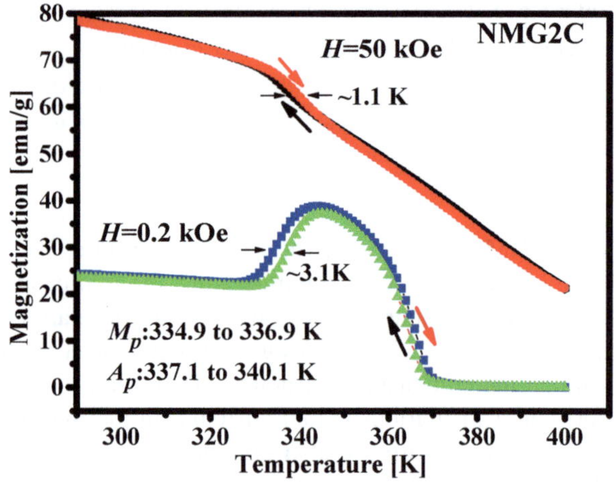

Fig. 5.21 NMG2C microwire M-T curve at 0.2 and 50 kOe

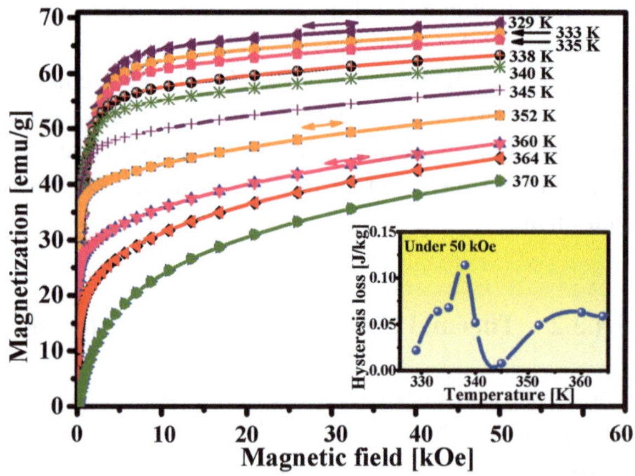

Fig. 5.22 Magnetization curves of NMG2C microwire at different temperatures and temperature dependence of hysteresis

order of magnitude smaller than that of NMG4C microwire and can be ignored. The small thermal hysteresis and negligible hysteresis of NMG2C lay the foundation for obtaining a high RC value.

5.4.3.3 Characteristics of Martensitic Transformation

It can be seen from Fig. 5.21 that the phase transformation temperature of NMG2C microwire in the magnetic structure partially coupled state moves to the high-temperature direction with the increase of magnetic field during the rising/cooling process. It can be seen that the magnetic-driven transformation from ferromagnetic austenite to ferromagnetic martensite occurs in the microwire. From the derivation

Table 5.7 Peak temperature of NMG2C microwire at 0.2 and 50 kOe

Microwire	A_p (0.2 kOe)	A_p (50 kOe)	M_p (0.2 kOe)	M_p (50 kOe)	ΔT_{0h}	ΔT_{0c}
	K					
NMG2C	337.1	340.1	334.9	336.9	3.0	2.0

of Fig. 5.21, the phase transition peak temperatures of NMG2C microwire at 0.2 and 50 kOe are shown in Table 5.7 [3, 9].

In the process of heating and cooling, when the magnetic field increases from 0.2 to 50 kOe, the changes of ΔT_{0h} and ΔT_{0c} are 3.0 and 2.0 K, respectively, that is, dA_p/dH = ~0.06 K/kOe, which is smaller than that of NMG4C. The enthalpy of phase transition Q is 4.6 J/g.

It can be seen that T_0 of microwire is 333.7 K, and dA_p/dH = ~0.06 K/kOe. If the three are brought into formula (5.2), we can get $\Delta M = 8.3$ emu/g, which is consistent with the difference of saturation magnetization between austenite and martensite near transformation at 50 kOe.

In order to study the effect of magnetic field on the magnetic properties of the microwire during temperature rise and fall, especially during phase transition, the field cooling (FC) and zero field cooling (ZFC) M-T curves of the microwire at 0.2 kOe were measured, as shown in Fig. 5.23. It can be seen from the curves of FC and FH that the magnetization of microwire martensite increases with the decrease of temperature when the magnetic field of 0.2 kOe drops to ~4 K. The ZFC curve coincides with FC curve at high temperature, but they are separated during cooling process, and the magnetization of ZFC curve is lower than that of FC. The explanation of this phenomenon is as follows: When cooling in the absence of magnetic field, the magnetic moment in the microwire is arranged along its own easy magnetization direction. In general, if there is no anisotropy in the microwire, the magnetic moment should be zero at low temperature. However, because NMG2C microwire has strong magnetic crystal anisotropy along the axis of easy magnetization axis, the magnetic

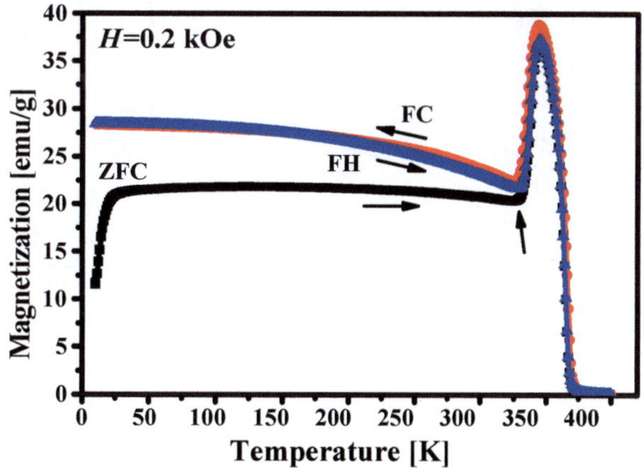

Fig. 5.23 Field cooling (0.2 kOe) and zero field cooling M-T curves of NMG2C microwire

moment is still >10 emu/g at ~4 K even when cooled at zero field. However, since the zero field drops to low temperature, when the magnetic field is applied at the beginning of the heating process, the magnetic moment cannot be turned in time, which is still a locked state. However, with the increase of temperature, more energy can disturb the locked system state, and the Zeeman effect is also driven, which makes the magnetic moment turn due to the action of external field. The magnetic moment tends to be arranged along the direction of the magnetic field and finally reaches the same level as FC at the threshold temperature. In addition, at high temperature, the magnetization of austenite decreases with the increase of temperature and becomes paramagnetic above the magnetic transition point, and the magnetization temperature drops to zero.

5.4.3.4 Magnetocaloric Properties

The phase transition peak point is shown in the embedded diagram in Fig. 5.15. Therefore, isothermal magnetization curves at different temperatures are made with 337 and 365 K as the dense points in the heating process. The results are shown in Fig. 5.24a [3, 9]. Similar to NMG4C and NMGC1C, the saturation magnetization of the microwire decreases with the increase of temperature. However, unlike the magnetic structure coupling state, NMG2C microwire has ferromagnetic austenite. Although austenite and martensite have the same magnetic state, the martensite phase is monoclinic, which has higher magnetocrystalline anisotropy than the cubic austenite phase. Therefore, it is difficult to be magnetized under low magnetic field, so it appears as shown in Fig. 5.24. For NMG2C microwire, the magnetic field is ~1.3 kOe, that is, when $h < 1.3$ the magnetic field is higher than that of martensite, and the magnetization of austenite is higher than that of martensite. However, when the magnetic field is larger than the magnetic field, the magnetization saturation of

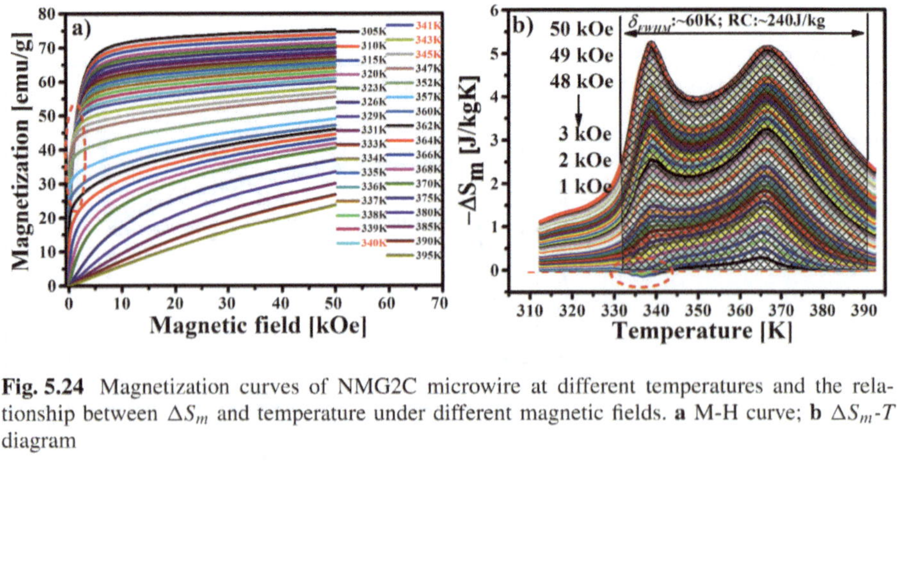

Fig. 5.24 Magnetization curves of NMG2C microwire at different temperatures and the relationship between ΔS_m and temperature under different magnetic fields. **a** M-H curve; **b** ΔS_m-T diagram

martensite is achieved through the turning of magnetic moment and even the movement of twin boundaries due to the Zeeman effect. In this process, the proportion of magnetic moments arranged along the external field in the unit volume of martensite increases rapidly, and the magnetization increases continuously.

Based on the isothermal magnetization curves at different temperatures, the relationship curves between ΔS_m of NMGC1C microwire and temperature under different magnetic fields are calculated according to formula (5.10), as shown in Fig. 5.24b. It can be seen from the figure that there are two extremum peaks of ΔS_m at 337 K and 365 K, corresponding to the structural phase transition and the magnetic phase transition, respectively. The values at 50 kOe are ~−5.2 J/kg K and ~−5.3 J/kg K, respectively. In addition, different from the magnetic structure coupling state, a positive magnetocaloric peak appears in the low field near the structural phase transition, which has a maximum value of ~0.14 J/kg K at 1 kOe, then decreases with the increase of magnetic field and finally becomes negative magnetocaloric at ~3 kOe. This phenomenon is related to the difficulty in the magnetization of martensite in the low field, which results in the intersection of magnetization curves of martensite and austenite in Fig. 5.24a. Marcos et al. reported on the sources of different MCEs and found that positive MCE comes from the coupling between magnetic moment and magnetic structure between martensite variants, which is also the source of SME [42]. This positive MCE generally occurs in the region where the magnetic field is less than the saturation magnetic field intensity and is proportional to the saturation magnetization of martensite. It is found that the saturation magnetization and magnetic field strength of Ni–Mn–Ga alloy have maximum values at the standard stoichiometric ratio of Ni_2MnGa alloy, and the value of positive MCE is the maximum [35, 42]. However, the saturation magnetic field intensity decreases with the increase of e/a, and when $e/a = 7.7$, the saturation magnetic field intensity of martensite near the transformation is the minimum and positive MCE does not appear. This also explains that there is no positive MCE in NMG4C due to $e/a \approx 7.7$, while NMG2C microwire has a smaller positive MCE due to $e/a \approx 7.67$.

Figure 5.25 shows the relationship between the maximum value of $\Delta S_m{}^{max}$ and the magnetic field near the structural phase transformation and magnetic transformation. For the structural phase transformation, the relationship between the two is linear, but for the magnetic phase transformation, the relationship between the $\Delta S_m{}^{max}$ and the magnetic field is shown in the figure. The value of index 0.74 is related to the composition, structure and magnetic state of the alloy. Generally speaking, $n > 2/3$ is due to the shape anisotropy, size effect and local inhomogeneity of small size microwires after heat treatment.

It can be seen that there is little difference between the two magnetic peaks at the temperature of Fig. 5.24 in the thermal region. Due to this phenomenon, the two peak regions combined with the transition region with higher magnetocaloric value formed a greatly broadened δT_{FWHM}, with 50 kOe reaching ~60 K. In addition, as described in Sect. 6.4.1.2, the hysteresis of NMG2C microwire is very small and can be ignored. Therefore, combined with the broadened δ_{FWHM} and minimal hysteresis, the RC_{net} value of NMG2C microwire under different magnetic fields is calculated as 240 J/kg according to Fig. 5.22. The comparison of RC_{net} values of NMG2C,

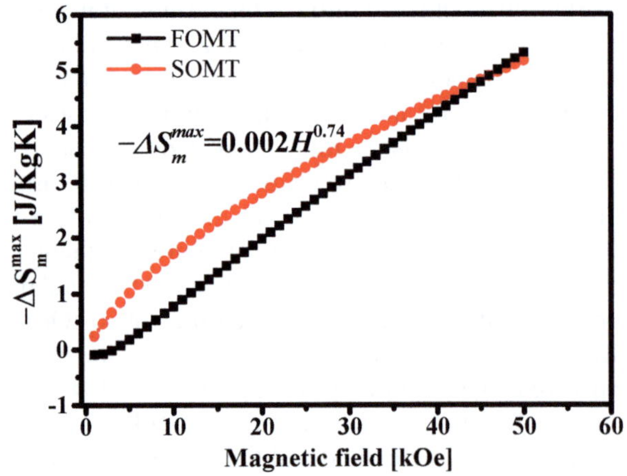

Fig. 5.25 NMG2C primary and secondary $\Delta S_m{}^{max}$ changes with magnetic field

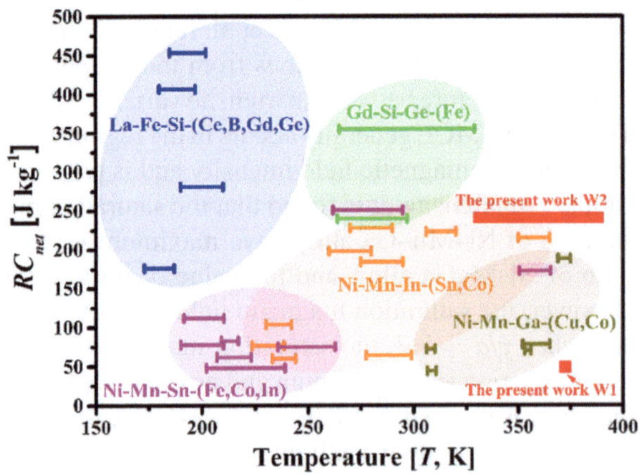

Fig. 5.26 Comparison of net RC_{net} values of different alloys at different temperatures

NMG4C microwire and other magnetic refrigeration materials is summarized in Fig. 5.26. As shown in the figure, the RC_{net} value of NMG2C at 50 kOe is ~240 J/kg. Although NMG4C has high ΔS_m, its RC_{net} value is very small due to its ~3 K δT_{FWHM}. Compared with rare earth compounds, NMG2C has a smaller RC_{net} value [43], but it is more economical and practical.

5.4.4 Effect of Cu Doping on Magnetocaloric Properties of Ni–Mn–Ga Microwires

The magnetic structure coupling state can also be obtained by adding a fourth element. For Mn-rich Ni–Mn–Ga alloy, its T_c remains basically unchanged [44], which is

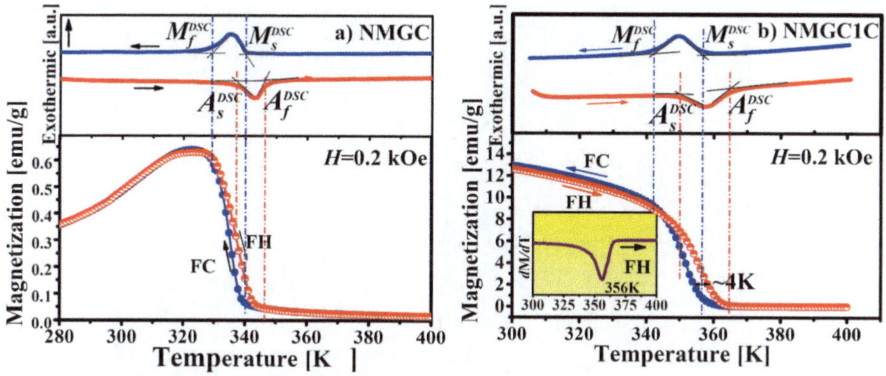

Fig. 5.27 DSC and M-T curves of NMGC1 microwire before and after the orderly heat treatment. **a** NMGC1; **b** NMGC1C

obviously too high for the application of MCE. The intrinsic brittleness of Ni–Mn–Ga alloy is large. Among the many choices of the fourth component, Cu has been widely concerned because of its low price, high efficiency in adjusting phase transition temperature and promoting the characteristics of MCE. In this section, Cu is used to replace Ga in Ni–Mn–Ga alloy to improve microwire plasticity and achieve the purpose of magnetic structure coupling. In addition, as the most expensive element in Ni–Mn–Ga alloy, replacing Ga with Cu can also reduces the cost [6, 9].

The DSC curves and M-T curves of as-prepared and ordered NMGC1 microwires are shown in Fig. 5.27. The results show that the martensitic transformation and magnetic transformation of the microwires occur simultaneously before and after the ordering, that is, during the heating process, the microwires directly change from ferromagnetic martensite to paramagnetic austenite, resulting in the first-order magnetic transformation. In Fig. 5.27b, the first derivative of temperature rising curve of ordered microwire is shown, which shows that T_c is 356 K. Compared with NMG4 microwire, T_c decreases by 16 K. Although it does not reach room temperature, the addition of Cu obviously reduces T_C. It can be seen from Fig. 5.27a that under the same external magnetic field, the magnetization of the as-prepared microwires is one order of magnitude lower than that of the ordered microwires. This phenomenon is related to the low atomic order, high defect density and internal stress caused by rapid solidification. Therefore, it is proved that the prepared microwires cannot be used as magnetic refrigeration materials due to their low magnetic properties. However, the magnetic properties of the microwires are significantly improved after the ordered treatment, and due to the existence of the magnetic structure coupling state, it is expected to obtain excellent magnetocaloric properties.

In order to obtain a large refrigeration working range under the condition of magnetic phase transition coupling, the vacuum Mn free ordering heat treatment of NMGC1 was also carried out. In order to simplify the test process, several characteristic lines are selected for EDs point and line energy spectrum analysis. The typical results are shown in Fig. 5.28.

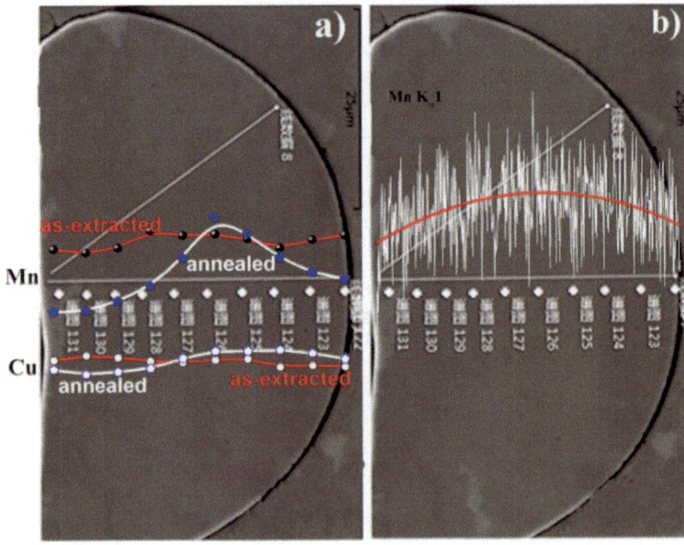

Fig. 5.28 Distribution of Mn element in cross section of NMGC1C microwire after vacuum Mn free ordering heat treatment. **a** point distribution; **b** line scan

The SEM in Fig. 5.28a shows the cross section of NMGC1C microwire, and the white and red curves are the energy spectrum results of ten points on the central line of the cross section of the microwire before and after ordering. It can be seen that Mn and Cu have little change, but Mn element in NMGC1C microwire increases gradually from microwire surface to inside, but Cu element has no significant change. Figure 5.28b shows the results of continuous line scanning at the same position. It can also be seen that the composition of Mn changes gradiently from the surface to the inside, which is consistent with the results of point scanning. To sum up, the same as NMG2C microwire, after vacuum Mn free ordering heat treatment, the NMGC1C microwire has a gradient distribution of Mn from the surface to the inside, but the change of Cu element composition is very small.

In order to determine the overall phase transformation characteristics of NMGC1 microwire after ordered heat treatment, the DSC curves of alloy ingot, as-prepared and ordered microwires during the rising/cooling process were tested. The results are shown in Fig. 5.29. All martensitic transformation temperatures obtained by the tangent method are shown in Table 5.8.

As shown in Table 5.8, the transformation width of microwires after ordered heat treatment is 14.8 and 13.6, which is higher than that of alloy ingots and as-prepared microwires, which proves that composition gradient plays a role in broadening the transformation width. Therefore, combined with the effect of the magnetic structure coupling state and the broadened phase transition width, it is expected to obtain large ΔS_m and wide δT_{FWHM} in NMGC1C microwires.

It can be seen from Fig. 5.27b that the extreme value of the M-T curve $\partial M/\partial T$ appears at 356 K. Therefore, a series of isothermal magnetization curves during the

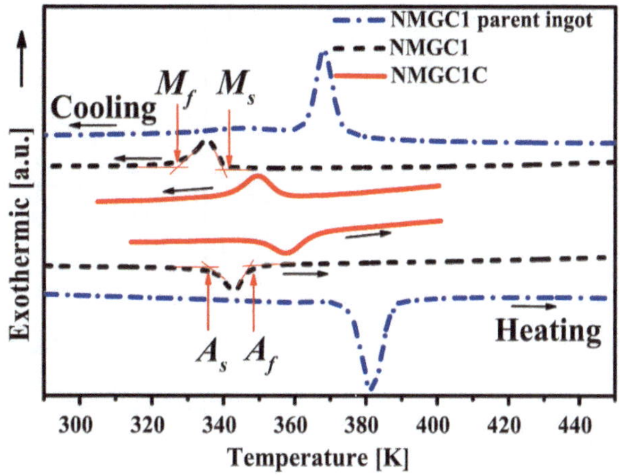

Fig. 5.29 DSC curves of NMGC alloy gold ingot, as-prepared and ordered microwires

Table 5.8 Phase transformation information of NMGC alloy ingots, as-prepared and ordered microwires

Objects	A_s	A_p	A_f	M_s	M_p	M_f	T_c	A_f-A_s	M_s-M_f
	K								
Parent ingot	376.5	381.4	387.4	373.0	368.2	363.6	348.3	10.9	9.4
NMGC1	338.0	342.9	347.3	340.0	334.9	328.8	337.1	9.3	11.2
NMGC1C	350.6	357.6	365.4	356.4	349.7	342.8	356.0	14.8	13.6

heating process is also tested near 356 K. The results are shown in Fig. 5.30a. It can be seen from the figure that the saturation magnetization of martensite decreases after Cu doping. In order to determine the T_c of the microwire, the isothermal magnetization curve near the phase transition temperature is also selected as the Arrott curve. The results are shown in Fig. 5.30b. It can be seen that the T_c point is between 356 and

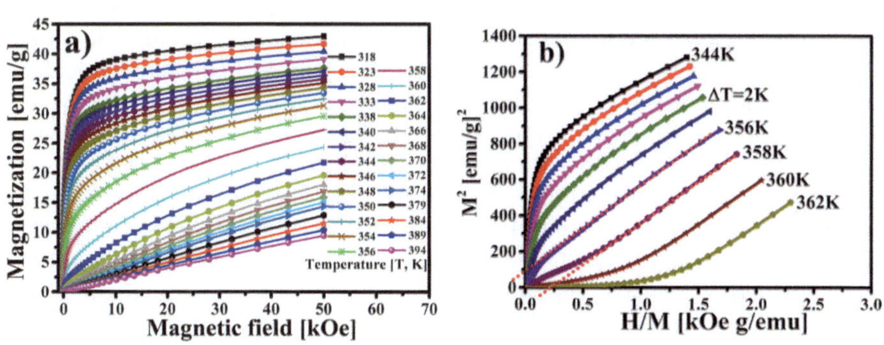

Fig. 5.30 Magnetization curves and Arrot curves of NMGC1C microwire at different temperatures. **a** *M-H* curve; **b** Arrot curve

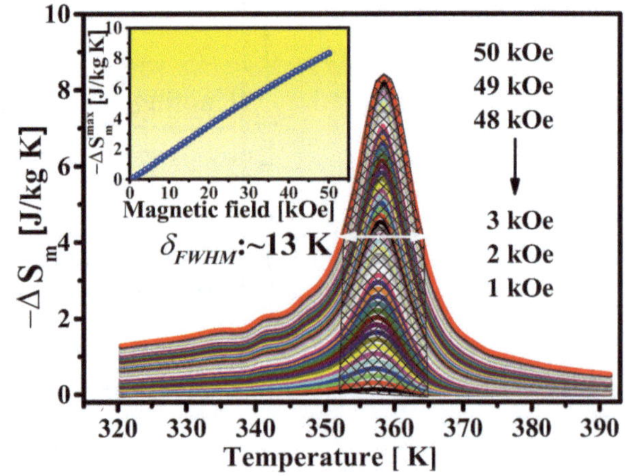

Fig. 5.31 Temperature-dependent ΔS_m in different magnetic fields of NMGC1C microwire

358 K, which is consistent with the results of the first derivative method. It can be seen from Table 5.8 that the A_s and A_f of the microwire are 350.6 and 365.4 K, respectively. Therefore, the magnetic transformation process and martensitic transformation occur at the same time as NMG4C microwire, which further confirms the state of magnetic structure coupling.

Based on the isothermal magnetization curves at different temperatures, the relationship curves between ΔS_m of NMGC1C microwire and temperature under different magnetic fields are calculated according to formula (5.8), as shown in Fig. 5.31. It can be seen from the figure that the maximum value of ΔS_m appears at 359 K, and the maximum value of ΔS_m of microwire is −3.6 J/kg K and −8.3 J/kg K at 20 and 50 kOe, which is much larger than that of Ni–Mn–In–Co [45] and Ni–Mn–Ga Fe [1] microwires. The curve of the relationship between the maximum ΔS_m and the magnetic field is shown in the embedded diagram. The results show that the microwire has the same approximate linear change trend as NMG4C microwire and does not reach saturation at 50 kOe.

It can be seen from Fig. 5.31 that the microwire δT_{FWHM} reaches ~13 K at 50 kOe, which is the maximum value obtained in Ni–Mn–Ga Cu alloy. According to formula (1.16), the RC values of the microwire under different magnetic fields are calculated. Because the hysteresis is very small and negligible, the RC value of the microwire is ~78 J/kg at 50 kOe. Compared with other Ni-Mn-based alloys, NMGC1C microwires have smaller ΔS_m (−8.3 J/kg K) in a magnetic field of 50 kOe, but its δT_{FWHM} (~13 K) is larger than that of other Ni–Mn-based ribbons or bulk single crystal or polycrystalline alloy materials; the RC value of the microwire is (~78 J/kg) equivalent to that of the remaining Ni–Mn-based alloys (70–115 J/kg) and higher than that of other reported Ni–Mn–Ga–Cu bulk polycrystalline and single-crystal alloys (72–75 J/kg); besides, compared with Gd and $LaFe_{13-x}Si_x$ alloys, NMGC alloys do not contain rare earth elements and are economically non-toxic [6, 9].

5.4.5 *Effect of Fe Doping on Magnetocaloric Properties of Ni–Mn–Ga Microwires*

In this section, the effects of Fe doping on martensitic transformation, magnetic transformation, phase transformation hysteresis, saturation magnetization and magnetocaloric properties of Ni–Mn–Ga microwires [1, 7, 46] are systematically studied.

5.4.5.1 Effect of Magnetic Field on Martensitic Transformation and Curie Temperature

For the ordered microwires, the *M-T* curves of the samples were measured under a constant magnetic field of 10 Oe, as shown in Fig. 5.32. The heating and cooling rates were 5 K/min. The *M-T* curves of microwires with different Fe contents have similar rules. During the heating process, all the *M-T* curves have martensite–austenite transformation and magnetic transformation of austenite phase. The former found that the existence of thermal hysteresis is a first-order reversible structural transformation, and the latter is a second-order ferromagnetic–paramagnetic transformation, and the

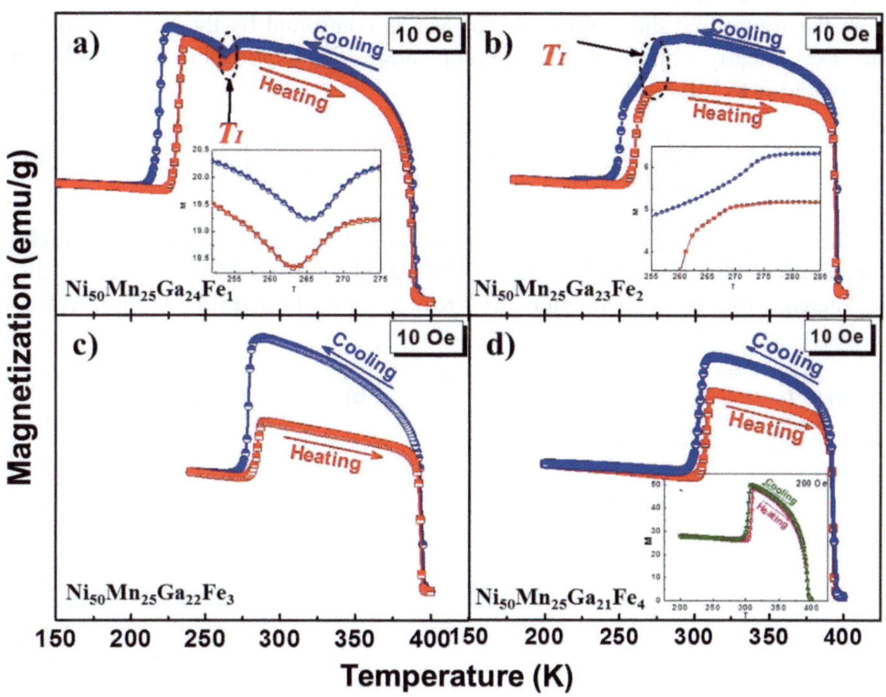

Fig. 5.32 M-T curve of $Ni_{50}Mn_{25}Ga_{26-x}Fe_x$ microwire. **a** $Ni_{50}Mn_{25}Ga_{24}Fe_1$; **b** $Ni_{50}Mn_{25}Ga_{23}Fe_2$; **c** $Ni_{50}Mn_{25}Ga_{22}Fe_3$; **d** $Ni_{50}Mn_{25}Ga_{21}Fe_4$

magnetic structure changes from order to disorder. In addition, a significant downward jump of magnetization occurs at ~270 K, corresponding to pre-MT [47]. The latent heat of phase transformation is small (about 2 J/g) in the microwires with Fe content of $x = 1$ and $x = 2$, respectively. The pre-MT occurs at 268 K and 272 K (Fig. 5.32a, b), respectively. There is no premartensitic phase in the microwires with $x > 2$. Previous studies have shown that in Ni2MnGa system, the premartensitic phase is caused by the magnetoelastic coupling between structure and magnetic degrees of freedom, and the TA2 branch (1/3 1/3 0) phonons in the alloy show obvious softening with temperature [47]. It can also be found from the thermomagnetic curve that the magnetization value of ferromagnetic austenite phase in cooling process is higher than that of heating process under 10 Oe magnetic field. In order to explore the cause of magnetization difference in heating and cooling process, *M-T* curve was measured under a slightly higher magnetic field of 200 Oe, as shown in Fig. 5.32d. The change of magnetization value of austenite phase under 200 Oe magnetic field is much smaller than that of 10 Oe, and MT temperature slightly moves to low temperature.

Figure 5.33a shows the phase diagram of MT temperature and Curie temperature of $Ni_{50}Mn_{25}Ga_{25-x}Fe_x$ microwire with Fe content. Pre-M, ferro-M, ferro-A and para-A represent premartensite phase, ferromagnetic martensite phase, ferromagnetic austenite phase and paramagnetic austenite phase, respectively. It can be seen from the figure that the temperature of MT is controlled by the composition and increases linearly with the change of composition. The relationship between M_s and Fe content is as follows:

$$\frac{\Delta T}{\text{Fe(at.\%)}} = \left|\frac{T_6 - T_1}{\text{Fe}_6 - \text{Fe}_1}\right| = \left|\frac{361.3 - 224.8}{6 - 1}\right| = 27.3 \tag{5.11}$$

The average increase rate of MT temperature with Fe is 27.3 k/at.%, which is very close to that measured by DSC. Compared with $Ni_{50}Mn_{25}Ga_{26-x}Cu_x$ alloy, the increase rate is much smaller (76.4 K/at.%) [48]. In addition, the pre-MT and

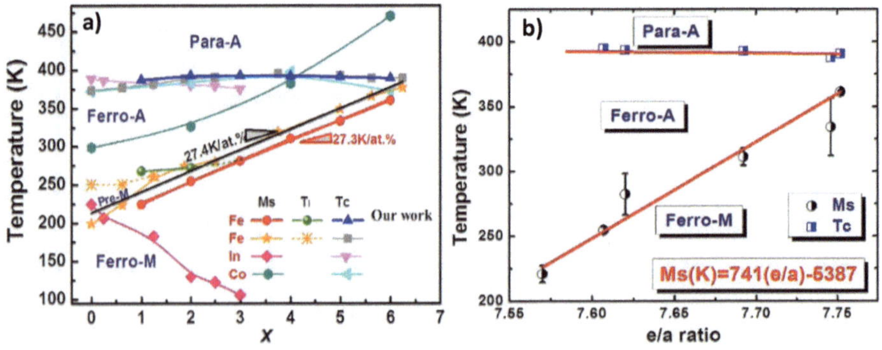

Fig. 5.33 Phase diagram of $Ni_{50}Mn_{25}Ga_{26-x}Fe_x$ microwire. **a** The temperature composition phase diagram; **b** the relationship between M_s, T_c and *e/a*

MT temperatures move toward high temperature with the increase of Fe and finally converge together. The temperature rise rate of pre-MT is lower than that of MT in $Ni_{50}Mn_{25}Ga_{23}Fe_2$ microwire, and the temperature of premartensite and MT is close, which may lead to interesting phenomena such as MCE. The Curie temperature of the microwire is stable at ~390 K, which is ~25 K higher than that of the ternary Ni2MnGa alloy because the magnetization value of Fe atom is higher than that of Ga atom [49]. The characteristic of the phase diagram is $T_c > M_s$, that is, MT occurs between two ferromagnetic phases. In addition, when $x > 4$, MT temperature is higher than room temperature, which is favorable for development and application at room temperature. In addition, the effects of alloying elements on M_s and T_c are different. Some common doped alloys such as Ni2MnGa1-xFex ($0 \leq x \leq 0.4$) and Ni2MnGa1-xInx ($0 < x < 0.25$) are also given in Fig. 5.33. For the change of M_s and T_c of ($Ni_{50.26}Mn_{27.30}Ga_{22.44}$) 100 xCox ($0 < x < 6$) alloys, the effect of different elements on T_c is weak, and the magnetic transformation of these alloys occurs in the range of 390 K or slightly higher than this temperature [7]. However, M_s strongly depends on the type of doping elements, such as Fe, In and Co. The substitution of Co for Ga leads to the rapid growth of M_S, while the substitution of In for Ga leads to the decrease of M_s, although the radius of In atom is larger than that of Ga atom. The M_s and T_c lines divide the phase diagram into several regions: ferro-M, ferro-A and para-A. The slope of linear fitting is 741 K/(*e*/*a*), which is lower than that of ternary Ni–Mn–Ga alloy (937 K/(*e*/*a*) [50], which also shows that *e*/*a* is a very convenient parameter to describe the structural stability and magnetic ordering degree in Heusler SMAS [47].

5.4.5.2 Effect of Fe Content on Phase Transformation Hysteresis

Thermal hysteresis is the standard to distinguish the first- and second-order phase transitions. The thermal hysteresis of structural phase transition is defined as A_p-M_p, and that of magnetic phase transition is defined as T_c^H-T_c^C. The curve of thermal hysteresis with Fe content is shown in Fig. 5.34. The thermal hysteresis of structural phase transition decreases from 12.5 K ($x = 1$) to 3.3 K ($x = 4$), which means that the average change of thermal hysteresis is 3.1 K/at.%, and the change of structural phase transition hysteresis is very small when $x > 4$. The decrease of hysteresis reflects the decrease of friction energy consumption at austenite–martensite interface. Compared with the hysteresis of structural transformation, the hysteresis of magnetic transformation is small.

Based on the friction theory that martensite thermal hysteresis originates from the phase interface [51], the energy dissipation of phase interface movement can be calculated. The friction model takes the percentage of martensitic transformation as the order parameter, and the free energy of the system has the following form:

$$F = a + bm + cm^2 \tag{5.12}$$

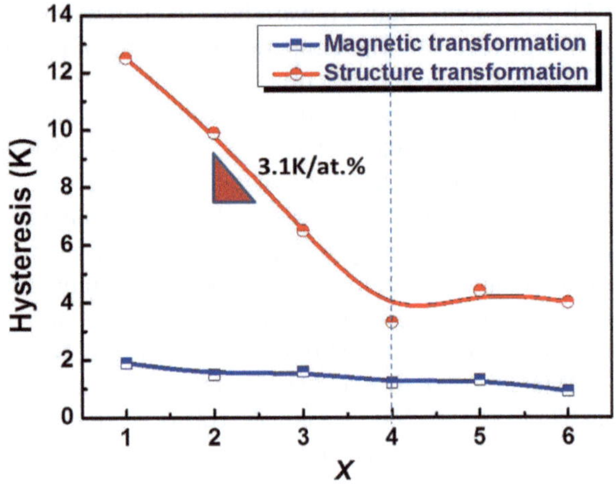

Fig. 5.34 Relationship between the thermal hysteresis of structure and magnetic phase transition and Fe content

where m is the percentage of martensitic transformation and a, b and c are functions of temperature independent of m. The friction force of interface motion can be written as follows:

$$F_r = H_r - T S_r \tag{5.13}$$

H_r and S_r are the friction enthalpy and entropy.

Figure 5.35 takes the cooling and heating curves of $Ni_{50}Mn_{25}Ga_{24}Fe_1$ microwire as an example to illustrate the calculation process [51]. It can be seen from the figure that the linear part of the phase transition loop is between 10 and 90%. The phase transition is a thermoelastic equilibrium state. The free energy function and friction function of the system can be expressed according to the state equation, $L_c = 0$ in the cooling process and $L_h = 0$ in the heating process:

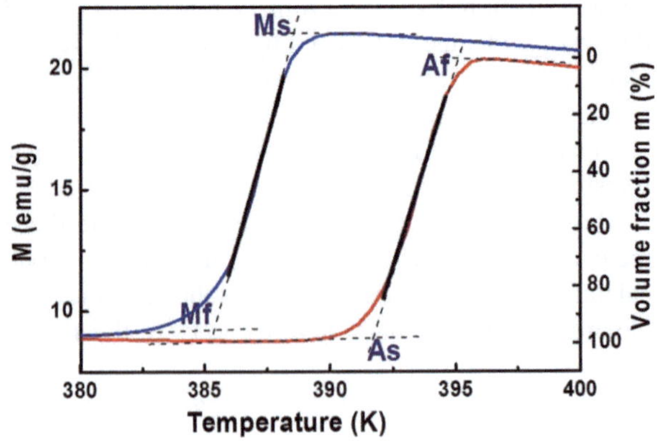

Fig. 5.35 Linear fitting of *M-T* curve of $Ni_{50}Mn_{25}Ga_{24}Fe_1$ microwire during heating and cooling

$$L_c = m + A_c T - B_c = 0, \quad L_h = m + A_h T - B_h = 0 \tag{5.14}$$

For $Ni_{50}Mn_{25}Ga_{24}Fe_1$ microwire:

$$L_c = m + 1.82T - 382.4 = 0, \quad L_h = m + 1.65T - 367.3 = 0$$

According to the fact that the equilibrium state has the minimum free energy, i.e., $\partial F/\partial m = 0$, when the system is unbalanced, the driving force of phase transition is, the resistance of phase boundary movement is, the positive MT boundary condition is and the inverse MT boundary condition is

$$F_r = -\Delta S \frac{(A_h - A_c)T + B_h - B_c}{A_h + A_c} m \tag{5.15}$$

$$F = (1.74T - 374.9)m + 0.5m^2 + C \tag{5.16}$$

$$F_r = (0.085T - 7.55)m \tag{5.17}$$

Based on Clausius–Clapeyron equation

$$\frac{\mathrm{d}\sigma}{\mathrm{d}T} = -\frac{\Delta S}{\varepsilon V} = -\frac{\Delta H}{\varepsilon T_0 V} \tag{5.18}$$

For Ni_2MnGa alloy, the ε is 0.06 and V is 32 cm^3/mol. F_r can be calculated.

As shown in Fig. 5.36, the relationship between interface friction energy consumption and Fe content is shown. The results show that the friction energy consumption of twin boundary decreases with Fe content, which is similar to the thermal hysteresis in structural transformation. The thermal hysteresis in the process of MT is due to

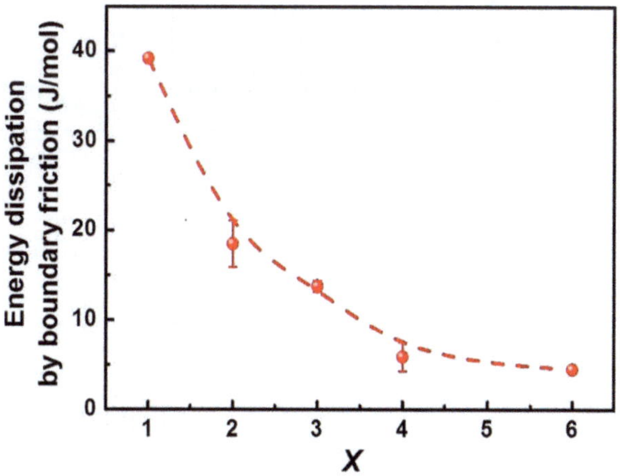

Fig. 5.36 Relationship between interfacial friction energy consumption and Fe content

the friction during the interface moving, which means that the resistance of austenite and martensite in MT decreases with the increase of Fe.

5.4.5.3 Effect of Fe Content on Saturation Magnetization

The magnetization curve is a very important means to characterize the magnetic properties of ferromagnetic materials. Figure 5.37 shows the hysteresis loops of different Fe contents in the as-prepared and ordered heat-treated microwires. The microwires in the two states show the characteristics of ferromagnetism and parromagnetism, respectively [52]. The as-prepared microwires are difficult to magnetize and cannot reach saturation under the magnetic field of 50 kOe. On the contrary, the microwires after heat treatment reach saturation under the magnetic field of ~0.2 kOe, which is due to the introduction of the internal stress, chemical ordering heat treatment reduces defects and residual stresses, resulting in homogenization of composition [28].

For ferromagnetic materials, the magnetization process can be roughly divided into the following four stages [53]: The first is the reversible rotation process of the magnetic domain. At this time, if the external magnetic field is removed, the domain wall will return to its original position; the second is the irreversible magnetization process. With the increase of the magnetic field, the magnetization curve rises rapidly, and the magnetization intensity increases sharply. At this time, after removing the external field, the position of the magnetic domain wall cannot be restored; the third is the magnetic moment of the domain, after the movement of the domain wall is basically completed, the magnetization value can only be increased by the rotation

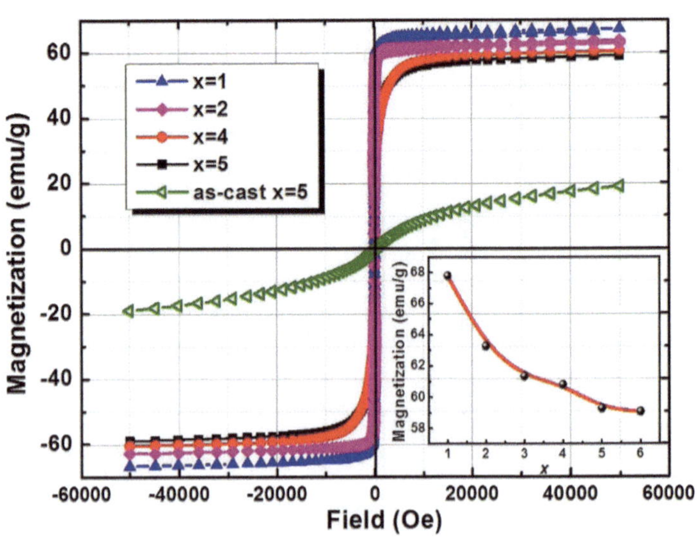

Fig. 5.37 Magnetization curve at room temperature

of the domain and magnetic moment; fourthly, the magnetization value increases little in the stage of approaching saturation, although the external field is large. For polycrystalline samples, the domain structure is very complex due to the disordered orientation of the crystal axis and the interaction between grains. The domain wall displacement process and the magnetization vector rotation process are not easy to separate in a low field. However, in the strong magnetic field, the process of domain wall displacement will stop completely, and the rotation of magnetization vector becomes the only way of magnetization (the fourth stage). In this case, because the magnetization of polycrystals comes from the rotation of magnetization vector, they have a common law, that is, they generally obey the law of approaching saturation. The law in the strong magnetic field can be expressed as follows [53, 54]:

$$M = M_S\left(1 - \frac{a}{H} - \frac{b}{H^2}\right) + \chi_P H \tag{5.19}$$

where M_s is the saturation magnetization; a is the inhomogeneity parameter; B is the anisotropy parameter; and χ_p is the susceptibility of paramagnetic magnetization under higher magnetic field. The a/H term only works in the low field, and the χ_p term is very small and can be ignored. Therefore, Eq. (5.19) in high field can be simplified as follows:

$$M = M_S\left(1 - \frac{b}{H^2}\right) \tag{5.20}$$

After derivation, the constant b satisfies the following relations:

$$b = \frac{8}{105}\frac{K_1^2}{M_S^2\mu_0^2} \tag{5.21}$$

where K_1 is the magnetocrystalline anisotropy constant and μ_0 is the vacuum permeability.

The saturation magnetization and magnetocrystalline anisotropy constant of the microwire can be obtained by using isothermal magnetization curve, formula (5.19) and Eq. (5.20). Taking $Ni_{50}Mn_{25}Ga_{19}Fe_6$ microwire as an example, the calculation process is illustrated as follows: The M-H curve is transformed into the M/H^2 curve, as shown in Fig. 5.38a, and then the points in the approaching saturation stage are selected for linear fitting, as shown in Fig. 5.38b, and the intercept and slope of the curve on the M-axis are calculated. The intercept is the saturation magnetization of the microwire.

Figure 5.37 shows the relationship between saturation magnetization and Fe content calculated by the law of approaching saturation. It can be seen that the saturation magnetization decreases with the increase of Fe content. When Fe content increases from 1 to 6 at.%, the saturation magnetization decreases from 66.44 emu/g to 58.36 emu/g, about 8.1 emu/g. The saturation magnetization of the ternary

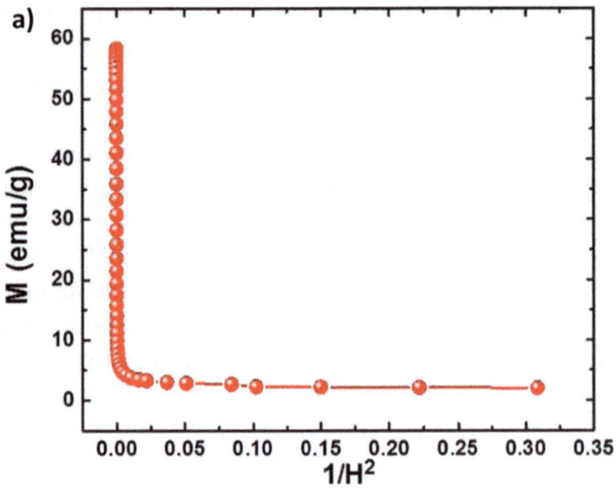

Fig. 5.38 Calculation of saturation magnetization of $Ni_{50}Mn_{25}Ga_{19}Fe_6$ microwire

Ni_2MnGa alloy with stoichiometric ratio is 66 emu/g [55], which is close to that of the microwire with Fe content $x = 1$. According to the principle of preferential number of valence electrons in ferromagnetic exchange [56], when the number of valence electrons is greater than Mn, Fe doping will preferentially occupy Mn site and make Ga vacancy. Therefore, the excess Mn atoms occupy the Ga vacancy to form antiferromagnetic exchange [56, 57], as shown in Fig. 5.39.

The occupation rule and the transition from ferromagnetism to antiferromagnetism lead to the change of exchange interaction, which leads to the decrease of saturation magnetization with Fe content. In addition, as mentioned earlier, there is no direct

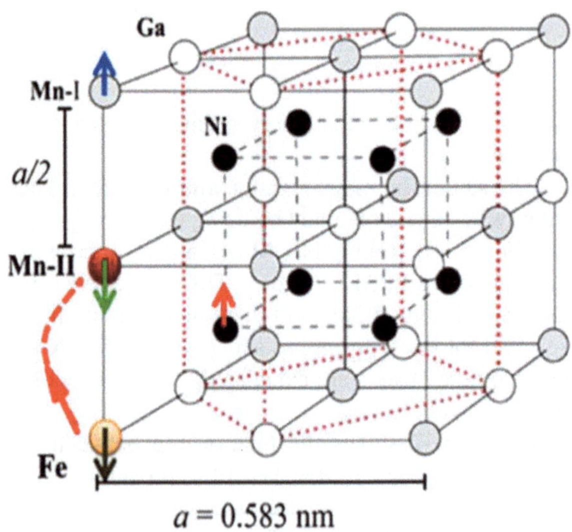

Fig. 5.39 Antiferromagnetic arrangement of Mn sites in Ni–Mn–Ga–Fe alloy

D-D coupling between Mn–Mn atoms in Ni_2MnGa crystal to form antiferromagnetic ordered arrangement of pure Mn materials. Instead, the ferromagnetic parallel ordered arrangement is formed by coupling exchange of itinerant electrons of Ni and Ga. It can be seen that the exchange of Mn–Mn atoms and the distance between Mn–Mn atoms will directly affect the magnetic properties of the materials. In Ni–Mn–Ga alloys, the magnetic coupling between Mn atoms is achieved by the itinerant electrons provided by Ni and Ga. The decrease of Ga content reduces the conduction electrons provided to Mn atoms, which weakens the exchange between Mn atoms [53], which is one of the reasons for the decrease of saturation magnetization.

Comprehensive analysis shows that the substitution of Fe can not only effectively change the MT temperature of the microwire, but also significantly increase the Curie temperature and significantly change the saturation magnetization of the material, which provides a new idea for the extensive research of small size materials in the future.

5.4.5.4 Magnetic Refrigeration

It is also shown in Fig. 5.32 that at MT, the magnetization value $\Delta M(H) = M_M - M_P$ (M_M and M_P are the magnetization values of martensite and parent phase) has an obvious change. In Fig. 5.40, the curves of ΔM and H at the transformation are given, and ΔM is closely related to the magnetic field. Figure 5.40a is a magnified view in low field, which clearly shows that ΔS_m is negative, its absolute value decreases with the increase of magnetic field and has a minimum value. Figure 5.40b is a high field diagram, which shows that ΔM reaches a positive saturation value ΔM_{sat} with the increase of magnetic field, which approximately follows an exponential relationship with the magnetic field when reaching saturation [35]. The saturation magnetization value depends on the difference of saturation magnetic

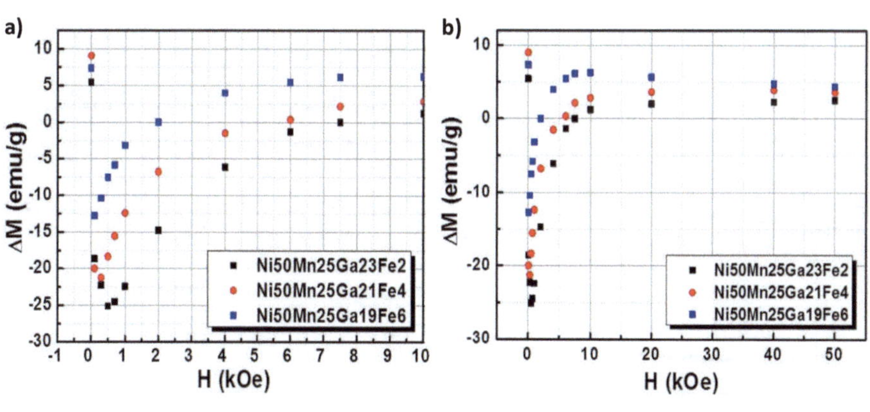

Fig. 5.40 Relationship between magnetization difference and magnetic field at MT. **a** In low field; **b** 0–50 kOe

moment between martensite and parent phases. In the figure, the magnetic fields of $Ni_{50}Mn_{25}Ga_{21}Fe_4$ and $Ni_{50}Mn_{25}Ga_{19}Fe_6$ microwires are 7.5, 5.0 and 2.0 kOe, respectively. $Ni_{50}Mn_{25}Ga_{23}Fe_2$ microwire has a minimum value of ΔM when $\Delta M < 0$, and $Ni_{50}Mn_{25}Ga_{19}Fe_6$ microwire has a maximum value of ΔM when $\Delta M > 0$.

In order to further explore the influence of magnetic field on the magnetic properties of microwires, the $M \sim T$ curves of 100 Oe and 50 kOe $Ni_{50}Mn_{25}Ga_{23}Fe_2$ microwires were tested. As shown in Fig. 5.41a, the magnetization value of austenite state increased with the decrease of temperature in the field cooling (FC) curve at 100 Oe, corresponding to the PM-FM phase transformation, the magnetization decreased sharply after reaching the peak, corresponding to the austenite to MT. The field heat (FH) curve is a process of inverse phase transition. In addition, the magnetization of zero field cooling (ZFC) curve is lower than that of FC and FH curves in the low-temperature region because of the twin boundary shift and magnetocrystalline anisotropy of martensite phase [58]. The $M \sim T$ curve of high field 50 kOe is shown in Fig. 5.41b. In the low-temperature region, the dependence of magnetization on temperature is opposite to that of low field. The illustration shows an enlarged view of MT.

Under zero or very low magnetic field, martensite is composed of a series of twin variants with different orientations. In each of the variants, there are magnetic domains with alternating magnetization and opposite directions. A small magnetic field is still enough to make some of the variants arrange along the direction of the easily magnetized axis. In an intermediate field, the magnetic domain tends to be consistent with the twin band. Due to the strong magnetocrystalline anisotropy of martensite phase, the variation fraction is increased by reducing the angle between the easy magnetization axis and the magnetic field, so that the difference of Zeeman energy is minimized. On the contrary, the single form of magnetic saturation has been formed in the austenite state when the intermediate magnetic field is applied, which has high magnetic symmetry $M_M < M_A$. Finally, the martensite crystal forms a magnetic saturation monovariant under high magnetic field, and $M_M > M_A$.

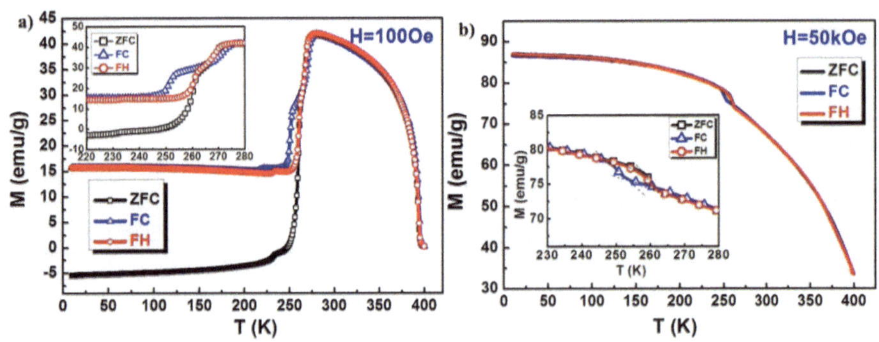

Fig. 5.41 Magnetization curves of ZFC, FC and FH of $Ni_{50}Mn_{25}Ga_{23}Fe_2$ microwire. **a** 100 Oe; **b** 50 kOe

Figure 5.42 shows the $M \sim T$ curves of $Ni_{50}Mn_{25}Ga_{23}Fe_2$, $Ni_{50}Mn_{25}Ga_{21}Fe_4$ and $Ni_{50}Mn_{25}Ga_{19}Fe_6$ microwires under the action of 200 Oe magnetic field. The illustrations show the dM/dT and T curves of FH so that the phase transition temperatures can be accurately determined. The peak temperatures A_p are 261.5 K, 309.2 K and 361 K, respectively. The curves of H/M and T at temperatures higher than Curie temperature are also shown in Fig. 5.42c (right). In this temperature range, the curves follow Curie–Weiss law [53]. In 1907, the French physicist Weiss found that the dielectric constant or magnetic susceptibility above the Curie temperature is the paraelectric phase or paramagnetic phase, that is, the transition temperature of ferromagnetic material is called Curie point. At this temperature, ferromagnetic material will lose ferromagnetism and present paramagnetism. Curie–Weiss law is expressed as follows:

$$\chi = \frac{C}{T - T_C} \tag{5.22}$$

In (5.22), T_c is the Curie point, and C is the Curie constant. The magnetic susceptibility can be written as follows:

$$\chi = \frac{AM}{QH} \tag{5.23}$$

A is the molecular weight, Q is the mass of matter, and H is the external magnetic field.

By comparing (5.22) with (5.23), it can be seen that after the Curie point, H/M is directly proportional to T. The intersection point between the linear curve and the temperature axis is the curiosities temperature θ_p of the microwire. In Fig. 5.42c, the constant C is 0.00878, and θ_p is 392.1 K.

In order to characterize MCE, $Ni_{50}Mn_{25}Ga_{23}Fe_2$ with transformation temperature lower than room temperature, near room temperature and higher than room temperature was selected. For $Ni_{50}Mn_{25}Ga_{21}Fe_4$ and $Ni_{50}Mn_{25}Ga_{19}Fe_6$ microwires, a series of isothermal magnetization curves were tested near the MT temperature. For the MT temperature, which is closest to the Curie temperature, the isothermal magnetization curves are also tested near the Curie point. The temperature interval

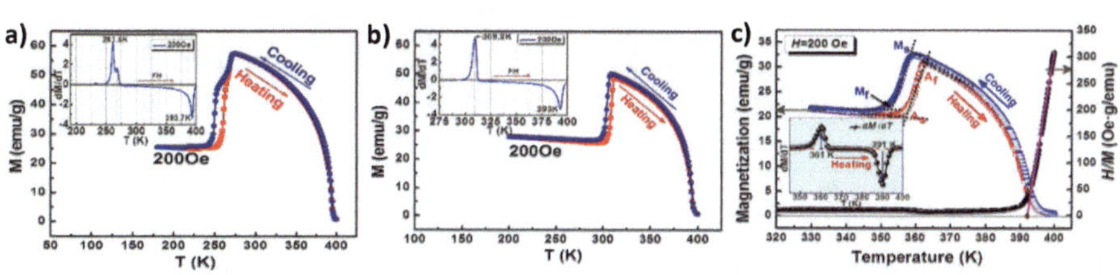

Fig. 5.42 $M \sim T$ curve of microwire under 200 Oe, illustration shows dM/dT and T curve. **a** $Ni_{50}Mn_{25}Ga_{23}Fe_2$; **b** $Ni_{50}Mn_{25}Ga_{21}Fe_4$; **c** $Ni_{50}Mn_{25}Ga_{19}Fe_6$

is 1–10 K, as shown in (a), (b) and (c) in Fig. 5.43. The effect of temperature and magnetic field on the magnetization curve is illustrated by taking Fig. 5.43b as an example. When the temperature is lower than the transformation temperature, the martensite phase has high magnetic anisotropy compared with the austenite phase due to its lower symmetry. The illustration shows that there is a cross point when the transformation from ferromagnetic martensite to ferromagnetic austenite exists. For $Ni_{50}Mn_{25}Ga_{21}Fe_4$ microwire, the magnetic field at this point is ~4.5 kOe. Below this cross field, the magnetization value of austenite is higher than that of martensite phase, that is, $M_A > M_M$. On the contrary, the magnetization higher than this cross field is $M_A < M_M$. For $Ni_{50}Mn_{25}Ga_{23}Fe_2$ and $Ni_{50}Mn_{25}Ga_{19}Fe_6$ microwires, the cross fields are ~7.6 and 1.5 kOe, respectively, which means that the critical cross field in Ni–Mn–Ga–Fe alloy is dependent on composition. In addition, the magnetization and demagnetization curves (358–364 K and 390 K) near the first-order MT and the second-order magnetic phase transition are also given in the illustration of Fig. 5.43c. The coincidence degree of the curves is good, and the hysteresis loss is close to 0, which is absolutely beneficial to magnetic refrigeration.

The relationship between the calculated magnetic entropy change and temperature and magnetic field is shown in Fig. 5.44a–c. The three groups of curves have

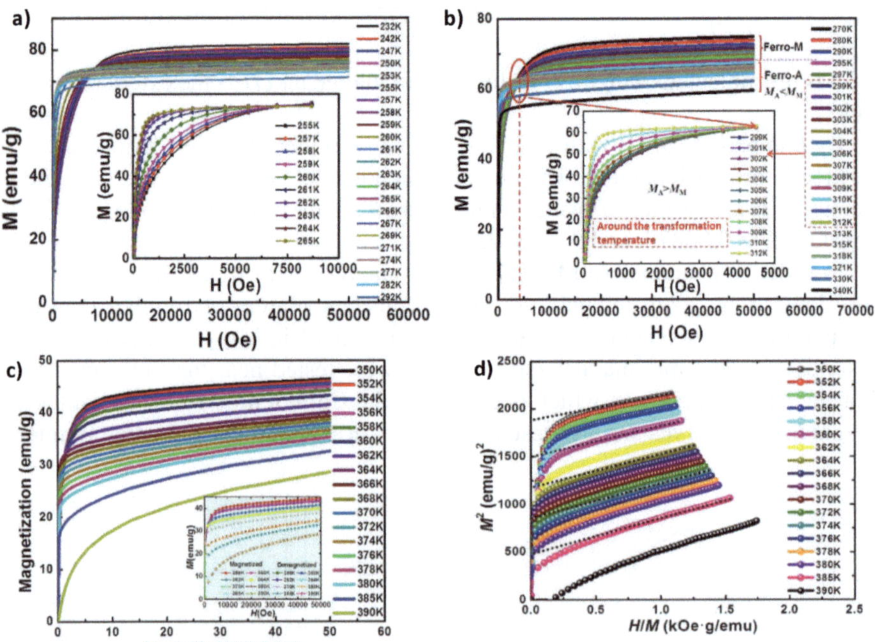

Fig. 5.43 Isothermal magnetization curve of the microwire, the illustration shows the $M \sim H$ curve and Arrott curve near the phase transition temperature. **a** $Ni_{50}Mn_{25}Ga_{23}Fe_2$; **b** $Ni_{50}Mn_{25}Ga_{21}Fe_4$; **c** $Ni_{50}Mn_{25}Ga_{19}Fe_6$; **d** Arrott ($M^2$ vs. H/M) plots for $Ni_{50}Mn_{25}Ga_{19}Fe_6$ microwires

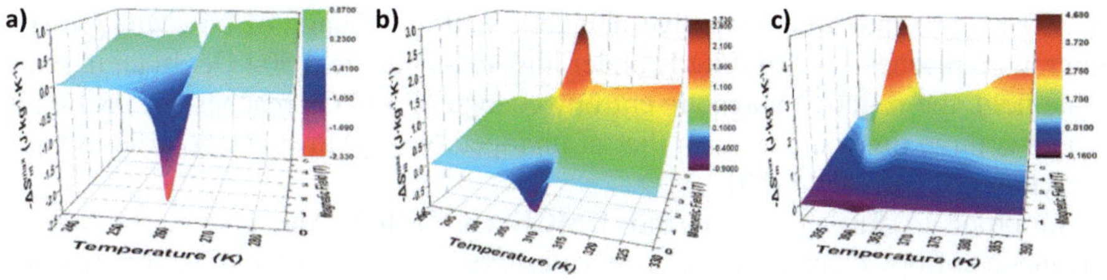

Fig. 5.44 Curves of ΔS_m, T and ΔH of microwires. **a** $Ni_{50}Mn_{25}Ga_{23}Fe_2$; **b** $Ni_{50}Mn_{25}Ga_{21}Fe_4$; **c** $Ni_{50}Mn_{25}Ga_{19}Fe_6$

similar characteristics: The magnetic entropy change at the structural phase transition shows a narrow peak, and the corresponding half-height width is about ~6 K. For FOMT, the peak appears at the phase transition temperature, and for SOMT, the peak appears at the Curie point. The ΔS_m of $Ni_{50}Mn_{25}Ga_{23}Fe_2$ microwire is positive at ~260.5 K, ΔH < 46 kOe. The maximum value of $-\Delta S_m$ is ~-0.3 J/kg K under high magnetic field 50 kOe, and the corresponding MCE is DMCE. On the contrary, the maximum value of ΔS_m is ~2.12 J/kg K under low magnetic field of 7.6 kOe. The corresponding MCE is IMCE. Large magnetic entropy change can be obtained in such a low magnetic field, which can be achieved by permanent magnets. The DMCE and IMCE of $Ni_{50}Mn_{25}Ga_{21}Fe_4$ microwire appear at 309.5 K near room temperature. The maximum value of ΔS_m is ~0.9 J/kg K at low field 4.5 kOe. When the magnetic field increases to 20 kOe, the magnetic entropy change varies from positive to negative and reaches the maximum negative value about -2.71 J/kg K under high magnetic field 50 kOe. With the increase of Fe doping content, for example, MT temperature of $Ni_{50}Mn_{25}Ga_{19}Fe_6$ microwire approaches Curie temperature, the magnetocrystalline anisotropy weakens, and the negative value dominates. DMCE is the traditional magnetocaloric behavior. There is only a small positive magnetic entropy change at 1.5 kOe. On the other hand, the maximum negative value of 360.5 K is ~-4.7 J/kg K at high magnetic field 50 kOe. In an intermediate field of 20 kOe, the maximum magnetic entropy change is ~-1.71 J/kg K.

The DSC curve shows that the enthalpy change of the structural phase transition is $\Delta H = 10.14$ J/g, and the entropy change can be calculated by the following formula:

$$\Delta S \approx \Delta H / T_0 \tag{5.24}$$

If the average transformation temperature $T_0 = (A_f + M_s)/2$ has the same Gibbs energy at the equilibrium transformation temperature of austenite and martensite, then T_0 and ΔS are calculated to be 360.5 K and 28.1 J/kg K, and the magnetic entropy change is obviously much lower than the transformation entropy, which can be further improved by increasing the magnetic field.

The results show that the peak temperature of DMCE and IMCE is close to MT temperature. Therefore, the magnetic entropy change has an important relationship with MT, and MT plays an important role in MCE. In addition, in Fig. 5.44c, in addition to a narrow peak centered at 360.5 K, there is also a broadened small peak at 390 K. The maximum negative magnetic entropy change of $Ni_{50}Mn_{25}Ga_{19}Fe_6$ microwire reaches ~−2.91 J/kg K under high magnetic field 50 kOe. In an intermediate field, the negative magnetic entropy change of 30 kOe reaches −2.05 J/kg K, which is three times higher than that of undoped $Ni_{50.95}Mn_{25.45}Ga_{23.6}$ microwire in the same magnetic field [59].

In order to further understand DMCE and IMCE, the change of sign of magnetic entropy change can be clearly observed from the curve of isothermal field change, magnetic entropy change and temperature. The magnetic entropy change of $Ni_{50}Mn_{25}Ga_{23}Fe_2$, $Ni_{50}Mn_{25}Ga_{21}Fe_4$ and $Ni_{50}Mn_{25}Ga_{19}Fe_6$ microwires at each temperature is selected as the function of magnetic field to draw the image, as shown in Fig. 5.45.

As shown in Fig. 5.45, each group of curves consists of two parts, one is that the magnetic entropy change is negative, and it decreases linearly with the increase of magnetic field; the other is that the maximum magnetic entropy change first reaches a positive maximum and then decreases linearly with the increase of magnetic field. The proportion of these two parts in different components is different as shown in Fig. 5.45a the linear decreasing part with negative magnetic entropy change is weak; (b) the two are equivalent; (c) the linear decreasing part with negative magnetic entropy change dominates. It is found that with the increase of magnetic field, a positive maximum magnetic entropy change is first reached, and then the maximum peak value appears near the phase transition temperature when the magnetic field decreases. For $Ni_{50}Mn_{25}Ga_{23}Fe_2$, $Ni_{50}Mn_{25}Ga_{21}Fe_4$ and $Ni_{50}Mn_{25}Ga_{19}Fe_6$ microwires, they are 260.5 K, 309.5 K and 360.5 K, respectively, which are just corresponding to the phase transition temperature.

The relationship between $\Delta S_M{}^{\max}$ and H is shown in Fig. 5.46. It is found that the initial increase stage of $\Delta S_M{}^{\max}$ of $Ni_{50}Mn_{25}Ga_{19}Fe_6$ microwire is weak. For $Ni_{50}Mn_{25}Ga_{23}Fe_2$, $Ni_{50}Mn_{25}Ga_{21}Fe_4$ and $Ni_{50}Mn_{25}Ga_{19}Fe_6$ microwires, the slope is 0.6, 0.87 and 1.0 J/kg K for each 10 kOe increase in magnetic field. The linear variation of $\Delta S_M{}^{\max}$ with H is very common in rare earth materials, such as pure Gd materials reported and some soft magnetic amorphous materials [60]. Only in Gd_2In alloy [61], the sign change of MCE sign induced by magnetic field at the same temperature is also found, but the positive and negative magnetic entropy changes are slightly different. This alloy is caused by the separation of antiferromagnetic and

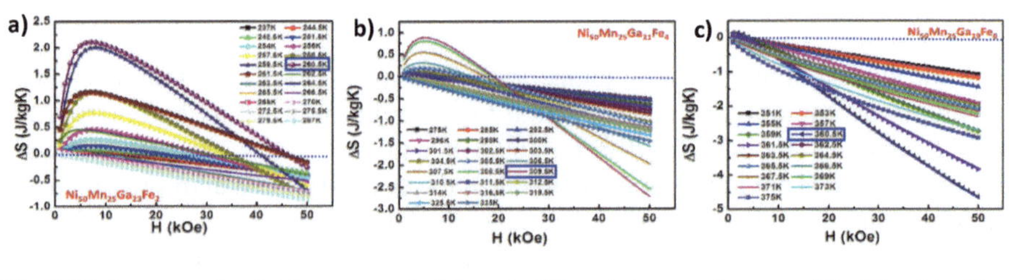

Fig. 5.45 Isomagnetic field curves of ΔS_m at different temperatures. **a** $Ni_{50}Mn_{25}Ga_{23}Fe_2$; **b** $Ni_{50}Mn_{25}Ga_{21}Fe_4$; **c** $Ni_{50}Mn_{25}Ga_{19}Fe_6$

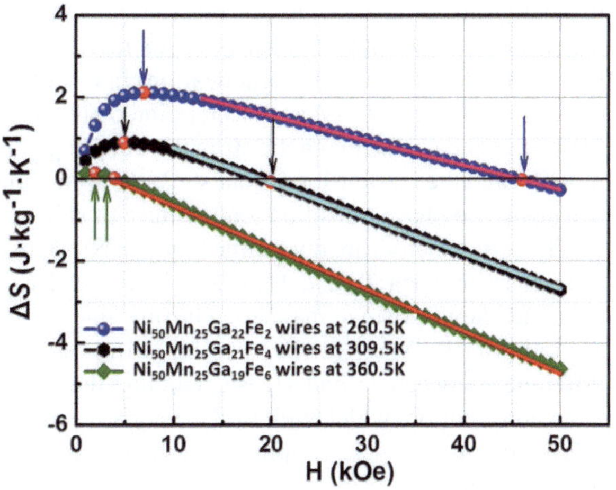

Fig. 5.46 Relationship curve of ΔS_m and magnetic field at MT temperature

ferromagnetic states caused by magnetic field-induced variable magnetic transformation. It is obvious that the micromechanism of Gd_2In alloy and Ni–Mn–Ga–Fe microwire is not the same. However, compared with Gd_2In alloy, the peak temperature of Ni–Mn–Ga–Fe microwire is increased by 170–270 K, which is more practical for room-temperature refrigeration.

More interesting analysis shows that the relationship between $\Delta S_M{}^{\max}$ and magnetic field varies with Fe content, that is, it is also regulated by *e*/*a*. As shown in Fig. 5.47, the maximum value of positive and negative magnetic entropy changes near the phase transition temperature and the relationship between the magnetic field corresponding to the maximum value of positive magnetic entropy change and *e*/*a* are given. With the increase of *e*/*a*, the maximum value of positive magnetic entropy change decreases, the maximum value of negative magnetic entropy change increases, and the magnetic field corresponding to the maximum value of positive

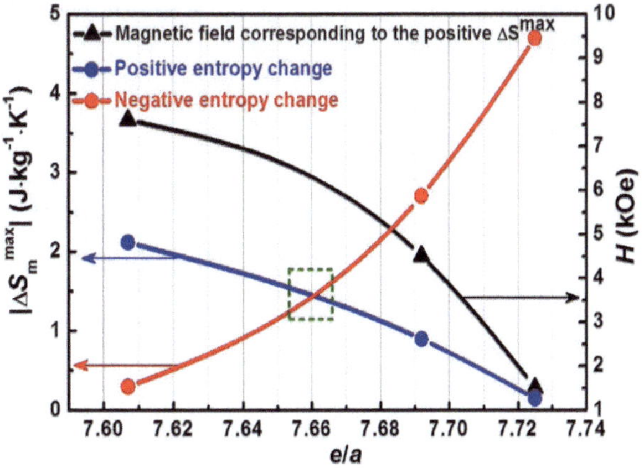

Fig. 5.47 Relation curve between magnetic field and *e*/*a* corresponding to $|\Delta S_m{}^{\max}|$ and maximum value of positive magnetic entropy change

magnetic entropy change also decreases. This magnetic field value is consistent with the position of the intersection point of martensite to austenite transformation in Fig. 5.43 of the isothermal magnetization curve and corresponds to the magnetic field of $\Delta M = 0$ in Fig. 5.40. Marcos [35] pointed out that the MCE near MT is mainly due to two factors: (1) the magnetic structure coupling between the mesoscopic magnetic moment and the martensitic variants; (2) the spin phonon coupling on the microscale can cause the phase transition temperature to shift with the magnetic field.

These two factors are determined by *e*/*a*. $Ni_{50}Mn_{25}Ga_{23}Fe_2$ microwire has a large difference between T_c and M_s ~ 139 K, *e*/*a* $\approx$ 7.6. Mesoscopic coupling is dominant and shows positive magnetic entropy change in low field. On the contrary, the *e*/*a* ratio of $Ni_{50}Mn_{25}Ga_{21}Fe_4$ and $Ni_{50}Mn_{25}Ga_{19}Fe_6$ microwires is about 7.7, the difference between T_c and M_s is small, and the corresponding microcoupling is the main contribution, which will produce negative magnetic entropy change. When *e*/*a* ~ 7.66, the mesoscopic coupling and microscopic coupling are equal, as shown in the box in Fig. 5.47.

Based on Fig. 5.44c, the relationship between $\Delta S_M{}^{max}$ at the structural and magnetic phase transitions of $Ni_{50}Mn_{25}Ga_{19}Fe_6$ microwire and magnetic field is plotted, as shown in Fig. 5.48. For the second-order magnetic phase change materials, at a given temperature value, $\Delta S_M{}^{max}$ and magnetic field conform to the following law [62]:

$$\Delta S_M^{\max} \propto H^n \tag{5.25}$$

The exponent *n* is related to the magnetic state and can be written as follows:

$$n = \frac{d \ln \Delta S_m^{\max}}{d \ln H} \tag{5.26}$$

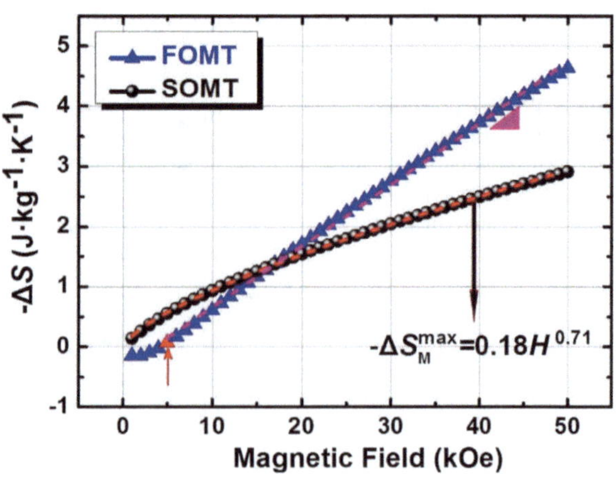

Fig. 5.48 Relation curve of $|\Delta S_M{}^{max}|$ at the primary and secondary phase transitions of $Ni_{50}Mn_{25}Ga_{19}Fe_6$ microwire and magnetic field

The fitting results are also plotted in Fig. 5.48. For $Ni_{50}Mn_{25}Ga_{19}Fe_6$ microwire around the magnetic phase transition, the index $n = 0.71$, which deviates from 2/3 calculated by mean field theory [62], which is close to the index of Gd-based and Fe-based amorphous wires ($n = 0.74$ and 0.75) [60, 63], but much lower than that of FePd ribbons ($n = 0.84$) [64]. Based on the 2/3 mean field theory, the microwire n value slightly higher than 2/3 is due to local component inhomogeneity, shape anisotropy and geometric size effect.

5.4.6 Cooling Temperature Broadening and Magnetic Cooling Capacity Design

In this section, the idea of broadening δT_{FWHM} and optimizing the magnetocaloric properties are mainly described in this section. The schematic diagram shown in Fig. 5.49 can be used for a brief explanation [3, 9]. Figure 5.49a shows that the temperature difference between structural transformation and magnetic transformation ($\Delta T_{\text{F-SOT}}$) is large. At this time, the coupling effect between magnetic and structural is small, and the two transformation peaks do not affect each other, that is,

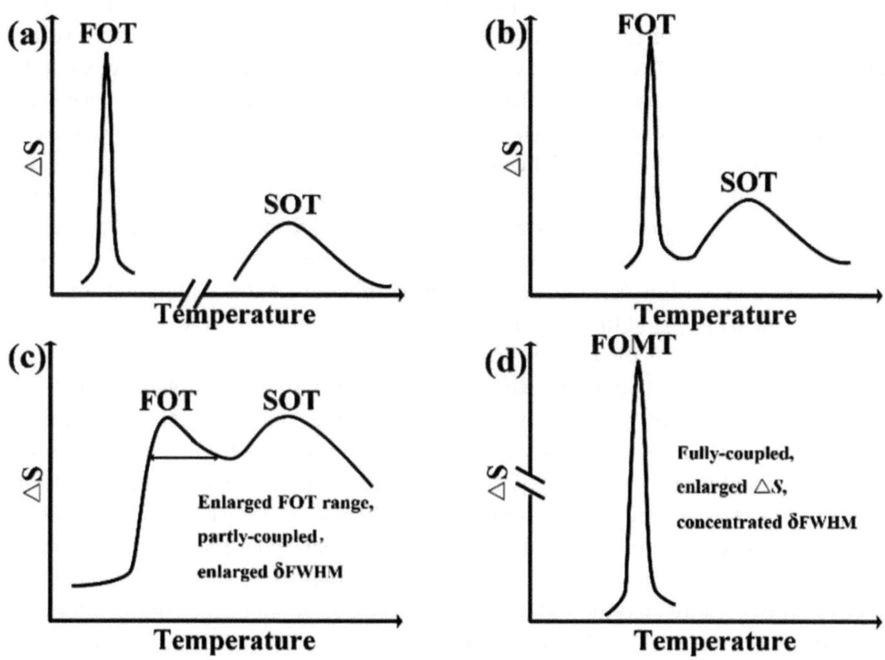

Fig. 5.49 Schematic diagram of ΔS_m evolution with different first- and second-order phase transition widths. **a** FoT is far away from SOT; **b** FOT approaches SOT slowly; **c** FOT is partially coupled with SOT; **d** FOT is fully coupled with SOT

two separated peaks appear as shown in the figure. Moreover, due to the small interaction between martensitic variants and magnetic moment, the ΔS_m value is generally small. With the decrease of $\Delta T_{\text{F-SOT}}$, the two-phase transition magnetocaloric peaks are close to each other, and finally a transition zone appears, as shown in Fig. 5.49b. However, in general, martensitic structural transformation is a non-diffusion type transformation. For the alloy with uniform composition, the transformation width is very narrow. Therefore, when $\Delta T_{\text{F-SOT}}$ decreases and two magnetocaloric peaks meet, the two magnetocaloric peaks are inconsistent as shown in (b). However, the practical application process, especially the Ericsson cycle refrigerator, needs the stable state of ΔS_m value in a wide temperature range, so the state shown in (b) has no practical value.

Therefore, considering that ΔS_m is higher and the temperature range is narrow at the structural phase transition than that at the magnetic phase transition, a method of broadening the temperature width of structural phase transition is designed to widen the width of magnetocaloric peak at the structural phase transition, so as to achieve the effective transition between the two peaks, and the ΔS_m value is numerically stable in a wide temperature range. Therefore, on the basis of Fig. 5.49b, based on the fact that the phase transition temperature of Ni–Mn–Ga microwire depends on composition and Mn element is easy to volatilize at high temperature, the composition distribution with gradient distribution from surface to interior is introduced into the microwire by heat treatment under vacuum condition, so as to widen the phase transition temperature range of structure. The schematic diagram of the results is shown in Fig. 5.49c. The effective transition of the two magnetocaloric peaks makes the transition region possess the ΔS_m value equivalent to the magnetocaloric peak, thus broadening the overall δT_{FWHM} and increasing the RC value of the alloy. In addition, when $\Delta T_{\text{F-SOT}}$ approaches 0, i.e., the state of magnetic structure transformation coupling, the situation as shown in (d) appears. Due to the magnetic structure coupling, the ferromagnetic martensite directly transforms into the paramagnetic austenite, which makes ΔM increase. If the width of structural transformation is not broadened, a very high ΔS_m value will appear in a very narrow temperature range. However, the δT_{FWHM} at this time is far less than that in Fig. 5.49c, so the RC value in Fig. 5.49d is far less than that in Fig. 5.49c.

In general, the magnetic structure coupled and partially coupled Ni–Mn–Ga–(Cu) microwires were prepared by a simple method, and the idea of optimizing the microwire MCE by adjusting the composition and heat treatment process was realized. In addition, it can be seen from the above that the microwires before and after ordering have no fracture after more than 100 cold and thermal cycles under low stress. This service state is much more severe than that under the effect of magnetic field. Therefore, the mechanical stability of microwires is sufficient to meet the needs of refrigeration. Combined with the advantages of one-dimensional microwire materials, such as good mechanical stability, easy heat dissipation and being used as subcomponents to form complex shape components, the design method in this paper makes Ni–Mn–Ga–(Cu) microwire have a good application prospect in magnetic refrigeration.

5.5 Damping Characteristics of Ni–Mn–Ga–X Alloy Microwires

The damping properties of materials are widely used in vibration reduction and noise reduction. At present, the research on Ni–Mn–Ga alloys is mainly focused on the magnetic-induced strain properties, but the research on damping properties is less. Since damping property is very sensitive to the microstructure of materials, damping is mainly used in the study of martensitic transformation in the early stage. It is found that internal friction peak is generated during martensitic transformation and reverse transformation, accompanied by the minimum value of modulus [65]; for the first type of Ni–Mn–Ga alloy, premartensitic transformation occurs, accompanied by damping maximum and minimum modulus [66]. Gavriljuk et al. [67] found that there is a damping peak in the range of martensite temperature, which is due to the reconstruction of martensite lattice. In addition to the above research on the relationship between damping performance and martensitic transformation of Ni–Mn–Ga alloy, the composition of martensitic transformation peak has also attracted great attention. Chang et al. [68] found that the transformation damping peak consists of isothermal and non-isothermal parts, the former has nothing to do with the heating/cooling rate, while the latter changes with the rising/decreasing temperature rate. This section mainly introduces the damping characteristics of Ni–Mn–Ga–X one-dimensional microwire materials [10, 12].

5.5.1 Ni–Mn–Ga Microwire

5.5.1.1 Effect of Vibration Frequency on Damping Properties of Microwires

Dynamic mechanical thermal analyzer DMA Q800 was used to test the damping performance of the microwire. Figure 5.50a shows the damping temperature spectrum of NMG1 microwire at different vibration frequencies during cooling. It can be seen from the figure that the damping change law is the same at different frequencies. In the beginning, before 30°C, with the decrease of temperature, the damping gradually decreases. When the temperature further decreases, the damping value increases rapidly, the damping peak appears at 10 °C, and the damping value oscillates after 0 °C. The tangent method can be used to determine the starting and ending points of damping peak, which is consistent with the M_s point of 21.6 °C and M_f point of 5.6 °C measured by DSC, indicating that the peak position corresponds to martensitic transformation, and the damping peak corresponds to the martensitic transformation process of microwire. It can also be seen from the figure that the peak position of the curve does not change with the decrease of frequency, indicating that the vibration frequency has no effect on the position of damping peak. However, with the decrease of vibration frequency, the fluctuation of the curve in the low-temperature

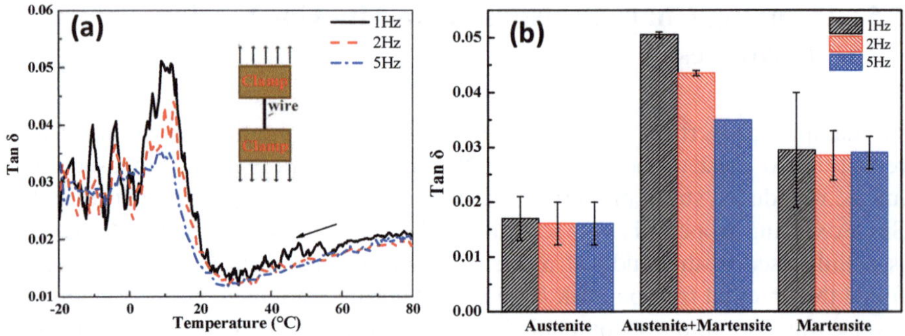

Fig. 5.50 Damping temperature spectra of NMG1 microwire. **a** At different frequencies and **b** comparison of damping values of austenite, martensite and their mixtures

range is more obvious, because there are many twin interfaces in martensite at low temperature, and anelastic motion occurs under the action of external dynamic load. At lower frequencies, the anelastic motion is more sufficient and can consume more energy, so the fluctuation of the curve is more obvious.

In Fig. 5.50a, the average value of martensite, austenite and their mixed state is selected to draw the histogram as shown in Fig. 5.50b, where the mixed state of the two is selected as the damping peak. It can be seen from the figure that the damping value of the material in the mixed state is the largest, followed by the martensite state, and the austenite state is the smallest. In the state of austenite and martensite, the vibration frequency has little effect on the damping, while in the mixed state, the damping value decreases with the increase of vibration frequency. In the Dejonghe model [69], the damping value of the mixed state is related to the vibration frequency as follows:

$$\text{Tan}\delta = \frac{\upsilon}{2\pi}\left\{\frac{\partial V_m}{\partial T}\cdot\frac{\partial T}{\partial t}\cdot\frac{1}{f}+\frac{4}{3}\sigma_0\frac{\partial V_m}{\partial \sigma}\left[1-\left(\frac{\sigma_c}{\sigma_0}\right)^3\right]\right\} \tag{5.27}$$

where υ is Poisson's ratio, $\frac{\partial V_m}{\partial T}$ is the amount of martensite formation at unit temperature, $\frac{\partial T}{\partial t}$ is the rate of rise/drop, f is the frequency, σ_0 is the stress amplitude, $\frac{\partial V_m}{\partial \sigma}$ is the martensitic transformation under unit stress, and σ_c is the critical stress. It can be seen from Eq. (5.27) that the damping value of the mixed state is proportional to the reciprocal of the vibration frequency, which is consistent with the experimental results in this paper.

5.5.1.2 Effect of Cooling Rate on Microwire Damping Properties

Figure 5.51a shows the damping temperature spectrum of Ni–Mn–Ga microwire at different cooling rates during the cooling process. It can be seen that with the

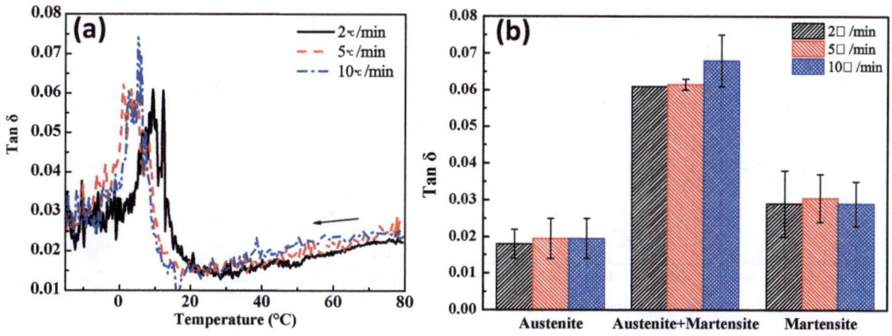

Fig. 5.51 Damping temperature spectra of NMG1 microwire. **a** At different cooling rates and **b** comparison of damping values of austenite, martensite and their mixed states

increase of cooling rate, the starting point of phase transformation peak moves to the low-temperature direction, which is because with the increase of cooling rate, the phase transformation hysteresis increases. The hysteresis from 2 to 5 °C/min is obvious, and the hysteresis changes little from 5 and 10 °C/min. Figure 5.51b shows the comparison of damping values of Ni–Mn–Ga microwires at different cooling rates under different conditions. It can be seen from the figure that in the mixed state, the damping increases with the rise/decrease rate. The damping peak in the phase transition process consists of three parts, namely the transient part IF_{Tr}, the phase transition part IF_{Pt} and the intrinsic part IF_{Int} [70]:

$$\mathrm{IF}(T) = \mathrm{IF}_{\mathrm{Tr}}(T) + \mathrm{IF}_{\mathrm{PT}}(T) + \mathrm{IF}_{\mathrm{Int}}(T) \tag{5.28}$$

Compared with the Dejonghe model, it can be seen that the transient part is directly proportional to the cooling rate, so the overall damping value increases. The same law as the vibration frequency, the rise/decrease rate has little effect on the damping of martensite state and parent phase state. It can also be seen from the figure that when the cooling rate is 10 °C/min, the damping value of mixed state is 0.075, which is three times of that in martensite state.

5.5.1.3 Effect of Strain Amplitude on Microwire Damping Properties

Figure 5.52a shows the damping temperature spectrum of NMG1 microwire with different strain amplitude during cooling process. It can be seen from the figure that the microwire shows more obvious fluctuation in martensitic state under 0.05% strain amplitude. Figure 5.52b is the damping comparison of microwires under different states and different strain amplitudes. It can be seen from the figure that the strain amplitude has no effect on the damping value in the mixed state. This may be because for the 0.05 and 0.1% strain amplitudes, the material is in a relatively small strain range, so the energy loss of the material is not large. In this small strain amplitude

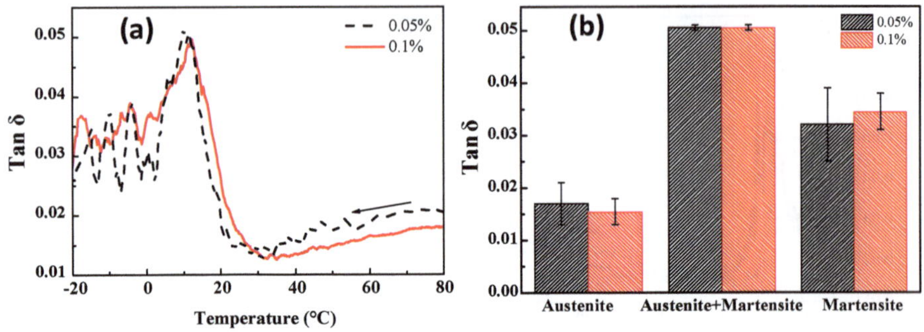

Fig. 5.52 NMG1 microwire. **a** Different strain amplitude damping temperature spectrum and **b** damping comparison under different states

range, the damping of mixed state is not affected by the strain amplitude influence. However, 0.1% strain amplitude impedance is slightly higher than 0.05% strain amplitude in martensite state, but it is opposite in austenite state. It can also be seen from the figure that the maximum damping of the microwire in martensitic state is 0.03, which is close to that of Ni–Mn–Ga single crystal [65] in martensitic state, while the high damping of Ni–Mn–Ga single crystal in martensitic state is due to the good mobility of martensitic twin boundary, which may indicate that the twin of Ni–Mn–Ga microwire in martensitic state also has good mobility.

To sum up, the damping value of NMG1 microwire decreases with the increase of vibration frequency in the mixed state of martensite and austenite, but increases with the increase of cooling rate; while in the state of martensite or austenite, it is not affected by vibration frequency and cooling rate. In martensitic state, damping increases with the increase of strain amplitude, while that in austenite decreases with the increase of strain amplitude.

5.5.2 Ni–Mn–Ga–Cu Microwire

The addition of Cu element can effectively improve the plasticity of the microwire, reduce the tendency of intergranular fracture and have the martensitic transformation characteristics of Ni–Mn–Ga ternary alloy. The high density dislocation and the increase of grain boundary area will promote the damping property of the microwire. This section describes the damping properties of as-prepared NMGC1 microwires [12]. Figure 5.53 shows the damping temperature spectra at different frequencies under the same strain amplitude. It can be seen from the figure that the damping property of the microwire is not sensitive to the amplitude, and the Tanδ values of martensite, austenite and peak value are 0.8, 0.4 and 1.2, respectively.

Figure 5.54 shows the damping temperature spectra of microwires with the same frequency and different strain amplitudes of 0.1, 0.2, 0.3 and 0.5%. It can be seen from the figure that the damping value of microwires increases with the increase of

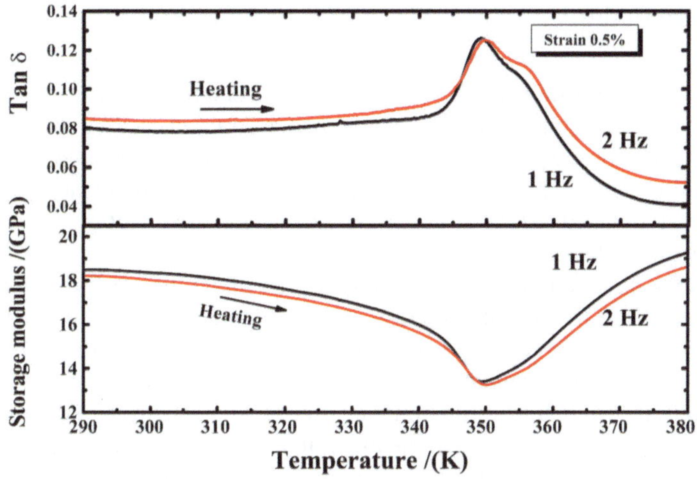

Fig. 5.53 Different frequency damping temperature spectrum of NMGC1 microwire

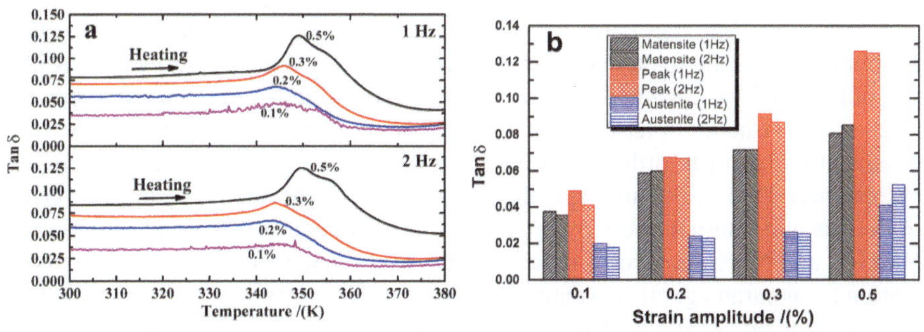

Fig. 5.54 NMGC1 microwire 1 and 2 Hz with different strain amplitude damping temperature spectrum and damping value statistics. **a** Different strain amplitude damping temperature spectrum; **b** damping value statistics

strain amplitude, showing strong strain amplitude dependence. At the same time, the martensitic transformation temperature moves to high temperature with the increase of strain amplitude, which indicates that the stability of martensite increases with strain amplitude. The damping value of martensite increases steadily with the increase of strain amplitude, but the damping value of austenite is almost unchanged below 0.3% strain amplitude, but increases rapidly at 0.5%, which is closely related to the generation principle of damping characteristics in different phases. The mechanism of martensite damping is the movement of twin boundary, and the increase of strain will make the twin boundary movement more intense, thus promoting the damping property. However, for austenite, the damping property is related to the movement of dislocations, and only when the critical stress is reached (0.5%), the damping property is improved significantly.

5.6 Conclusions

Polycrystalline Ni–Mn–Ga wires have a fully reversible two-way shape memory effect under tensile conditions. The phase change strain exceeding the 3–6 times of polycrystalline wires can be obtained by 3–6 times at lower stress. Both stress heat cycle training and superelastic training can improve grain boundary mobility and reduce stress-induced critical stress of martensite, and the effect of stress heat cycle training is more obvious.

The stress–strain curves of polycrystalline wire under the condition of martensite tensile are related to the critical stress of twin boundary movement of martensite in different grains, and the strain amplitude of different steps is different, which is related to the different orientation and size of twins in different grains. The critical stress of initial twin boundary motion under the tensile condition of oligomeric wire is in the range of 20–55 MPa, and the maximum strain can reach 4.55%. The stress heat cycle training can effectively reduce the critical stress of wire twin boundary motion from 20 to 9 MPa, which lays a foundation for the acquisition of MFIS.

After Regulation of Ni–Mn–Ga wire composition to $e/a = 7.7$ and chemical ordering, the magnetic structure phase transition coupling state can be obtained in $Ni_{50}Mn_{25}Ga_{25\text{-}x}Cu_x$ (x = 3.8 at.%) alloy. To achieve magnetostructural coupling state, gradient composition distribution of Mn along the surface to the interior of the wire was obtained by heat treatment. By using Ni–Mn–Ga wire with $T_C{}^A = 28$ K, the structural phase transition range is broadened (15.2–20.6 K) after heat treatment without Mn ordering. A partial magnetostructure coupling state can also be obtained. Due to the high specific surface area of the wire and the elimination of defects and internal stress by ordering heat treatment, the thermal hysteresis of NMG4C and NMG2C wires is reduced. As a result, magnetic field induced transformation from paramagnetic austenite to ferromagnetic martensite occurrs in NMG4C wire, and ferromagnetic austenite to ferromagnetic martensite in NMG2C wire.

The maximum magnetic entropy change in NMG4C (magnetostructural coupled state) is −24.9 J/kg K under 50 kOe, better than Ni–Mn–Ga single crystal. The refrigeration temperature range, however, is concentrated in a range of 3 K and greatly reduced the RC_{net} of refrigeration capacity (50 kOe, 48.3 J/kg). Under magnetostructure partial coupled state, the NMG2C wire shows two magnetic entropy peaks at ~337 K and 365 K, corresponding to magnetic entropy change about −5.2 J/kg K and −5.3 J/kg K, respectively. The magnetocaloric value of the transition zone of two peaks is ~4.0 J/kg K. Due to the wide working temperature interval, the wire has a large RC_{net} up to ~240 J/kg.

References

1. Y.F. Liu, X.X. Zhang, D.W. Xing et al., Magnetocaloric effect (MCE) in melt-extracted Ni-Mn-Ga-Fe Heusler microwires. J. Alloys Compd. **616**, 184–188 (2014)

2. M.F. Qian, X.X. Zhang, L.S. Wei et al., Effect of chemical ordering annealing on martensitic transformation and superelasticity in polycrystalline Ni-Mn-Ga microwires. J. Alloys Compd. **645**, 335–343 (2015)
3. M.F. Qian, X.X. Zhang, L.S. Wei et al., Tunable magnetocaloric effect in Ni-Mn-Ga microwires. Sci. Rep. **8**, 16574 (2018)
4. M.F. Qian, X.X. Zhang, C. Witherspoon et al., Superelasticity and shape memory effects in polycrystalline Ni-Mn-Ga microwires. J. Alloys Compd. **577**, S296–S299 (2013)
5. Y.F. Liu, X.X. Zhang, J.S. Liu et al., Superelasticity in polycrystalline Ni-Mn-Ga-Fe microwires fabricated by melt-extraction. Mater. Res.-Ibero-Am. J. Mater. **18**, 61–65 (2015)
6. X.X. Zhang, M.F. Qian, Z. Zhang et al., Magnetostructural coupling and magnetocaloric effect in Ni-Mn-Ga-Cu microwires. Appl. Phys. Lett. **108**(5), 52401 (2016)
7. Y.F. Liu, *Transformation Behavior and Properties of $Ni_{50}Mn_{25}Ga_{25-x}Fe_x$Shape Memory Alloy Microwires* (Harbin Institute of Technology, Harbin, 2015)
8. M.F. Qian, *Research on Fabrication and Stress-Induced Strain Properties of Ni-Mn-Ga Ferromagnetic Shape-Memory Microwires* (Harbin Institute of Technology, Harbin, 2011)
9. M.F. Qian, Investigation of the Microstructure, Thermal-Induced and Magnetocaloric Characteristics in Ni-Mn-Ga Microwires (Harbin Institute of Technology, Harbin, 2016)
10. L.S. Wei, X.X. Zhang, L. Geng et al., Damping properties of Ni-Mn-Ga microwires. J. Harbin Univ. Sci. Technol. **18**(02), 7–10 (2013)
11. Y. Xu, *Influence of Diameter and Grain Size on the Superelasticity and Shape Memory Effect of Ni-Mn-Ga Fiber* (Harbin Institute of Technology, Harbin, 2012)
12. X.X. Zhang, M.F. Qian, G.W. Wang et al., High damping capacity of Ni–Mn–Ga–Cu microwires prepared by melt-extraction technique. Rare Met. **1**, 1–6 (2017)
13. C. Segui, E. Cesari, J. Pons et al., Internal friction behaviour of Ni-Mn-Ga. Mater. Sci. Eng., A **370**(1–2), 481–484 (2004)
14. Y.T. Cui, L. Wu, S.Q. You, et al., Strain characteristics and superelastic response of $Ni_{53.2}Mn_{22.6}Ga_{24.2}$ single crystals. Solid State Commun. **149**(37–38), 1539–1542 (2009)
15. L.S. Wei, *Research on Hot Deformation Behavior and Microstructure of Polycrystalline Ni-Mn-Ga Alloys* (Harbin Institute of Technology, Harbin, 2013)
16. C. Segui, J. Pons, E. Cesari et al., Low-temperature behaviour of Ni-Fe-Ga shape-memory alloys. Mater. Sci. Eng., A **438**, 923–926 (2006)
17. R.F. Hamilton, H. Sehitoglu, Y. Chumlyakov et al., Stress dependence of the hysteresis in single crystal NiTi alloys. Acta Mater. **52**(11), 3383–3402 (2004)
18. V.A. Chernenko, V. L'Vov, J. Pons et al., Superelasticity in high-temperature Ni-Mn-Ga alloys. J. Appl. Phys. **93**(5), 2394–2399 (2003)
19. K. Oikawa, R. Saito, K. Anzai et al., Elastic and superelastic properties of NiFeCoGa microwires grown by micro-pulling-down method. Mater. Trans. **50**(4), 934–937 (2009)
20. H.E. Karaca, I. Karaman, A. Brewer, et al., Shape memory and pseudoelasticity response of NiMnCoIn magnetic shape memory alloy single crystals. Scripta Mater. **58**(10), 815–818 (2008)
21. H.E. Karaca, I. Karaman, D.C. Lagoudas et al., Recoverable stress-induced martensitic transformation in a ferromagnetic CoNiAl alloy. Scripta Mater. **49**(9), 831–836 (2003)
22. H.Y. Kim, H. Satoru, J.I. Kim et al., Mechanical properties and shape memory behavior of ti-nb alloys. Mater. Trans. **45**(7), 2443–2448 (2004)
23. Y. Ganor, D. Shilo, T.W. Shield, et al., Breaching the work output limitation of ferromagnetic shape memory alloys. Appl. Phys. Lett. **93**(12), 122509 (2008)
24. M.F. Qian, X.X. Zhang, L.S. Wei et al., Structural, magnetic and mechanical properties of oligocrystalline Ni-Mn-Ga shape memory microwires. Mater. Today: Proc. **2S**, S577–S581 (2015)
25. U. Gaitzsch, M. Potschke, S. Roth et al., A 1% magnetostrain in polycrystalline 5M Ni-Mn-Ga. Acta Mater. **57**(2), 365–370 (2009)
26. D.C. Dunand, P. Müllner, Size effects on magnetic actuation in Ni-Mn-Ga shape-memory alloys. Adv. Mater. **23**(2), 216–232 (2011)

27. N. Scheerbaum, O. Heczko, J. Liu, et al., Magnetic field-induced twin boundary motion in polycrystalline Ni-Mn-Ga fibres. New J. Phys. **10**, 073002 (2008)
28. Z.B. Li, Y.D. Zhang, C.F. Sanchez-Valdes, et al., Giant magnetocaloric effect in melt-spun Ni-Mn-Ga ribbons with magneto-multistructural transformation. Appl. Phys. Lett. **104**(4), 044101 (2014)
29. A. Smaili, R. Chahine, Composite materials for Ericsson-like magnetic refrigeration cycle. J. Appl. Phys. **81**(2), 824–829 (1997)
30. J. Liu, T. Gottschall, K.P. Skokov et al., Giant magnetocaloric effect driven by structural transitions. Nat. Mater. **11**(7), 620–626 (2012)
31. Z. Li, K. Xu, Y. Zhang, C. Tao, et al., Two successive magneto-structural transformations and their relation to enhanced magnetocaloric effect for $Ni_{55.8}Mn_{18.1}Ga_{26.1}$ Heusler alloy. Sci. Rep. **5**(1), 1–7 (2015)
32. B.J. Korte, V.K. Pecharsky, K.A. Gschneidner, The correlation of the magnetic properties and the magnetocaloric effect in $(Gd_{1-x}Er_x)$ NiAl alloys. J. Appl. Phys. **84**(10), 5677–5685 (1998)
33. V.A. Chernenko, E. Cesari, V.V. Kokorin et al., The development of New ferromagnetic shape-memory alloys in Ni-Mn-Ga system. Scr. Metall. Mater. **33**(8), 1239–1244 (1995)
34. F.X. Hu, B.G. Shen, J.R. Sun, Magnetic entropy change in $Ni_{51.5}Mn_{22.7}Ga_{25.8}$ alloy. Appl. Phys. Lett. **76**(23), 3460–3462 (2000)
35. J. Marcos, L. Manosa, A. Planes, et al., Multiscale origin of the magnetocaloric effect in Ni-Mn-Ga shape-memory alloys. Phys. Rev. B **68**(9), 094401 (2003)
36. G.J. Liu, J.R. Sun, J. Shen, et al., Determination of the entropy changes in the compounds with a first-order magnetic transition. Appl. Phys. Lett. **90**(3), 032507 (2007)
37. V.A. Chernenko, R.L. Anton, M. Kohl et al., Magnetic domains in Ni-Mn-Ga martensitic thin films. J. Phys.-Condens. Matter **17**(34), 5215–5224 (2005)
38. W. Li, S.T. Jiang, *Condensed Matter Magnetic Physics* (Science Press, Beijing, 2002)
39. A. Arrott, Criterion for ferromagnetism from observations of magnetic isotherms. Phys. Rev. **108**(6), 1394–1396 (1957)
40. Z.B. Li, J.L.S. Llamazares, C.F. Sanchez-Valdes et al., Microstructure and magnetocaloric effect of melt-spun $Ni_{52}Mn_{26}Ga_{22}$ ribbon. Appl. Phys. Lett. **100**(17), 17410217 (2012)
41. N.V.R. Rao, R. Gopalan, V. Chandrasekaran et al., Microstructure, magnetic properties and magnetocaloric effect in melt-spun Ni-Mn-Ga ribbons. J. Alloy. Compd. **478**(1–2), 59–62 (2009)
42. X. Jin, M. Marioni, D. Bono et al., Empirical mapping of Ni-Mn-Ga properties with composition and valence electron concentration. J. Appl. Phys. **91**(10), 8222–8224 (2002)
43. R.D. Shull, V. Provenzano, A.J. Shapiro et al., The effects of small metal additions (Co, Cu, Ga, Mn, Al, Bi, Sn) on the magnetocaloric properties of the $Gd_5Ge_2Si_2$ alloy. J. Appl. Phys. **99**(8), 08k908 (2006)
44. X.Z. Zhou, H. Kunkel, G. Williams et al., Phase transitions and the magnetocaloric effect in Mn rich Ni-Mn-Ga Heusler alloys. J. Magn. Magn. Mater. **305**(2), 372–376 (2006)
45. X.X. Zhang, S.P. Miao, J.F. Sun, Magnetocaloric effect in Ni-Mn-In-Co microwires prepared by Taylor-Ulitovsky method. Trans. Nonferrous Met. Soc. China **24**(10), 3152–3157 (2014)
46. Y.F. Liu, L. Luo, X.X. Zhang, et al., Magnetostructural coupling induced magnetocaloric effects in Ni-Mn-Ga-Fe microwires. Intermetallics **112**, 106538 (2019)
47. D. Soto, F.A. Hernandez, H. Flores-Zuniga, et al., Phase diagram of Fe-doped Ni-Mn-Ga ferromagnetic shape-memory alloys. Phys. Rev. B **77**(18), 184103 (2008)
48. Y.Y. Li, J.M. Wang, C.B. Jiang, Study of Ni-Mn-Ga-Cu as single-phase wide-hysteresis shape memory alloys. Mater. Sci. Eng. A **528**(22–23), 6907–6911 (2011)
49. P.J. Brown, J. Crangle, T. Kanomata et al., The crystal structure and phase transitions of the magnetic shape memory compound Ni_2MnGa. J. Phys.-Condens. Matter **14**(43), 10159–10171 (2002)
50. V.A. Chernenko, Compositional instability of β-phase in Ni-Mn-Ga alloys. Scripta Mater. **40**(5), 523–527 (1999)
51. W.H. Wang, J.L. Chen, Z.H. Liu, et al., Thermal hysteresis and friction of phase boundary, motion in ferromagnetic $Ni_{52}Mn_{23}Ga_{25}$ single crystals. Phys. Rev. B **65**(1), 012416 (2002)

52. Y. Hayasaka, S. Aoto, H. Date et al., Magnetic phase diagram of ferromagnetic shape memory alloys $Ni_2MnGa_{1-x}Fe_x$. J. Alloy. Compd. **591**, 280–285 (2014)
53. S. Tian, *Physical Properties of Materials* (Beijing University of Aeronautics and Astronautics Press, Beijing, 2004)
54. L. Gao, *Microstructure and Martensitic Transformation and Mechanical Properties of Ni-Mn-Ga-Re Magnetic Shape Memory Alloys* (Harbin Institute of Technology, Harbin, 2007)
55. P.J. Webster, K.R.A. Ziebeck, S.L. Town, M.S. Peak, Magnetic order and phase transformation in Ni_2MnGa. Phil. Mag. B **B49**, 295–310 (1984)
56. S.K. Ayila, R. Machavarapu, S. Vummethala, Site preference of magnetic atoms in Ni–Mn–Ga–M (M= Co, Fe) ferromagnetic shape memory alloys. Phys. Status Solidi (b) **249**(3), 620–626 (2012)
57. C.M. Li, H.B. Luo, Q.M. Hu, et al., Site preference and elastic properties of Fe-, Co-, and Cu-doped Ni_2MnGa shape memory alloys from first principles. Phys. Rev. B **84**(2), 024206 (2011)
58. A. Planes, L. Manosa, M. Acet, Magnetocaloric effect and its relation to shape-memory properties in ferromagnetic Heusler alloys. J. Phys.-Condens. Matter **21**(23), 233201 (2009)
59. R. Varga, T. Ryba, Z. Vargova et al., Magnetic and structural properties of Ni-Mn-Ga Heusler-type microwires. Scripta Mater. **65**(8), 703–706 (2011)
60. H.X. Shen, H. Wang, J.S. Liu et al., Enhanced magnetocaloric and mechanical properties of melt-extracted $Gd_{55}Al_{25}Co_{20}$ micro-microwires. J. Alloy. Compd. **603**, 167–171 (2014)
61. A. Bhattacharyya, S. Giri, S. Majumdar, Field induced sign reversal of magnetocaloric effect in Gd_2In. J. Magn. Magn. Mater. **324**(6), 1239–1241 (2012)
62. H. Oesterreicher, F.T. Parker, Magnetic cooling near Curie temperatures above 300 K. J. Appl. Phys. **55**(12), 4334–4338 (1984)
63. F.X. Qin, N.S. Bingham, H. Wang et al., Mechanical and magnetocaloric properties of Gd-based amorphous microwires fabricated by melt-extraction. Acta Mater. **61**(4), 1284–1293 (2013)
64. V.M. Prida, V. Franco, V. Vega et al., Magnetocaloric effect in melt-spun FePd ribbon alloy with second order phase transition. J. Alloys Compd. **509**(2), 190–194 (2011)
65. Z.M. Peng, X.J. Jin, Y.Z. Fan, et al., Internal friction and modulus changes associated with martensitic and reverse transformations in a single crystal $Ni_{48.5}Mn_{31.4}Ga_{20.1}$ alloy. J. Appl. Phys. **95**(11), 6960–6962 (2004)
66. V.A. Chernenko, C. Segui, E. Cesari et al., Some aspects of structural behaviour of Ni-Mn-Ga alloys. J. Phys. IV **7**(C5), 137–141 (1997)
67. V.G. Gavriljuk, O. Soderberg, V.V. Bliznuk et al., Martensitic transformations and mobility of twin boundaries in Ni_2MnGa alloys studied by using internal friction. Scripta Mater. **49**(8), 803–809 (2003)
68. S.H. Chang, S.K. Wu, Low-frequency damping properties of near-stoichiometric Ni_2MnGa shape memory alloys under isothermal conditions. Scripta Mater. **59**(10), 1039–1042 (2008)
69. D. Feng, *Metal Physics, vol. 3: Mechanical Properties of Metals* (Science Press, Beijing, 1999), pp. 190–203
70. R.B. Perez-Saez, V. Recarte, M.L. Nó et al., Anelastic contributions and transformed volume fraction during thermoelastic martensitic transformations. Phys. Rev. B **57**(10), 5684 (1998)

Chapter 6
Preparation and Properties of Magnetic Shape Memory Alloy Particles

6.1 Introduction

Chapters 2–6 introduce three-, two- and one-dimensional Ni–Mn-based alloy. This chapter discusses a zero-dimensional Ni–Mn-based micro-particle for preparation, phase transition and property. There are two main methods to prepare Ni–Mn-based alloy particles. The first is electric spark corrosion, that is, the target alloy material working as the electrode, and then partially melt local electrode through the discharge between the electrodes. The produced droplets in the low temperature liquid electrolyte rapidly cool to form particles. Tang [1–4] uses this method to prepare micron-scale spherical $Ni_{49}Mn_{30}Ga_{21}$ alloy particles. The second method is mechanical ball grinding [5, 6]. Because the Ni–Mn-based polycrystalline alloy is brittle, it is easy to crush the bulk alloy by mechanical ball grinding and sieving to obtain the alloy particles. Compared with the EDM method, the mechanical ball-grinding method is economical and practical, but the test parameters need to be explored to obtain alloy particles of different sizes.

Ni–Mn–Ga and Ni–Mn–In–Co alloys particles are discussed in this chapter. The different sizes of alloy particles are selected by an artificial mechanical grinding method. The isothermal magnetic entropy change of single particle and mixed particle, working temperature range and refrigeration ability are studied, and the advantages of mixed particle as magnetic refrigeration material are revealed [7–9].

6.2 Preparation of MSMA Micron-Particles

Ni–Mn–Ga and Ni–Mn–In–Co bulk alloys are prepared by induction melting under Ar atmosphere. The bulk alloy is crushed with a universal testing machine and then ground in an agate mortar. Eight different sizes of particles are obtained by screening the ground particles through standard sample sieve. The different sized particles are marked P1 (880 ~ 1000 μm), P2 (550 ~ 650 μm), P3 (325 ~ 380 μm), P4 (212 ~

X. Zhang and M. Qian, *Magnetic Shape Memory Alloys*,
https://doi.org/10.1007/978-981-16-6336-9_6

Fig. 6.1 Morphology of Ni–Mn–Ga alloy particle with different sizes. **a** 600 ~ 710 μm; **b** 63 ~ 96 μm; **c** 50 ~ 63 μm; **d** 45 ~ 50 μm; **e** 38.5 ~ 45 μm

230 μm), P5 (125 ~ 150 μm), P6 (61 ~ 96 μm), P7 (38 ~ 45 μm), P8 (13 ~ 18 μm), some of which are shown in Figs. 6.1 and 6.2. When the particle size is large, the aspect ratio is large. The particle of 600 ~ 710 μm has an average value of the particle aspect ratio ~3. As the particle size decreases, the aspect ratio decreases gradually and the particles tend to be spherical.

6.3 Heat Treatment of Ni–Mn–Ga Alloy Particles

In Ni–Mn–Ga alloy, Curie temperature T_c and martensitic transformation temperature change on a large scale by composition regulation, and the change of composition can be expressed by the outer layer electron concentration *e*/*a*. Generally, when *e*/*a* < 7.7, the martensitic transformation occurs below T_c; when *e*/*a* > 7.7, the martensitic transformation occurs above T_c; when the *e*/*a* approaches 7.7, the two phase transitions tend to occur at the same time, which leads to the coupling of the magnetostructure phase transition, so the magnetostructure coupling may be obtained by adjusting the Ni–Mn–Ga alloy composition to approach 7.7. Here, we try to prepare four kinds of particles, i.e. $Ni_{54}Mn_{21}Ga_{25}$ (NMG32), $Ni_{54.5}Mn_{20.5}Ga_{25}$ (NMG33), $Ni_{55}Mn_{20}Ga_{25}$ (NMG34) and $Ni_{54.75}Mn_{20.25}Ga_{25}$ (NMG35) [9].

The ordering heat treatment is carried out at 998 K for 2 h, 973 K for 10 h and 773 K for 20 h, followed by furnace cooling; the stress-relief heat treatment is 773 K for 20 h followed by furnace cooling. The chemical ordering refers to the full diffusion near the disorder–order transition temperature, making the re-arrangement of Mn and Ga elements to re-occupy the ordered state position in the solid solution lattice and thus improve the order degree. The purpose of 998 K for 2 h and 973 K for 10 h is to enable sufficient diffusion of atoms, while 773 K for 20 h is for stress-relief annealing (release internal stress and defects formed during grinding).

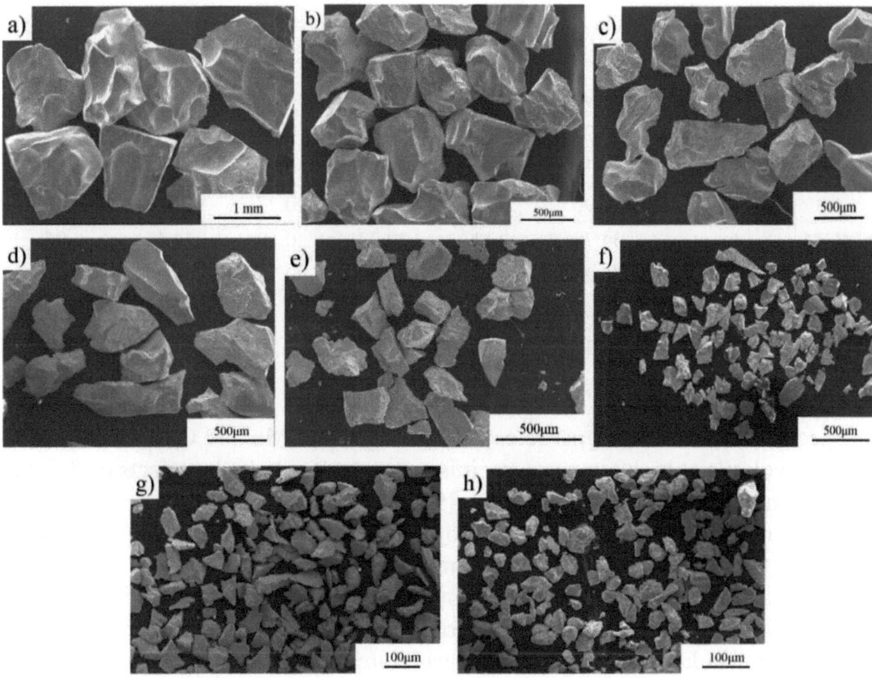

Fig. 6.2 Morphology of Ni–Mn–In–Co alloy particle with different sizes. **a** 880 ~ 1000 μm; **b** 550 ~ 650 μm; **c** 325 ~ 380 μm; **d** 212 ~ 230 μm; **e** 125 ~ 150 μm; **f** 61 ~ 96 μm; **g** 38 ~ 45 μm; **h** 13 ~ 18 μm

During the sealing of the sample in a quartz tube, the tube is washed repeatedly three times with Ar gas and then backed filled with 0.4–0.5 atm high purity Ar gas to inhibit the volatilization of alloy elements, and then the quartz tube is sealed. The stress-relief heat treatment adopts two different environmental atmosphere, one is 0.4–0.5 atm high purity Ar gas; the other is to vacuum the quartz tube to 0.5 Pa.

NMG33 and NMG34 alloy particles with size 38.5–45 μm are selected for ordering heat treatment and two kinds of stress-relief heat treatment to study the effect of different heat treatment processes on particle composition and martensitic transformation. The letters "PV" and "PA" are used to represent annealing state of the particles in vacuum and Ar, respectively, and the ordered state of the particles is represented by "PC".

6.3.1 Effect of Heat Treatment on Particle Composition

Phase transition temperature of Ni–Mn–Ga alloy strongly depends on the composition. The phase transition temperature affects the structure and properties of the martensite. So the regulation of particle composition is extremely important. Since

Table 6.1 Composition of bulk alloys and particles after different heat treatments (at.%)

Sample	Status	Ni		Mn		Ga	
NMG33	Ingot	54.00	0.16	21.16	0.13	24.84	0.13
	PV	54.27	0.21	21.30	0.18	24.43	0.19
	PA	54.37	0.22	21.24	0.18	24.39	0.16
	PC	53.96	0.31	21.86	0.27	24.18	0.46
NMG34	Ingot	54.78	0.11	20.73	0.10	24.49	0.05
	PV	55.20	0.34	20.40	0.18	24.40	0.31
	PA	55.42	0.32	20.54	0.36	24.04	0.21
	PC	54.33	0.30	21.37	0.20	24.30	0.38

the specific surface area of particles is very large, and Mn and Ga elements in the alloy are volatile at high temperature, it is necessary to study the effects of different heat treatments on particle composition [9].

In order to reveal the change of alloy composition in different heat treatment processes, the bulk alloy and particle composition after heat treatment are analyzed and summarized. Each component is obtained by averaging the composition of at least 10 particles. The uniformity of particle composition can be expressed by the composition standard deviation, and the results are shown in Table 6.1. Comparing the composition of particles after stress annealing in vacuum and argon atmosphere, the difference between the two is small, which indicates that the long-range diffusion ability of atoms is poor and the volatilization of elements is small under 773 K. Compared with bulk alloys, the homogeneity of the particles after stress-relief heat treatment decreases, and the composition uniformity of the particles decreases further after ordering, which may be due to the longer time and higher temperature of ordering heat treatment, and the change of composition between particles is more obvious.

6.3.2 *Effect of Heat Treatment on Particle Phase Change*

To determine the overall phase transition characteristics of NMG33 and NMG34 alloy micron particles after ordering and stress-relief heat treatment, the curves of the lifting/cooling process of NMG33 and NMG34 bulk alloys and alloy particles after different heat treatment are tested as shown in Fig. 6.3 [9]. All martensitic transformation temperatures obtained by the tangent method are shown in Table 6.2. From Fig. 6.2, it can be seen that the phase transition temperature of the particles moves to low temperature after heat treatment, and the phase change peak becomes narrower and smaller. It can be seen that the phase transition width of the particles after heat treatment is obviously widened relative to the bulk alloy. This may be due to the inhomogeneous composition distribution of alloy particles, in which the phase

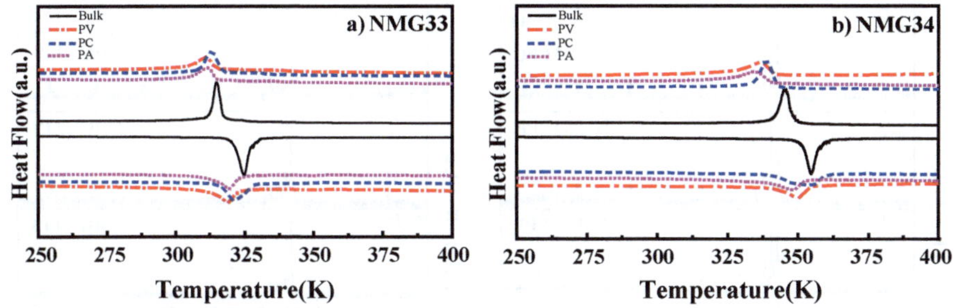

Fig. 6.3 Heating/cooling DSC curves of micron particles in bulk alloys after different heat treatments. **a** NMG33; **b** NMG34

Table 6.2 Phase transition temperature of particles at different heat treatment states (K)

Sample	Status	M_s	M_f	A_s	A_f	T_c cooling	T_c heating	$A_f - A_s$	$M_s - M_f$
NMG33	Ingot	317.5	312.2	320.0	327.6			7.6	5.3
	PV	315.0	302.6	308.7	322.6	339.8	351.7	13.9	12.4
	PA	315.1	304.5	309.5	324.6	338.6	353.1	15.1	10.6
	PC	318.1	307.4	316.4	328.8	333.6	348.7	12.4	10.7
NMG34	Ingot	348.9	340.8	350.9	358.0			7.1	8.1
	PV	341.7	326.3	341.6	354.6			13.0	15.4
	PA	340.4	323.9	337.6	353.2			15.6	16.5
	PC	343.6	332.4	341.2	359.5	310.3	315.1	18.3	11.2

transition temperature in various particles is different, leading to increased phase transition width.

For determining the specific phase of the particles after heat treatment, the XRD diffraction analysis of the two alloy particles under different heat treatment processes is carried out, as shown in Fig. 6.4. NMG33 bulk alloy contains NM martensite and a small amount of parent austenite phase after homogenization heat treatment, NMG34 the bulk alloy has a single phase NM martensite state. The two alloy particles contain 7M and NM martensite after stress-relief annealing and ordering treatment. The increase of martensite type in alloy particles is also the reason for the widening of martensitic transformation temperature range.

6.4 Magnetic and Magnetocaloric Effect of Micron Ni–Mn–Ga Particle

Small size materials have great advantages in heat conduction efficiency and can widen the phase change temperature width, reduce the hysteresis of materials, and

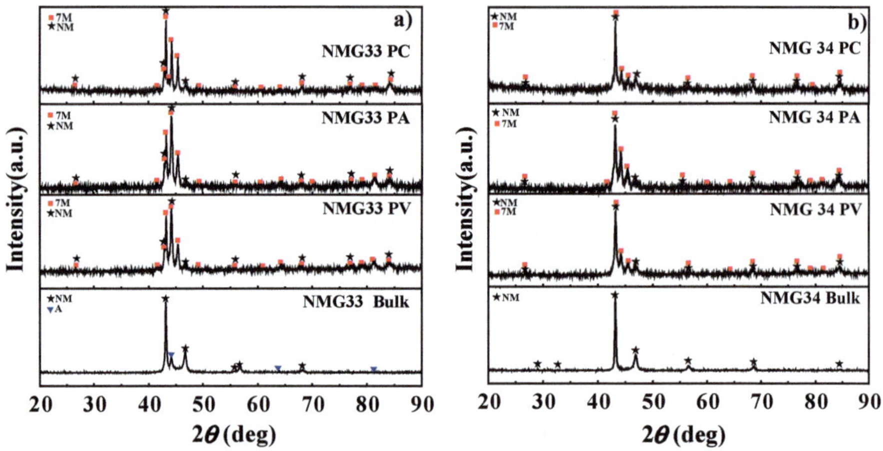

Fig. 6.4 XRD of particles under different heat treatment states. **a** NMG33; **b** NMG34

improve the magnetic refrigeration ability. NMG33 and NMG34 alloy particles (size 38.5–45 μm) are subjected to stress-relief heat treatment and ordering heat treatment. NMG33 and NMG34 alloy micron particles are mixed according to mass 1:1. The ΔS_m, ΔT_{FWHM} and RC values of the mixed particles are increased. This proves an effective strategy to obtain wide ΔT_{FWHM} in Ni–Mn–Ga alloys [7, 9].

It can be seen that the composition of the two alloy particles of "PV" and "PA" heat treatment is basically the same, and the martensitic transformation temperatures are similar. So it can be considered that the two are in the same state, and only the alloy particles of "PV" and "PC" heat treatment are studied later.

6.4.1 Martensite and Magnetic Transitions of Single-Sized Ni–Mn–Ga Particle

Figure 6.5 is M–T curves of the heating/cooling process of NMG33 and NMG34 alloy micron particles at 100 Oe and 50 kOe under stress annealing. The phase transition hysteresis of NMG33 alloy particles is ~1.3 K at 50 kOe and less than that of bulk alloy (~4.6 K); NMG34 alloy particles have hysteresis ~3.4 K at 50 kOe and also less than that of bulk alloy. After ordering, the phase transition hysteresis of NMG33 alloy particles is ~3.5 K at 50 kOe and less than that of the bulk alloy (~4.6 K), and the same lower hysteresis ~3.4 K at 50 kOe is found in NMG34 particle as compared to the bulk alloy (4.9 K).

The thermal hysteresis of NMG33 and NMG34 alloy particles is smaller than that of bulk alloy. Small size materials have a high proportion of grain free surface, and the alloy particles are single crystals. So, there is no bonding of adjacent grains in

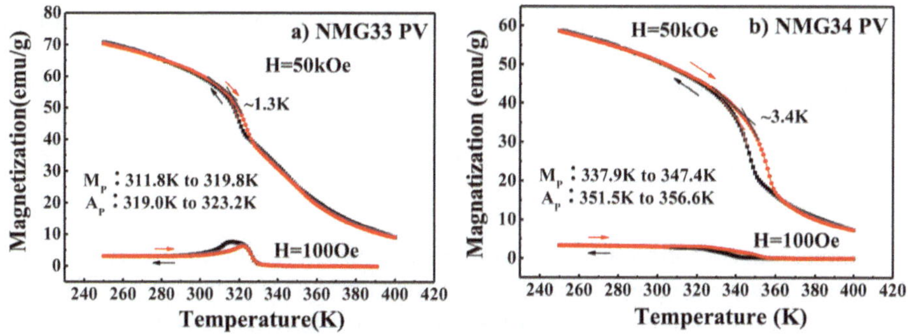

Fig. 6.5 M–T curve of particles under 100 Oe and 50 kOe. **a** NMG33PV; **b** NMG34PV

the phase transition process, and the volume change can be effectively released. This means the thermal hysteresis of the phase transition process is very small.

The hysteresis of magnetic refrigeration materials under magnetic field is an important factor affecting the magnetocaloric properties. Figures 6.6a and 6.7a are

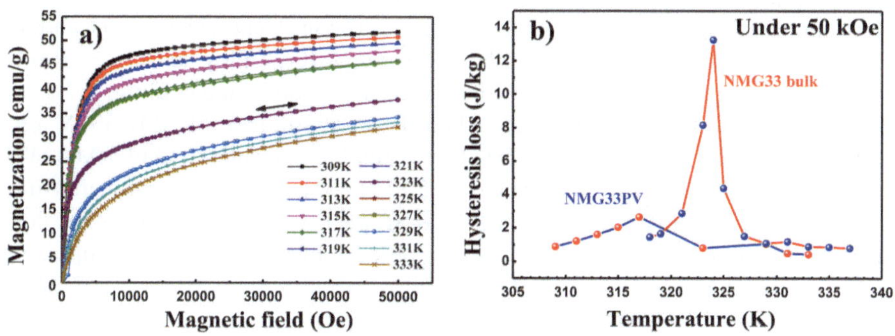

Fig. 6.6 NMG33PV alloy particle. **a** Isothermal magnetization curves; **b** hysteresis

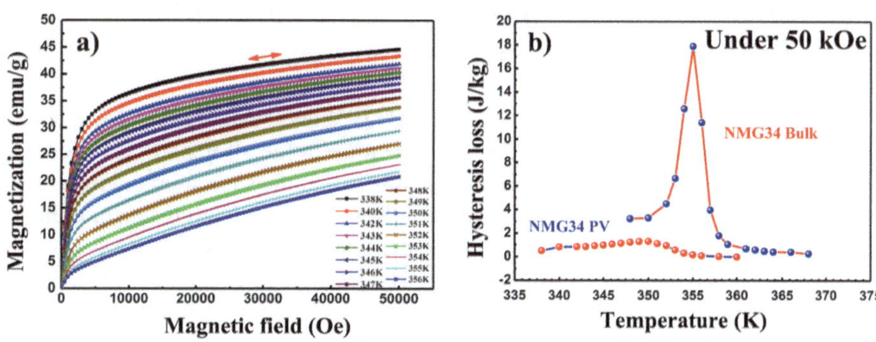

Fig. 6.7 NMG34PV alloy particle. **a** Isothermal magnetization curves; **b** hysteresis

Table 6.3 Peak temperature of phase transition for NMG33PV and NMG34PV alloy particles at 100 Oe and 50 kOe (K)

Sample	A_p (100 Oe)	A_p (50 kOe)	M_p (100 Oe)	M_p (50 kOe)	ΔT_{0h}	ΔT_{0c}
NMG33PV	319.0	323.2	311.8	319.8	4.2	8.0
NMG34PV	351.5	356.6	337.9	347.4	5.1	9.5

isothermal magnetization curves of NMG33 and NMG34 particles. Figures 6.6b and 6.7b show that the average magnetic hysteresis of NMG33PV and NMG34PV alloy particles at 50 kOe is only 1.97 and 1.04 J/kg, which is much smaller than the average hysteresis of bulk alloy. As with thermal hysteresis, the decrease of hysteresis also depends on the high surface area of particles and the decrease of defects and internal stress after heat treatment, which leads to the decrease of the resistance of magnetic domain wall motion under the action of magnetic field. From the above analysis, it can be seen that after stress-relief, the alloy micron particles have small thermal hysteresis and almost negligible magnetic hysteresis, which is beneficial to improve the magnetocaloric effect of the material.

The magnetization curves (Fig. 6.5) show that the phase transition temperature of the NMG33 and NMG34 alloy particles moves to high temperature with the increase of magnetic field during the heating/cooling. This is consistent with Figs. 6.6 and 6.7. The phase transition temperatures of NMG33 and NMG34 alloy particles at 100 Oe and 50 kOe are listed in Table 6.3. In the process of heating and cooling, ΔT_{0h} and ΔT_{0c} are 4.2 and 8.0 K, respectively when the magnetic field of NMG33PV alloy particles increased from 100 Oe to 50 kOe. When the magnetic field of NMG34PV alloy particles increased from 100 Oe to 50 kOe, the change of phase transition temperature ΔT_{0h} and ΔT_{0c} are 5.1 and 9.5 K, respectively. Since the magnetocaloric effect only takes into account the heating process in this paper, the temperature change rate of NMG33PV and NMG34PV alloy particles is $dA_p/dH = \sim 0.08$ K/kOe and $dA_p/dH = \sim 0.10$ K/kOe, respectively.

6.4.2 *Magnetocaloric Properties Single-Sized Ni–Mn–Ga Particle*

The isothermal magnetization curves of NMG33PV and NMG34PV alloy particles are shown in Fig. 6.8. The saturation magnetization of the alloy particles decreases with the increase of temperature. Figure 6.8 shows that the saturation magnetization of NMG33 alloy particles decreases significantly at 315 K, indicating that martensitic transformation begins at 315 K, which is higher than that of DSC test ~308.1 K. The saturation magnetization of NMG34 alloy particles decreases significantly at 347 K, showing that the martensitic transformation starts at 347 K, which is higher than the starting temperature of the martensitic transformation tested DSC ~341.6 K.

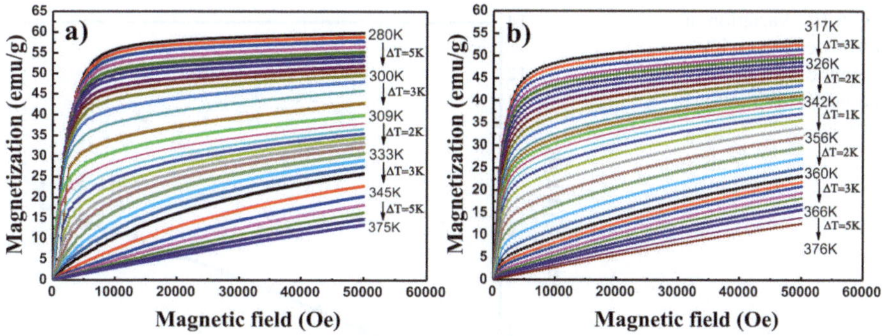

Fig. 6.8 Temperature-dependent magnetization curves of particles. **a** NMG33PV; **b** NMG34PV

The relationship between the ΔS_m and temperature of NMG33PV and NMG34PV alloy particles under different magnetic fields is calculated based on isothermal magnetization curves at different temperatures, as shown in Fig. 6.9. The maximum values ΔS_m NMG33PV alloy particles appear at 318 K, and at 20 kOe and 50 kOe − 2.82 J/(kgK) and −6.74 J/(kgK), respectively. The maximum values ΔS_m NMG34PV alloy particles appear at 351 K, and at 20 kOe and 50 kOe −5.29 J/(kgK) and −12.96 J/(kgK). The magnetocaloric values of both alloy micron particles are larger than those of $Ni_{50.7}Mn_{29.7}Ga_{19.6}$ [4] (−2 J/(kgK)) and $Ni_{54.5}Mn_{20.5}Ga_{25}$ [10] (−1.5 J/(kgK)) reported in the literature at 20 kOe.

In Fig. 6.9, the ΔT_{FWHM} of magnetocaloric peak of NMG33PV alloy particles is 20.3 K, 10 times that of bulk alloy, and the ΔT_{FWHM} of NMG34PV alloy particles is ~8.9 K, which is also larger than that of bulk alloy. The magnetocaloric effect is comprehensively evaluated by calculating the RC value of the material according to the ΔS_m–T curve. The RC_{net} value of NMG33PV alloy particles at 50 kOe is 113.13 J/kg, and the RC_{net} value of alloy particles at 50 kOe is 87.24 J/kg, which is much larger than the RC_{net} value of corresponding blocks.

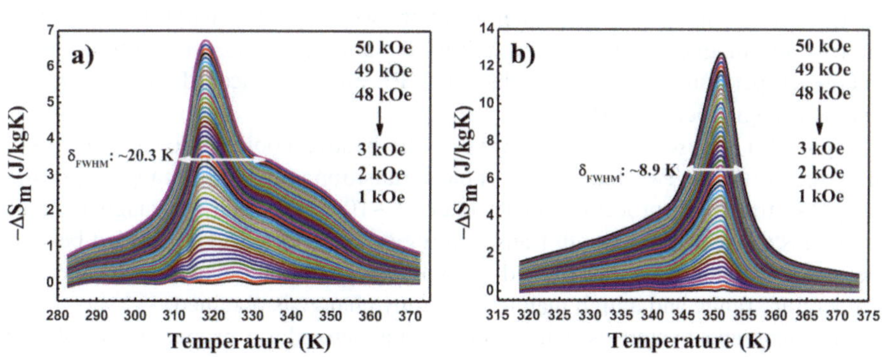

Fig. 6.9 ΔS_m under different magnetic fields of alloy particles. **a** NMG33PV; **b** NMG34PV

Fig. 6.10 Variation of entropy change with magnetic field in NMG33PV and NMG34PV alloy particles

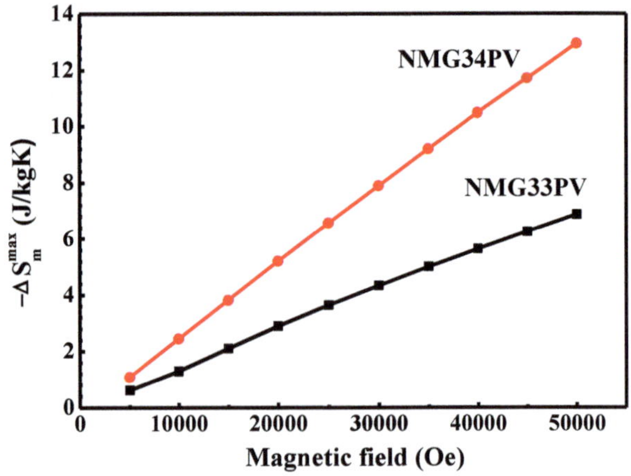

The relationship between the phase transition of NMG33PV and NMG34PV alloy particles and the magnetic field is shown in Fig. 6.10. The two lines are basically linear, and the slope is ~1.4 J/(kgKT) and ~2.6 J/(kgKT), respectively, which is smaller than the slope change value of single crystal $Ni_{52.6}Mn_{23.1}Ga_{24.3}$ reported in [11], ~4.0 J/(kgKT). Under the same magnetic field, the maximum isothermal magnetic entropy change of NMG34PV alloy particles is larger than that of NMG33PV, which is just opposite to the corresponding bulks.

6.4.3 Magnetic and Thermal Properties Ni–Mn–Ga Mixed Particle Size

From the above analysis, the hysteresis of micron particles decreases, the half-height width ΔT_{FWHM} of magnetocaloric peak increases, and the magnetic refrigeration ability is greatly improved. To further widen the working interval of refrigeration and improve the magnetic refrigeration ability, the alloy particles with different phase change temperature NMG33PV and NMG34PV are mixed according to mass 1:1 to test their magnetocaloric properties.

Figure 6.11 presents the M–T curve of the heating/cooling process for 50 kOe of mixed particles. Two independent phase transition temperature ranges appear in the phase transition process with an interval ~10.0 K. When the magnetocaloric properties are tested, a smooth transition region is expected to be formed between the two magnetocaloric peaks, and a favorable combination of a wide operating temperature range and a high magnetic entropy change is obtained.

The magnetic hysteresis can affect the magnetocaloric properties, so 50 kOe is applied for different temperature magnetization/demagnetization curves. The test temperature range contains the width of the two magnetocaloric peaks in

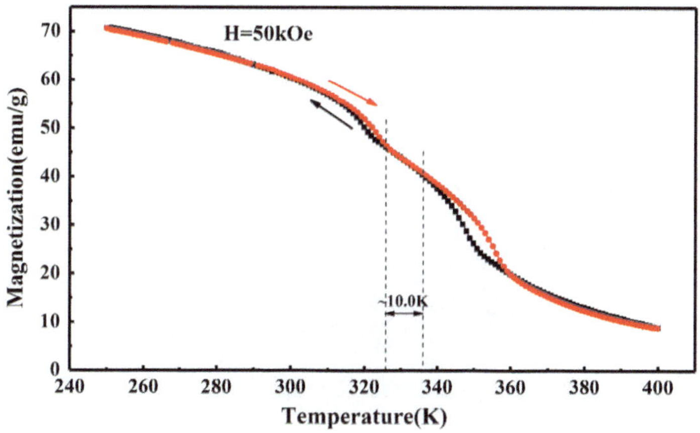

Fig. 6.11 M–T curve of the mixed particles under a field of 50 kOe

the M–T curve, and the results are shown in Fig. 6.12. The mixed particle addition/demagnetization curve basically coincides, indicating that the hysteresis is very small and negligible.

Based on the isothermal magnetization curve at different temperatures, the relationship between the ΔS_m and temperature of alloy mixed particles under different magnetic fields is calculated, as shown in Fig. 6.13. The two peaks appeared in ΔS_m near 318 K and 349 K, corresponding to the phase transition temperature of NMG33PV and NMG34PV alloy particles, respectively. The values at 50 kOe are

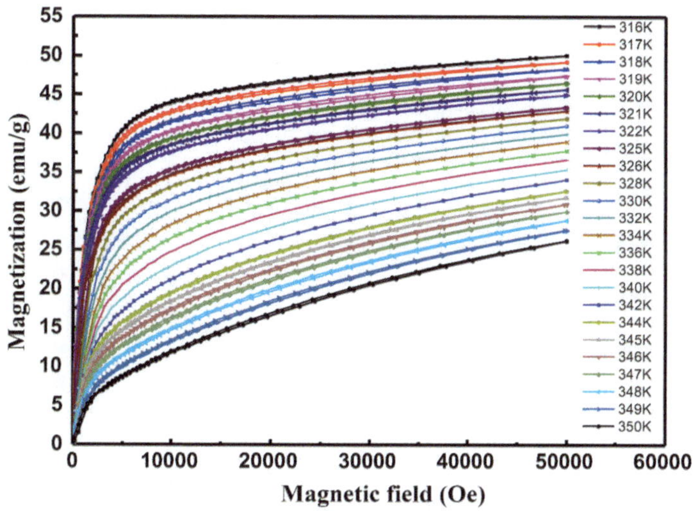

Fig. 6.12 Isothermal magnetization curves of mixed particles

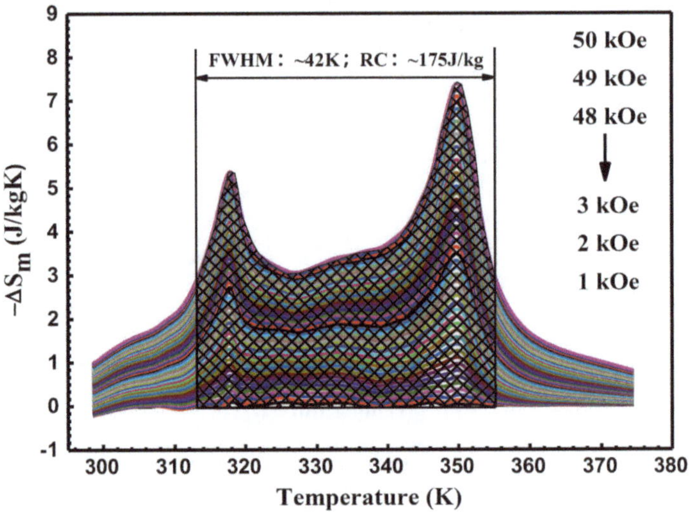

Fig. 6.13 Temperature-dependent ΔS_m of mixed alloy particles under various magnetic fields

5.44 J/kg and 7.46 J/kgK, respectively. Using the minimum value of the transition zone of the two magnetocaloric peaks as the magnetocaloric peak half-height ΔT_{FWHM}, reached ~42 K. RC_{net} value of the mixed particles at 50 kOe is much larger than that of the single particle size, as shown in Fig. 6.14.

The RC_{net} values of the mixed particles at 20 kOe and 50 kOe are ~70.0 J/kg and ~175 J/kg, respectively, as shown in Fig. 6.14. Although the NMG34PV alloy particle has a high ΔS_m, their ΔT_{FWHM}, ~8.9 K, is smaller than that of NMG33PV

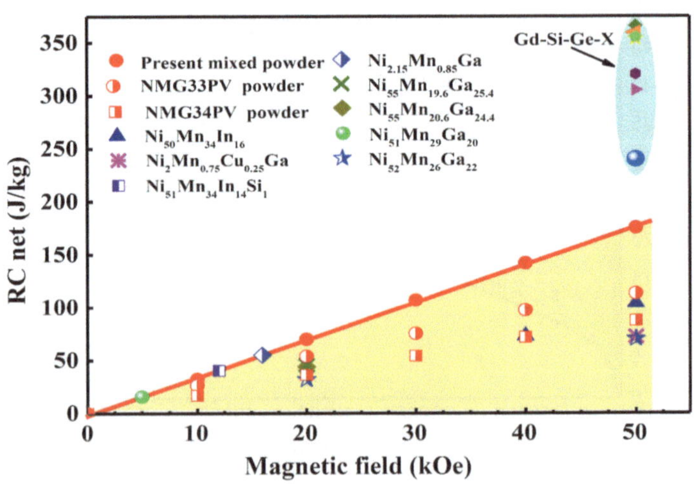

Fig. 6.14 Comparison of RC_{net} values of different alloys

alloy particles, ~20.3 K, resulting in a small RC_{net} value. RC_{net} values of mixed particles under different magnetic fields are higher than other Ni–Mn-based alloys [12–18]. Compared with rare earth compounds, mixed particles have smaller RC_{net} values [12, 19], but are more economical and practical.

6.4.4 *Optimization of Mixed Proportion of Different Particle Size*

When the NMG33PV and NMG34PV alloy particles are mixed at a mass ratio of 1:1, the two magnetocaloric peaks of the mixed particles are:

$$\Delta S_{\mathrm{m}}^{\mathrm{mix}} = w_{\mathrm{A}} \Delta S_{\mathrm{m}}^{\mathrm{A}} + w_{\mathrm{B}} \Delta S_{\mathrm{m}}^{\mathrm{B}} \tag{6.1}$$

where $\Delta S_{\mathrm{m}}{}^{\mathrm{mix}}$ represents the isothermal magnetic entropy change of mixed particles, $\Delta S_{\mathrm{m}}{}^{\mathrm{A}}$ and $\Delta S_{\mathrm{m}}{}^{\mathrm{B}}$ represent the isothermal magnetic entropy change of NMG33PV and NMG34PV alloy particles, respectively, and W^{A} and W^{B} represent the mass fraction of the alloy particles, respectively.

The magnetic entropy change ΔS_{m} of mixed particles by mass mixing law is shown in Fig. 6.15. The calculated results agree well with the actual test results, indicating that the isothermal magnetic entropy change of the mixed particles ΔS_{m} conforms to the mass mixing law, which can be used to optimize the mass ratio of the mixed particles and obtain the results to meet the actual demand.

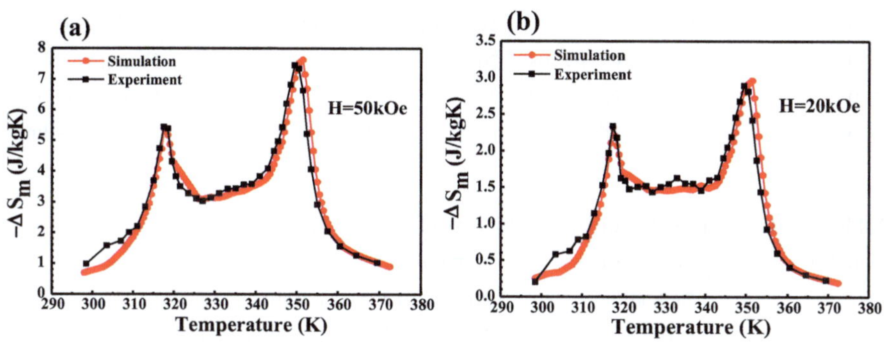

Fig. 6.15 ΔS_{m} of mixed particle by power law and comparison of experimental results. **a** $H = 50$ kOe; **b** $H = 20$ kOe

6.5 Heat Treatment of Ni–Mn–In–Co Alloy Particles

6.5.1 Effect of Stress-Relief Annealing on Particle Structure and Composition

The internal defects and internal stresses exist in the as-prepared particles obtained by mechanical grinding. Therefore, all particles are subjected to stress-relief annealing heat treatment to improve the particle structure and properties [8]. Considering that the defects and stresses introduced by different particles in the grinding process will be different, the stress-relief heat treatment process is shown in Table 6.4.

In order to reflect the change of the particle composition of the sample during different heat treatments, the bulk alloy after homogenization heat treatment and the particle composition after different heat treatment are tested. Each particle size component is averaged over at least 10 different regions, and the homogeneity of the particle composition is represented by the composition standard deviation, as shown in Tables 6.5, 6.6 and 6.7. The results show that the change of In and Co elements with the size is not obvious under the same heat treatment condition, but the change of Ni and Mn elements is obvious. When the particle size is more than 45 μm, the composition of alloy particles is basically unchanged, but when the particle size is less than 45 μm, the content of Mn decreases obviously, with the increase of Ni. The uniformity of particle composition decreases. This may be due to the less volatilization of large particle components under the protection of 0.5 atm

Table 6.4 Heat treatment process for Ni–Mn–In–Co alloy particles

Temperature (°C)	Object	Time (h)	Environment (atm Ar)	Cooling method
450	P2, P3, P5, P7, P8	20	0.5	Furnace cooling
700	P2, P3, P5, P7, P8	20	0.5	Furnace cooling
900	P2, P3, P5, P7, P8	20	0.5	Furnace cooling

Table 6.5 Heat treatment of different particle sizes at 450 °C/20 h

Number	Size (μm)	Composition
B	Bulk alloy	$Ni_{44.0\pm0.10}Mn_{38.5\pm0.08}In_{12.4\pm0.03}Co_{5.1\pm0.10}$
P1	880–1000	$Ni_{44.1\pm0.31}Mn_{38.2\pm0.35}In_{12.6\pm0.13}Co_{5.1\pm0.16}$
P2	550–650	$Ni_{44.0\pm0.19}Mn_{38.3\pm0.28}In_{12.5\pm0.16}Co_{5.2\pm0.16}$
P3	325–380	$Ni_{43.9\pm0.06}Mn_{38.3\pm0.19}In_{12.5\pm0.06}Co_{5.3\pm0.20}$
P4	212–230	$Ni_{44.0\pm0.15}Mn_{38.4\pm0.05}In_{12.5\pm0.16}Co_{5.1\pm0.08}$
P5	125–150	$Ni_{44.0\pm0.23}Mn_{38.5\pm0.07}In_{12.5\pm0.09}Co_{5.0\pm0.17}$
P6	61–96	$Ni_{44.0\pm0.14}Mn_{38.2\pm0.37}In_{12.6\pm0.15}Co_{5.2\pm0.09}$
P7	38–45	$Ni_{44.3\pm0.04}Mn_{38.2\pm0.10}In_{12.4\pm0.20}Co_{5.1\pm0.10}$
P8	13–18	$Ni_{44.6\pm0.18}Mn_{38.0\pm0.08}In_{12.5\pm0.15}Co_{4.9\pm0.06}$

Table 6.6 Composition of different particle sizes in heat treatment at 700 °C/20 h

Number	Size (μm)	Composition
B	Bulk alloy	$Ni_{44.0\pm0.10}Mn_{38.5\pm0.08}In_{12.4\pm0.03}Co_{5.1\pm0.10}$
P1	880–1000	$Ni_{43.9\pm0.17}Mn_{38.5\pm0.09}In_{12.5\pm0.07}Co_{5.1\pm0.10}$
P2	550–650	$Ni_{43.9\pm0.22}Mn_{38.4\pm0.14}In_{12.5\pm0.08}Co_{5.2\pm0.13}$
P3	325–380	$Ni_{44.0\pm0.17}Mn_{38.4\pm0.13}In_{12.4\pm0.17}Co_{5.3\pm0.16}$
P4	212–230	$Ni_{44.0\pm0.17}Mn_{38.4\pm0.15}In_{12.4\pm0.10}Co_{5.2\pm0.13}$
P5	125–150	$Ni_{44.0\pm0.15}Mn_{38.4\pm0.18}In_{12.4\pm0.08}Co_{5.2\pm0.10}$
P6	61–96	$Ni_{44.0\pm0.24}Mn_{38.3\pm0.25}In_{12.5\pm0.02}Co_{5.2\pm0.05}$
P7	38–45	$Ni_{44.4\pm0.23}Mn_{38.2\pm0.25}In_{12.3\pm0.08}Co_{5.1\pm0.13}$
P8	13–18	$Ni_{44.5\pm0.30}Mn_{38.0\pm0.45}In_{12.4\pm0.00}Co_{5.1\pm0.14}$

Table 6.7 Composition of different particle sizes in heat treatment at 900 °C/20 h

Number	Size (μm)	Composition
B	Ingot	$Ni_{44.0\pm0.10}Mn_{38.3\pm0.08}In_{12.4\pm0.03}Co_{5.1\pm0.10}$
P2	550–650	$Ni_{44.0\pm0.26}Mn_{38.3\pm0.20}In_{12.4\pm0.12}Co_{5.2\pm0.10}$
P8	13–18	$Ni_{44.6\pm0.40}Mn_{37.9\pm0.63}In_{12.4\pm0.45}Co_{5.1\pm0.32}$

argon atmosphere. Furthermore, the standard deviation of alloy particle composition becomes larger with the increase of heat treatment temperature, which indicates that the higher the temperature, the more volatilization of the alloy elements, the worse the composition uniformity.

6.5.2 *Effect of Stress-Relief Annealing on Martensitic Transformation of Particles*

To reveal the overall phase transition characteristics and the corresponding magnetic properties of different particle size samples after heat treatment with stress-relief annealing 450 °C/20 h above, all samples are tested with M–T curves under an external magnetic field of 200 Oe and 50 kOe [8]. Through the analysis of the M–T curve by the tangent method, we can get the martensitic transformation temperatures and thermal hysteresis ($\Delta T = \frac{A_S+A_f}{2} - \frac{M_s+M_f}{2}$). The result is shown in Table 6.8. With the increase of the applied magnetic field, the phase transition temperature points of each alloy particle decrease, and the phase transition process hysteresis increases. Meanwhile, the hysteresis of martensitic transformation process after 450 °C/20 h heat treatment increases with the decrease of alloy particle size.

The phase transition hysteresis is a key factor affecting the magnetocaloric effect of materials, which is mainly related to the resistance during the martensitic transformation. Although the phase transition hysteresis of the material is greatly reduced

Table 6.8 Martensite temperature of different particles after heat treatment at 450 °C/20 h

Number	Magnetic field (Oe)	A_S	A_f	M_S	M_f	ΔT
		K				
P1	200	358	374	353	347	16
	50,000	343	363	344	332	15
P2	200	344	354	344	335	9.5
	50,000	318	339	327	305	12.5
P3	200	343	364	347	331	14.5
	50,000	315	345	330	298	16
P5	200	346	370	349	327	20
	50,000	325	350	328	297	25
P7	200	349	371	347	318	27.5
	50,000	333	353	325	273	44
P8	200	352	372	346	317	30.5
	50,000	332	353	323	270	46

after the bulk alloy is prepared into small size particles, the phase transition process hysteresis increases when the particle size is too small. It is speculated that this may be caused by the 450 °C thermal treatment, where the temperature is too low to completely effectively remove the internal stress and defects. M–T curves of the particles after high temperature stress-relief heat treatment show that the alloy particles still maintain the martensitic transformation after increasing the annealing temperature.

Unlike the low-temperature treatment, the phase transition hysteresis of alloy particles change after increasing the temperature: For particle samples with relatively large size ($D > 325$ μm), the phase transition hysteresis is largely unaffected; for particle samples with small size ($D < 325$ μm), the phase transition hysteresis is greatly reduced. The hysteresis does not change when the annealing temperature continues to increase to 900 °C, which indicates that 700 °C/20 h heat treatment has basically removed the internal stress.

Different heat treatment processes and particle size will inevitably lead to the change of phase transition temperature, which is the key factor affecting the final magnetocaloric effect of the material. The phase transition temperature of each composite gold particle sample after high temperature annealing is shown in Tables 6.9 and 6.10. With the increase of the applied magnetic field, the phase transition temperature moves towards the low temperature as a whole. This phenomenon also indicates that the alloy martensitic phase transition process will be affected by the magnetic field.

Since the phase transition hysteresis is one of the key factors affecting the final magnetocaloric properties of the materials, the thermal hysteresis of alloy particles of various sizes after different stress-relief heat treatment is summarized and analyzed as shown in Fig. 6.16. For alloy particles of the same size, when the particle size

Table 6.9 Martensite temperature of different particles after heat treatment at 700 °C/20 h

Number	Magnetic field (Oe)	A_s	A_f	M_s	M_f	ΔT
		K				
P2	200	345	355	346	334	10
	50,000	322	339	325	309	13.5
P3	200	347	355	345	335	11
	50,000	323	336	323	309	13.5
P5	200	346	354	341	332	14.5
	50,000	326	337	320	307	18
P7	200	351	361	349	331	16
	50,000	331	347	329	305	22
P8	200	340	366	351	325	15
	50,000	316	352	332	297	19.5

Table 6.10 Martensite temperature of different particles after heat treatment at 900 °C/20 h

Number	Magnetic field (Oe)	A_s	A_f	M_s	M_f	ΔT
		K				
P2	200	344	357	348	334	9.5
	50,000	325	341	329	308	14.5
P8	200	343	367	351	325	17
	50,000	316	350	333	290	21.5

is more than 325 μm, the hysteresis is basically not affected by the heat treatment temperature; while when the particle size is less than 325 μm, the phase transition hysteresis decreases obviously with the increase of heat treatment temperature. This phenomenon suggests that the hysteresis may be related to the internal defects and stress state of the material. Combined with the above results of microstructure and phase, the small and medium size particles ($D < 325$ μm) have more stress than the large size particles in the grinding process, which leads to the internal stress accumulation. The stress-relief heat treatment process for large size particles ($D > 325$ μm) is 450 °C/20 h, while for small size particles ($D < 325$ μm) it requires a higher temperature of 700 °C/20 h.

Heat treatment of 450 °C/20 h of heat-treated alloy particles and 900 °C/20 h at higher temperatures reduced the hysteresis of small size particles by annealing 700 and 900 °C but did not affect large size particles. Moreover, comparing the two heat treatments 700 and 900 °C found no change occur even for small size particle hysteresis, which indicates that 700 °C/20 h heat treatment has basically removed the stress and defects inside the particles.

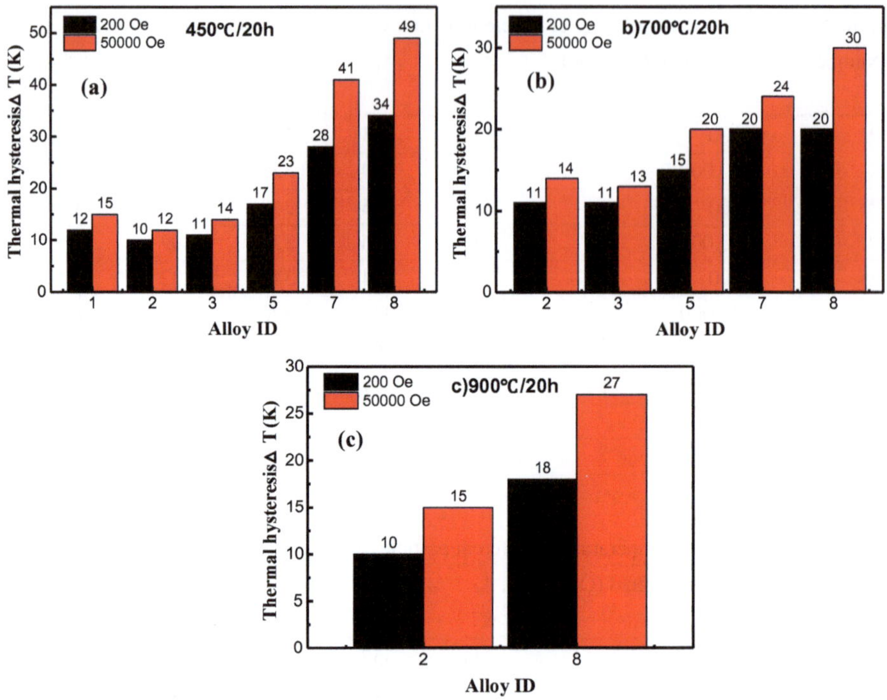

Fig. 6.16 Phase change hysteresis of the particles of different sizes under different annealing states. **a** 450 °C/20 h; **b** 700 °C/20 h; **c** 900 °C/20 h

6.6 Magnetic and Thermal Properties of Ni–Mn–In–Co Particles

In order to verify the practical application of the alloy particles studied in this experiment, the magnetocaloric effect of Ni–Mn–In–Co alloy particles with single particle size and mixed particle size is measured by designing the composition of the particles [8].

6.6.1 Composition Ni–Mn–In–Co Mixed Particles

Phase transition temperature of Ni–Mn–In–Co alloy strongly depends on the composition, and the phase transition temperature affects the structure and properties of the alloy martensite because the specific surface area and free surface of particles are very large, and Mn elements in the alloy are easy to volatilize at high temperature. The composition of particle samples in different size ranges is basically unchanged when 0.5 atm high purity argon is filled in the glass tube. Therefore, in order to

Table 6.11 Heat treatment process for Ni–Mn–In–Co particle for composition adjustment

Temperature (°C)	Object	Time (h)	Environment	Cooling method
900	P2, P5, P7	5	Vacuum	Furnace cooling
900	P2, P5, P7	10	Vacuum	Furnace cooling
900	P2, P5, P7	40	Vacuum	Furnace cooling
900	P7	2	0.2 atm Ar	Furnace cooling
900	P7	5	0.2 atm Ar	Furnace cooling
900	P7	10	0.2 atm Ar	Furnace cooling

mix alloy particles with different phase change temperatures to broaden the working temperature range of alloy materials, the argon concentration and heat treatment time are adjusted. Heat treatment process is listed in Table 6.11.

In order to reveal the effect of the above volatile heat treatment on the regulation of alloy particle composition, the components of all samples are tested as shown in Table 6.12, and at least ten particle samples of this component are taken for average at each component point. The component deviation of each group of particles after heat treatment is large, which indicates that the temperature can effectively control the phase transition temperature of alloy particles. The minimum content of Mn elements decreased with the increase of holding time, but the content of Mn elements is slightly higher than that of 40 h after 5 h treatment. This may be because the temperature treatment has more influence on the volatilization of elements in small size particles, the particle composition is more uneven, and the number of ten samples taken in the test composition is less. The alloy particles filled with 0.2 atm argon gas can be seen that the content of In elements in the particles decreases, the content of Co elements increases but the content of Mn elements does not change.

Table 6.12 Composition of different samples after heat treatment

Size	Heat treatment process	Composition
P7 (38–45 μm)	① 900 °C/2 h, 0.2 atm Ar	$Ni_{43.4\pm0.69}Mn_{38.4\pm0.58}In_{11.6\pm0.66}Co_{6.6\pm0.90}$
	② 900 °C/5 h, 0.2 atm Ar	$Ni_{43.4\pm0.75}Mn_{38.4\pm0.64}In_{11.7\pm0.80}Co_{6.5\pm1.19}$
	③ 900 °C/10 h, 0.2 atm Ar	$Ni_{44.1\pm0.43}Mn_{38.3\pm0.37}In_{12.1\pm0.86}Co_{5.5\pm0.88}$
	④ 900 °C/5 h, vacuum	$Ni_{45.2\pm0.34}Mn_{37.3\pm0.74}In_{12.3\pm0.58}Co_{5.2\pm0.28}$
	⑤ 900 °C/10 h, vacuum	$Ni_{45.1\pm0.31}Mn_{36.9\pm0.29}In_{12.9\pm0.27}Co_{5.1\pm0.09}$
	⑥ 900 °C/40 h, vacuum	$Ni_{45.0\pm0.44}Mn_{37.7\pm0.25}In_{12.2\pm0.29}Co_{5.1\pm0.45}$
P5 (125–150 μm)	⑦ 900 °C/5 h, vacuum	$Ni_{44.0\pm0.26}Mn_{38.0\pm0.40}In_{13.0\pm0.20}Co_{5.0\pm0.13}$
	⑧ 900 °C/10 h, vacuum	$Ni_{43.8\pm0.33}Mn_{38.2\pm0.20}In_{12.9\pm0.20}Co_{5.1\pm0.11}$
	⑨ 900 °C/40 h, vacuum	$Ni_{39.7\pm0.40}Mn_{32.0\pm0.27}In_{23.6\pm0.50}Co_{4.7\pm0.11}$
P2 (550–650 μm)	⑩ 900 °C/5 h, vacuum	$Ni_{44.0\pm0.32}Mn_{38.4\pm0.30}In_{12.5\pm0.15}Co_{5.1\pm0.07}$
	⑪900 °C/10 h, vacuum	$Ni_{44.0\pm0.25}Mn_{38.4\pm0.21}In_{12.5\pm0.15}Co_{5.1\pm0.11}$
	⑫900 °C/40 h, vacuum	$Ni_{44.0\pm0.25}Mn_{38.4\pm0.17}In_{12.5\pm0.16}Co_{5.1\pm0.08}$

Compared with bulk alloy, the hysteresis of small size Ni–Mn–In–Co alloy particles is greatly reduced and the performance is improved. In order to further broaden the working temperature range of the material, several different heat-treated alloy particles in the previous section are selected for 1:1 equal mass mixing to test their properties.

6.6.2 Magnetic and Thermal Properties of Single Particle Size Ni–Mn–In–Co Particles

Here, 0.2 atm Ar environment and vacuum environment are selected for the heat treatment of P5 (125–150 μm) and P7 (38–45 μm) particles. The M–T curves are tested for the above groups of particles separately, as shown in Fig. 6.17. Compared with alloy particles with the same particle size range after stress-relief heat treatment, the temperature $|A_f - A_s|$ of martensitic transformation increases, which is due to the uneven composition distribution caused by the volatilization of the particles after heat treatment at high temperature for a long time.

In addition, comparing the particle samples of different heat treatment processes of the same size, it can be found that the phase transition temperature moves to the high temperature direction with the extension of the holding time, as shown in Fig. 6.18. The phase transformation temperature of P5 (125–150 μm) alloy particles after 5 and 40 h holding is ~27 K, and that of P7 (38–45 μm) alloy particles after 5 and 10 h treatment is ~10 K. This change is smaller than that of P5 particles because of the addition of 0.2 atm Ar during the heat treatment to suppress the volatilization of the components, and because of the reduction of the heat treatment time leading to a smaller change of the compositions.

The above analysis shows that P7 (38–45 μm) sample single size alloy particles have large compositional inhomogeneity, and the change of phase transition temperature of P5 (125–150 μm) alloy particles is more obvious after 900 °C holding

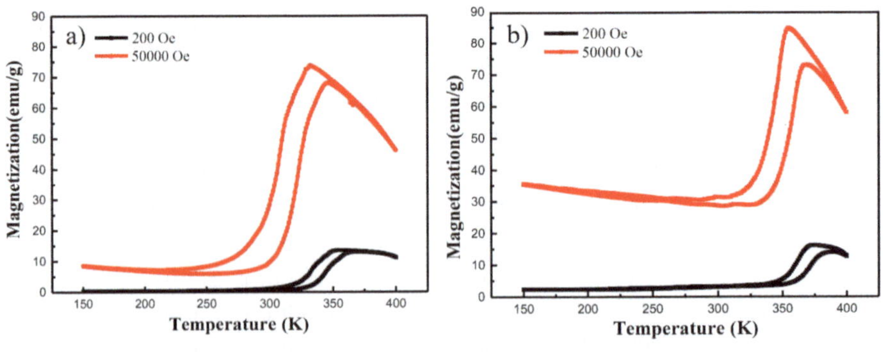

Fig. 6.17 M–T curve of P5 (125–150 μm) particles after different heat treatments. **a** 900 °C/5 h; **b** 900 °C/40 h

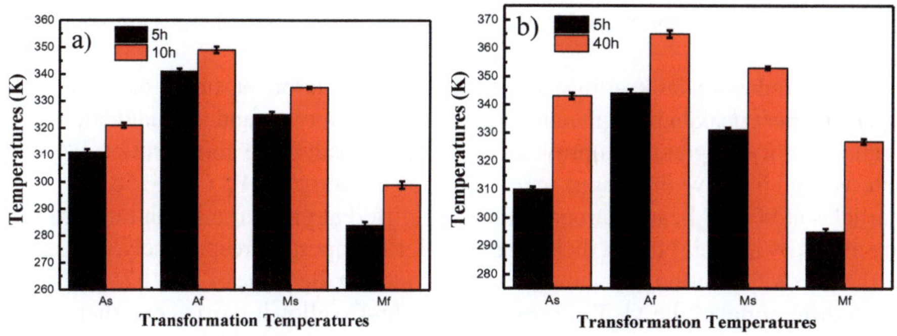

Fig. 6.18 Phase change temperature of Ni–Mn–In–Co alloy particle under 50 kOe. **a** P7 particles annealed at 0.2 atm argon, 900 °C/5 h and 900 °C/10 h; **b** P5 particle annealed at vacuum environment, 900 °C/5 h and 900 °C/40 h

different time, so the P5 particle samples of heat treatment 5 and 40 h are mixed at 1:1 mass ratio to widen the phase transition temperature range of the alloy. M–T curve test for mixed particles is shown in Fig. 6.19. The M–T curves show two peaks P1, P2, corresponding to the phase transition processes of the two particles. Compared with single size alloy particles, the phase transition temperature of mixed particles is obviously widened, which increases from 50 K of single particles to 100 K; and the magnetization is improved, while the phase transition hysteresis remains unchanged. It is expected that a smooth transition zone is formed between the two magnetocaloric peaks when the magnetocaloric properties are tested to obtain

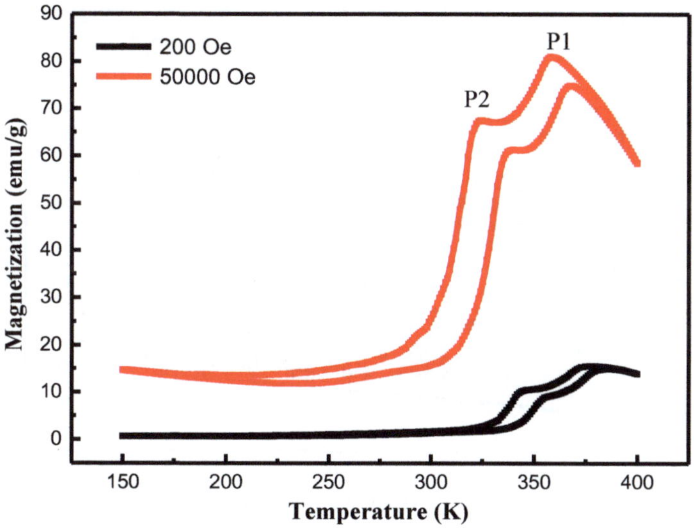

Fig. 6.19 M–T curve of mixed particles (P5 at 900 °C/5 h and 900 °C/40 h) after heat treatment

a favorable combination of a wide operating temperature interval and a high magnetic entropy change.

Compared with bulk alloys, the hysteresis of micron particles decreases, the magnetic heat peak half height and width ΔT_{FWHM} increase and the magnetic refrigeration ability is greatly improved. To further broaden the refrigeration working interval and improve the magnetic refrigeration capacity, the P5 (125–150 μm) alloy particles at 900 °C heat treatment 5 and 40 h different phase transition temperatures are mixed by mass 1:1 to test their magnetocaloric properties, respectively. According to the Maxwell relation, the value of ΔS_m depends on the value of $\partial M/\partial T$, so the first order differential of the M–T curve of Ni–Mn–In–Co alloy is carried out, therefore, a series of isothermal magnetization curves are measured near 323 and 364 K, that is, the test temperature range is 280–390 K, and the temperature interval of isothermal magnetization curves is 2, 3 and 5 K, the M–H curve is shown in Fig. 6.20. The saturation magnetization of the alloy particles increases with the increase of temperature, indicating that more and more martensite is transformed into ferromagnetic austenite with the increase of temperature. At 326 K, the saturation magnetization increases obviously, which indicates that a particle with a higher phase transition temperature starts to occur at 326 K. The magnetization of the mixed particles between 335 and 353 K shows the law of up–down fluctuation, which is due to the decrease of the magnetization of the particles with lower temperature of the previous phase transition near the end of the inverse martensitic transformation. The particles with higher temperature of the latter phase transformation are still in the phase of martensite to austenite transformation. The superposition of these two factors leads to the fluctuation of the overall magnetization of the mixed particles.

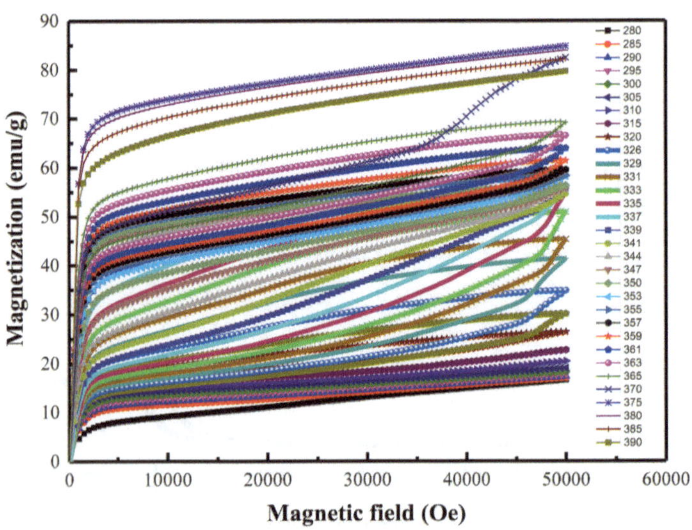

Fig. 6.20 M–H curve of mixed P5 at 900 °C/5 h and 900 °C/40 h particles after heat treatment

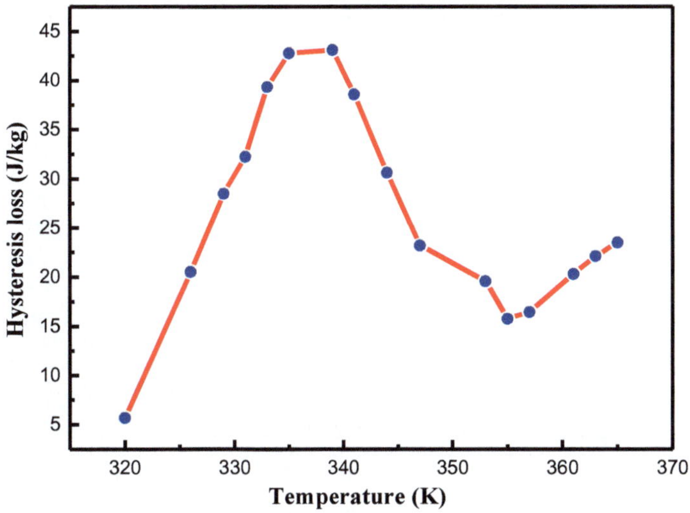

Fig. 6.21 Magnetic hysteresis–temperature curve of mixed P5 (125–150 μm) particles

Magnetic hysteresis produced by magnetic refrigeration materials under the action of magnetic field is also an important factor affecting magnetic thermal properties. According to the calculation of temperature magnetization/demagnetization curves near the phase transition when the maximum magnetic field of the mixed particles is 50 kOe in Fig. 6.20, the area surrounded by each temperature closed curve is after hysteresis, and the curve with hysteresis is obtained by calculation as shown in Fig. 6.21. The results show that the average hysteresis of mixed particles is 26.45 J/kg.

On the basis of isothermal magnetization curves at different temperatures, ΔS_m is calculated by Maxwell equation. The relationship curves of ΔS_m and temperature under different magnetic fields are obtained is shown in Fig. 6.22. Two peaks P1 and P2 appear in ΔS_m near 335 and 371 K, corresponding to the phase transition temperature of the alloy particles after two different heat treatments. The values at 50 kOe are obtained 5.27 J/(kg) and 10.67 J/(kg), respectively. When the minimum value of the transition zone of the two magnetocaloric peaks is taken as the magnetocaloric peak half-height width ΔT_{FWHM}, the working temperature range of the refrigeration reaches ~54 K, which is much larger than that of the Ni–Mn–In–Co alloy reported in the literature. For evaluating the magnetocaloric properties of magnetic refrigeration materials, it is necessary to consider not only the values of ΔS_m and refrigeration temperature ranges, but also their temperature dependence. The magnetocaloric effect is evaluated synthetically by the RC value of the material, expressed by subtracting the hysteresis value from the integral in the ΔT_{FWHM} range (T_1–T_2) under the ΔS_m–T curve. The calculated RC value of the mixed particles is ~180 J/kg.

It can be seen from Fig. 6.22 that the magnetic entropy change intensity of mixed particles is low in the transition region, so there is still much room for improvement. By adjusting the phase transition temperature of particles of the same size, the phase

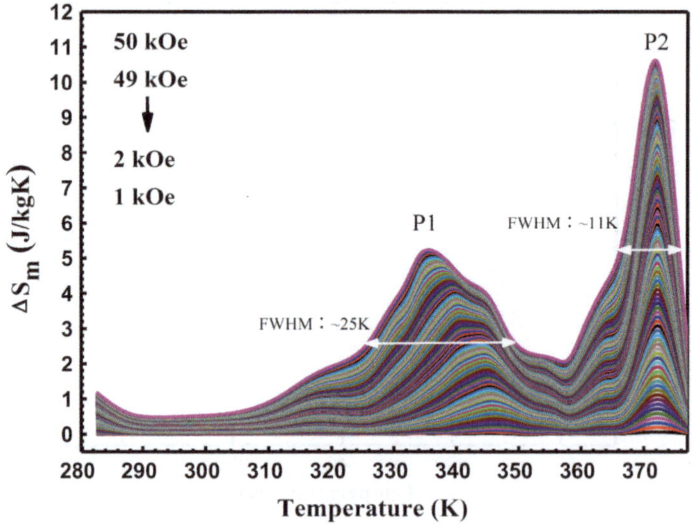

Fig. 6.22 Temperature-dependent ΔS_{m} in different magnetic fields of mixed P5 (125–150 μm) particles

transition temperature is near 356 K, and the entropy change and temperature curves are shown in Fig. 6.23. The results show that the actual operating temperature range ~53 K, refrigeration capacity RC value of ~290 J/kg can be obtained to achieve the giant magnetocaloric effect.

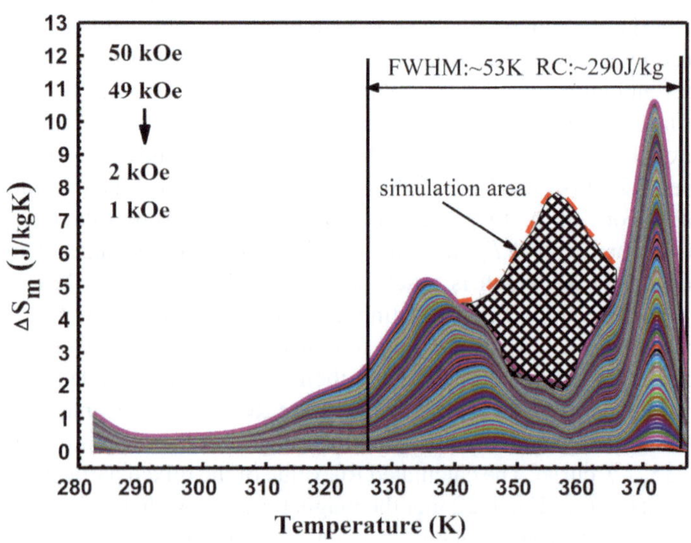

Fig. 6.23 ΔS_{m} of mixed particles under the changes of temperature in different magnetic field

6.7 Conclusions

Magnetic shape memory alloys are usually brittle, making preparation of micron-particles via mechanical crushing, grinding and sieving possible. Other methods such as electric spark corrosion can also be used, but the composition must be controlled during melting and electric spark corrosion of the alloy. Heat treatment of the particle is also necessary for chemical ordering or reducing the internal stress caused by mechanical crushing. The composition of the particles is tailorable.

The heat-treated particle exhibits well martensitic transformation, whose temperature is related to the heat treatment temperature and time, as well as the size of particles. The hysteresis of particles is smaller than the bulk alloy.

The magnetocaloric effect of particles is tailorable by mixing two or more particles, with successive martensitic transformation temperature. The working temperature interval can be greatly enhanced by this strategy.

References

1. Y.J. Tang, D.J. Smith, H. Hu et al., Structure and phase transformation of ferromagnetic shape memory alloy $Ni_{49}Mn_{30}Ga_{21}$ fine particles prepared by spark erosion. IEEE Trans. Magn. **39**(5), 3405–3407 (2003)
2. V.C. Solomon, D.J. Smith, Y. Tang et al., Microstructural characterization of Ni-Mn-Ga ferromagnetic shape memory alloy powders. J. Appl. Phys. **95**(11), 6954–6956 (2004)
3. V.C. Solomon, J.I. Hong, Y. Tang et al., Electron microscopy investigation of spark-eroded Ni-Mn-Ga ferromagnetic shape-memory alloy particles. Scripta Mater. **56**(7), 593–596 (2007)
4. Y.J. Tang, V.C. Solomon, D.J. Smith et al., Magnetocaloric effect in NiMnGa particles produced by spark erosion. J. Appl. Phys. **97**(10) (2005)
5. B. Tian, F. Chen, Y. Liu et al., Structural transition and atomic ordering of $Ni_{49.8}Mn_{28.5}Ga_{21.7}$ ferromagnetic shape memory alloy powders prepared by ball milling. Mater. Lett. **62**(17–18), 2851–2854 (2008)
6. B. Tian, F. Chen, Y. Liu et al., Effect of ball milling and post-annealing on magnetic properties of $Ni_{49.8}Mn_{28.5}Ga_{21.7}$ alloy powders. Intermetallics **16**(11–12), 1279–1284 (2008)
7. M.F. Qian, X.X. Zhang, Z.G. Jia et al., Enhanced magnetic refrigeration capacity in Ni-Mn-Ga micro-particles. Mater. Des. **148**, 115–123 (2018)
8. K.Y. Gao, *Research on the Martensitic Transformation and Magnetocaloric Properties of Ni-Mn-In-Co Powders* (Harbin Institute of Technology, Harbin, 2017)
9. X.H. Wan, *Research on the Phase Transformation and Magnetocaloric Properties of Ni-Mn-Ga Micron-Particles* (Harbin Institute of Technology, Harbin, 2016)
10. Y.V.B. de Santanna, M.A.C. de Melo, I.A. Santos et al., Structural, microstructural and magnetocaloric investigations in high-energy ball milled $Ni_{2.18}Mn_{0.82}Ga$ powders. Solid State Commun. **148**(7–8), 289–292 (2008)
11. F.X. Hu, B.G. Shen, J.R. Sun et al., Large magnetic entropy change in a Heusler alloy $Ni_{52.6}Mn_{23.1}Ga_{24.3}$ single crystal. Phys. Rev. B **64**(13), 13241213 (2001)
12. T.B. Zhang, Y.G. Chen, Y.B. Tang, The magnetocaloric effect and hysteresis properties of melt-spun $Gd_5Si_{1.8}Ge_{1.8}Sn_{0.4}$ alloy. J. Phys. D Appl. Phys. **40**(18), 5778–5784 (2007)
13. Z.B. Li, Y.D. Zhang, C.F. Sanchez-Valdes et al., Giant magnetocaloric effect in melt-spun Ni-Mn-Ga ribbons with magneto-multistructural transformation. Appl. Phys. Lett. **104**(4), 44101 (2014)

14. Y.P. Zhang, R.A. Hughes, J.F. Britten et al., Magnetocaloric effect in Ni-Mn-Ga thin films under concurrent magnetostructural and Curie transitions. J. Appl. Phys. **110**(1), 13910 (2011)
15. N.V.R. Rao, R. Gopalan, V. Chandrasekaran et al., Microstructure, magnetic properties and magnetocaloric effect in melt-spun Ni-Mn-Ga ribbons. J. Alloys Compd. **478**(1–2), 59–62 (2009)
16. S. Stadler, M. Khan, J. Mitchell et al., Magnetocaloric properties of $Ni_2Mn_{1-x}Cu_xGa$. Appl. Phys. Lett. **88**(19), 192511 (2006)
17. L. Pareti, M. Solzi, F. Albertini et al., Giant entropy change at the co-occurrence of structural and magnetic transitions in the $Ni_{2.19}Mn_{0.81}Ga$ Heusler alloy. Eur. Phys. J. B **32**(3), 303–307 (2003)
18. V.K. Sharma, M.K. Chattopadhyay, R. Kumar et al., Magnetocaloric effect in Heusler alloys Ni(50)Mn(34)In(16) and Ni(50)Mn(34)Sn(16). J. Phys. Condens. Matter **19**(49), 49620749 (2007)
19. R.D. Shull, V. Provenzano, A.J. Shapiro et al., The effects of small metal additions (Co, Cu, Ga, Mn, Al, Bi, Sn) on the magnetocaloric properties of the $Gd_5Ge_2Si_2$ alloy. J. Appl. Phys. **99**(8), 08K908 (2006)

Chapter 7
Application of Magnetic Shape Memory Alloys

7.1 Introduction

Magnetic shape memory alloy (MSMA) has excellent functional properties and has a wide application prospect. The temperature-controlled shape memory effect and superelasticity have been used for temperature control sensors, actuators or damper. For MSMA, the high magnetic-field-induced strain (MFIS) can be applied for high frequency actuators, while the magnetocaloric effect can be applied for room temperature refrigeration, which is promising to replace the traditional gas compression refrigeration technology.

By preparing small-sized MSMA alloys, the working frequency can be further enhanced with little hysteresis and high reversibility. In addition, the working temperature interval can be increased, which is favorable for the cooling material. This chapter discusses the application of temperature- and magnetic-controlled shape memory effect and magnetocaloric effects.

7.2 Temperature-Controlled Shape Memory Effect

Traditional shape memory alloys (such as nickel–titanium alloys) are widely used in aerospace, electronic energy, civil and medical devices because of their excellent shape memory effect and superelasticity, combined with high specific strength, corrosion resistance and good biocompatibility. Functional materials with sensing and actuating capacity at multiple signals (i.e. heat, magnetic, light, electricity, etc.) are required. Unlike traditional shape memory alloys that work at fairly low frequency, MSMAs may work at very high frequency up to 5000 Hz.

X. Zhang and M. Qian, *Magnetic Shape Memory Alloys*,
https://doi.org/10.1007/978-981-16-6336-9_7

7.2.1 Aerospace Field

Application of shape memory alloy in aerospace field mainly includes deployable antenna, drive/separation mechanism, deformable intelligent device and intelligent robot, so as to provide heat control, active position control and so on [1].

7.2.1.1 Lunar Spacecraft Antenna

In order to collect information on the moon and send it back to earth, the lunar spacecraft needs a half-moon antenna with a diameter of several meters. For the convenience of transportation, the shape memory alloy is used to fold the martensite antenna into a small volume at low temperature. After arriving at the moon, the antenna restores the unfolded state by means of sunlight irradiation heating which drives backward martensitic transformation. This is firstly realized in 1970 when the United States used nickel–titanium alloy to prepare the antenna of the Apollo spacecraft. When the antenna is launched, it is folded into a sphere of only 5 cm diameter. In recent years, the development of in-orbit printing (4D printing) is expected to provide a similar effect in sunlight.

7.2.1.2 Powerless Drive

Memory alloys can be used as powerless driving because of their driving force during phase recovery. In the field of aerospace, the simple action of opening or closing can be completed by making shape memory alloy into an adaptive powerless driver with temperature change. The technology has been used in the "Martian Pathfinder" spacecraft test unit shield open action, driving satellite solar panels and others [2]. Compared with magnetostrictive or piezoelectric material, SMA has the advantages of large driving stroke, high working stress, large driving force and sufficient solar radiation resistance in space. Solar memory alloy without power driver can directly use solar radiation energy to change the temperature field and use the phase change memory effect to obtain the required driving force and displacement.

7.2.1.3 Separation Device

Shape memory alloy has been used in satellite separation mechanisms [3]. The separation device is divided into pyrotechnic and non-fire devices. The explosive device is driven by the explosion of gunpowder or the combustion of high pressure gas to push the function mechanism to complete the predetermined function. At present, the shape memory driver is one of the main non-fire release devices.

Shape memory alloy separation devices have the following types: (1) Shape memory alloy brittle bolt. The shape memory alloy driver is used to break the titanium screw with the weakening groove in the tension state. When the shape memory alloy is heated to the phase transition temperature by the built-in heater, the shape memory driver is elongated. Bolts with weakening grooves extend until failure breaks. This process reduces preload and impact. The brittle bolt can be used to release 22,000 N of load in less than 25 s. Shape memory alloy driver can be reused after cooling and returning to martensite state [3]. (2) Release device. Air Force Research Laboratory and Lockheed Martin have developed two release devices, LFN separation nuts [4, 5]. The connection force of the LFN separation nut and the TSN separation nut can reach 1300 and 2500 kg, respectively. Both of them use redundant shape memory alloy to drive the steel ball brake with a responding time <50 ms.

The U.S. Air Force Research Laboratory and Starsys have worked on a release device based on shape memory alloys QWKNUT [5]. QWKNUT works like a LFN separation nut, which uses shape memory alloy triggers to release bolts from segmented nuts. Its volume and mass are similar to the LFN separation nut. The friction force has to be reduced in order to reduce the electric energy consumption and increase release speed for QWKNUT. QWKNUT successfully released the Falcon Sat I spacecraft in 2000.

Starsys developed a FASSN separation nut [5–7]. It uses the shape memory alloy torsion drive device to start the rotation separation of the flywheel nut so that 90% of the energy is stored in the flywheel nut after the separation is completed, and the energy is converted into the rotational kinetic energy of the flywheel nut. The separation impact is very small. FASSN separation nut also has fast separation speed (50 ms), small mass, compact volume and high reliability.

7.2.1.4 Deformable Aircraft Components

In order to reduce energy consumption or vibration of aircraft, the aircraft components have to adapt quickly to flight environment, such as jet flow. By using shape memory alloy deformation, the torsional angle of helicopter rotor blade can be changed, which can adapt to the torsion of rotor blade under different flight conditions and reduce the damage [8].

By installing the shape memory alloy driver at the engine nozzle and adjusting the driver by heating, the nozzle can achieve 20% expansion and contraction, meeting the requirements of reducing noise and saving energy under complex flight conditions [9]. Aircraft mobility depends on the mobility of the rear flap of the aircraft wing. At present, most aircraft uses hydraulic system to realize complex electromechanical transmission. In some cases, multiple hydraulic systems are required to operate at the same time in order to ensure reliability, making the aircraft heavy and costly. The choice of shape memory alloy can make the aircraft flap achieve the same effect as the hydraulic system by external stimulation, thus saving energy and reducing weight and cost [10]. The wire bents according to the set way by temperature control, which can achieve the purpose of swinging. In this case, if the magnetic drive of ferromagnetic shape memory alloy is used, the frequency of swing action speed can be greatly increased.

Intelligent skinning technology is to implant intelligent structures into aircraft skins, including detection elements (sensors), micro-processing controller (signal processors), driving elements (microbrakes) and interconnected circuits, to form a neural network of aircraft, which can not only perceive its own physical situation, but also remain sensitive to the external environment. This endows aircraft materials, components with the functions of self-detection, monitoring, correction, adaptation and memory, thinking, judgment and response. Similarly, the detection elements are pasted on the surface of the body structure or embedded in the body structure to monitor the health of the body.

7.2.1.5 Intelligent Robots

In the field of aerospace, robot substitution is usually used. As shown in Fig. 7.1, the fingers and wrists of a robot can bend and close by the expansion of the spiral

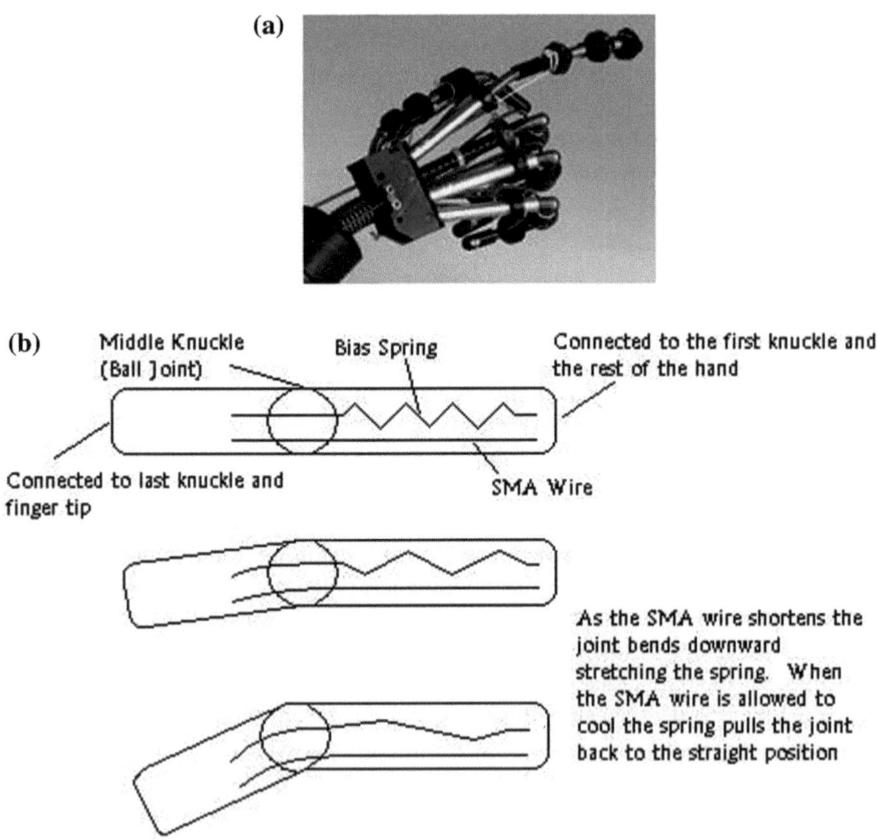

Fig. 7.1 Intelligent manipulator/robot muscle [10, 11]. **a** Physical map; **b** schematic

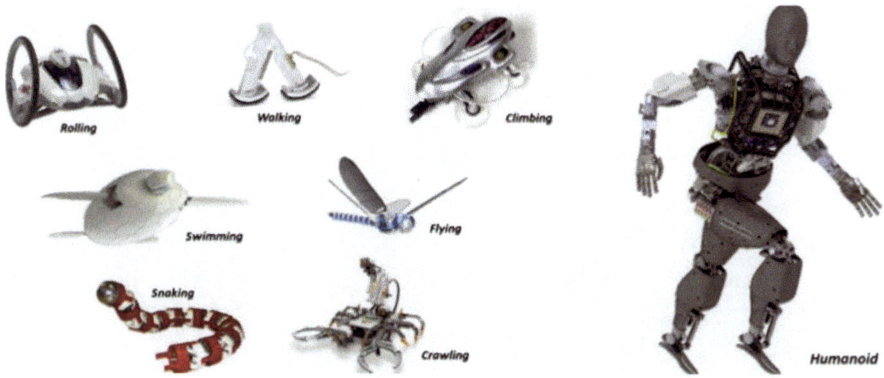

Fig. 7.2 Multiple actions realized by actuation of shape memory alloy in robots [11]

spring; the elbows and shoulders can bend by the expansion of the straight wire. Each shape memory alloy element is controlled by a direct pulse width adjustable current or magnetic field [10, 11]. In addition, a variety of robot actions can be realized, e.g. walking, turning, climbing/creeping, swimming/flying, snake swimming, and even human-like motion, as shown in Fig. 7.2 [11].

7.2.2 Automobile Component

Shape memory alloy is widely used in automobiles (Fig. 7.3), in the interior and engine compartment, such as radiator, fan clutch, clutch starting device, piston up and down motion device. It can also be used in the body and exterior parts, such as lamp, wiper, car shell, rearview mirror and others [11].

7.2.3 Medical Field

For medical application, good biocompatibility is necessary. Shape memory alloys can be used in dental braces, orthodontic appliances, bone healing [11]. They are also widely used in orthopedics [12], and have been successfully used in patellar [13], clavicle, carpal [14], femoral head [15] and spinal [16].

7.2.4 Daily Life

A temperature regulator based on a two-way shape memory effect can be used to adjust the change of water temperature by controlling the length of the shape memory

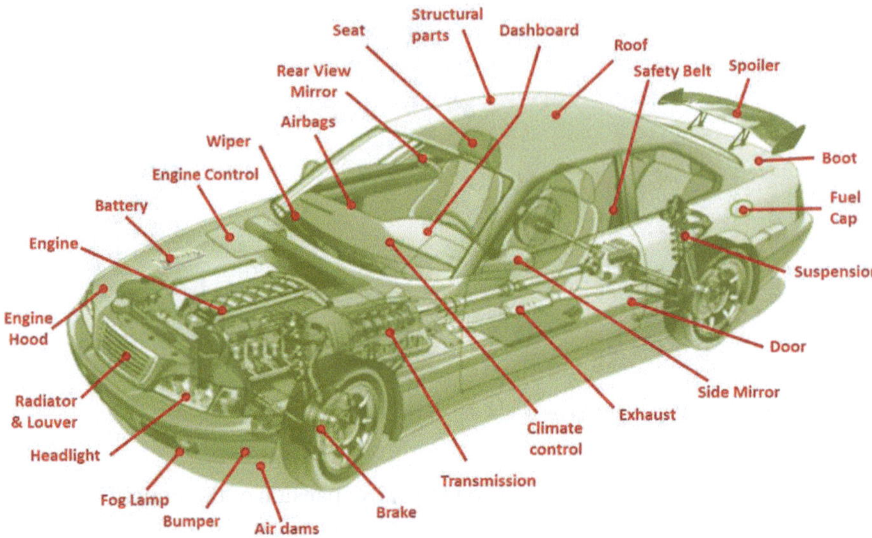

Fig. 7.3 Sample of application of shape memory alloys in automobile [11]

alloy spring [17]. Shape memory alloys can also be used as an electronic fastener, high pressure device plug, memory glasses frame and anti-ironing valve. Shape memory alloy microwire has been used in fire protection and heat insulation fabric [18] and temperature control structure fabric. A shape memory alloy for damping shock resistance is applied to the Alaskan Road Viaduct replacement bridge [19] in order to reduce the damage caused by the earthquake. After an earthquake of magnitude 7.5, the pier can still restore the original shape.

7.3 Magnetic Shape Memory Effect

Magnetic shape memory alloys, especially Ni–Mn–Ga alloy, have the ability to produce a large magnetic-field-induced strain. MSMAs can be driven quickly as actuators as well as sensors by taking advantage of the change of magnetization under external load. Compared with the traditional temperature-controlled memory alloy, the magnetron shape memory alloy has faster response speed, which is suitable for high frequency sensing and driving elements.

In the hydraulic field [20], piezoelectric materials and magnetostrictive materials have been widely used in the past few decades. The ferromagnetic shape memory alloy has the characteristics of high response speed and large strain compared to the above two materials. It is especially suitable for driving high speed and high frequency reversing valve [21–23]. In a series of ferromagnetic shape memory alloy actuators, the working frequency is as high as 400 Hz, i.e. operating response time is only 1.3 ms for a high speed switch valve designed by $Ni_{50}Mn_{28}Ga_{22}$ alloy [24]. It has the advantages of small forward flow pressure, reverse cutoff and small leakage. When the coil 3 generates a magnetic field perpendicular to the ferromagnetic shape memory bar 15, magnetic shape memory alloy rod elongates and pushes rod 13, and thus opens the coning spool 9. The pressure oil flows from the inlet 12 to the outlet 8. When power is off, reset spring 10 closes the spool and preloading spring 14 restores the shape of the ferromagnetic shape memory alloy rod.

An energy collector [25] is designed by the Adaptamat Company. The working principle is to repeatedly apply tension and compression load to deform the ferromagnetic shape memory alloy and generate voltage in the coil, which can be used for small load power supply ~150 kJ/m^3.

An actuator based on a ferromagnetic shape memory alloy film is shown in Fig. 7.4. The overall size is 7 mm × 2 mm × 5 mm [26]. The working principle is to place a curved film driver in an inhomogeneous magnetic field to control the precise movement of the mirror by MFIS. The magnetic and shape recovery forces are opposite in direction. Repeated motion can be realized in a single component driver. By applying alternating current, the periodic oscillation of the excitation signal is excited, and the reciprocating motion of the alloy is used to control the deflection of the micro mirror in the actuator. A prototype of a micro scanner has been developed based on this drive mechanism.

A mini-pump developed by Boise State University in the United States that can be used to transport sub microliter drugs directly to specific areas of the brain. This small volume micro pump can be placed in the head of a rat for drug delivery and monitoring of brain activity as the rat moves [27].

Magnetic shape memory alloys can also be used to detect the recovery of broken bones without ray imaging to detect healing and recovery state. The broken bone is fixed with a ferromagnetic shape memory alloy. A harmless magnetic field is applied

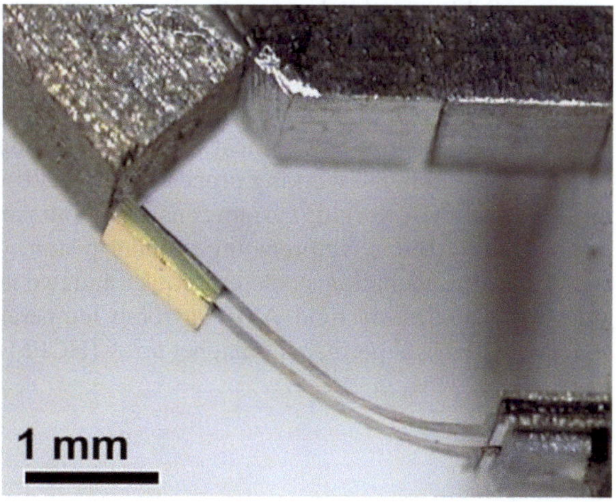

Fig. 7.4 A micro actuator prototype based on Ni–Mn–Ga alloy film [26]

to the alloy sheet to determine the healing by detecting the strain feedback produced in the alloy sheet [28].

7.4 Application of Magnetocaloric Effect

7.4.1 *Traditional Magnetic Refrigerator*

Usually, the room temperature magnetic refrigeration machine is mainly composed of magnetic field system, cooler (including magnetic refrigeration working fluid), cold and hot end heat exchanger, heat exchange fluid, driving mechanism and controller. The first prototype of magnetic refrigeration machine is prepared by General Electric Company of the United States. Room temperature magnetic refrigeration machine is divided into reciprocating and rotating types. Reciprocating machine is composed of permanent magnet field system and driving part, magnetic refrigeration material reheater, cold and hot end heat exchanger and driving part, motion control parts, data acquisition parts and machine control interface, etc. [29].

The performance of a refrigeration cooler is affected by many factors, including fluid flow rate, specific heat, thermal conductivity, size of magnetic particle size, magnetic field intensity, working speed and so on. Active magnetic Regenerator (AMR) can effectively reduce the lattice heat loss in the magnetic working fluid, and is a key component of the room temperature magnetic refrigeration system. The performance of the AMR has an important effect on the efficiency of the magnetic refrigerator. It is of great significance to study the performance between the internal magnetocaloric material and the heat exchange fluid in the AMR to improve the performance of the magnetic refrigerator.

A room temperature porous rotary AMR is divided into four regions: The 1st magnetic field area, the demagnetization area, the 2nd magnetic field area and the transition area [30]. Both magnetic field areas are cooled by water pipeline, and the demagnetization area is cooled by the chilled water. The whole AMR is divided into 36 small units, each of which is separated by a metal plate.

The AMR consists of two heat exchangers at high and low temperature ends, which are stored in the working process of the field supercooled flow and field heat flux [31]. An AMR working system consists of four parts [32]: two NdFeB permanent magnets; two AMR; a reciprocating cylinder-piston system driven by motor; passive cooler systems, including a cold end cooler and two hot end coolers. The permanent magnet magnetization field AMR at room temperature is usually provided by a concave cylindrical permanent magnet array (HCPMA).

7.4.2 *New Magnetic Refrigerator*

Room temperature magnetic refrigerators can be divided into an electromagnetic magnet, superconductor and permanent magnet according to the magnetic sources. According to the magnetic working fluids, it can be divided into magnetic fluid and solid magnetic working fluid. According to the movement mode of magnetic working fluid and magnetic field, it can be divided into static, reciprocating and rotating mode. These magnetic refrigeration prototypes have the following advantages and disadvantages:

(1) In stationary magnets mode, the magnetic working fluids remain stationary and are absent of moving parts. But it is rarely studied because electromagnet is necessary. However, the electromagnet can reach a high magnetic field strength only when a large volume of coil is used. This requirement makes it difficult for the stationary room temperature magnetic refrigerator to be commercially acceptable.
(2) In a reciprocating magnetic refrigerator, magnetic working fluid flows in and out of the magnetic field, and the working material is subject to magnetization and demagnetization to realize refrigeration. The structure of this refrigerator is simple and easy to realize. But it has the disadvantages of low motion frequency and low refrigeration efficiency, which seriously affects the refrigeration efficiency.
(3) Rotating magnetic refrigerator has compact structure, high motion frequency and high refrigeration efficiency. At the same time, it has complex structure. The circulation of fluid pipelines is complex, so it is often necessary to design complex special flow distribution valve group, and the working disc is subjected to unilateral force and so on.

7.4.2.1 New Composite Room Temperature Magnetic Refrigerator

Application of reciprocating magnetic refrigerators is limited because of their low cycle frequency. It cannot achieve higher refrigeration efficiency. A rotating magnetic refrigerator can achieve a very high frequency. But its heat exchange system is complex and has a high failure rate. To avoid these shortcomings, new composite refrigerator uses a dual rotating permanent (magnet) magnetic field system instead of fixed permanent magnetic field. The heat exchange is still reciprocating. This can ensure that the room temperature magnetic refrigerator heat exchange system is simple and reliable [33].

A composite permanent magnet magnetic refrigerator is mainly composed of seven parts [34]: magnetic field, active magnetic accumulator, power and transmission system, heat exchanger drive and circulation system, control system, cold and hot end heat exchanger and refrigerator. The magnetic field is composed of two nested cylindrical Halbach arrays, two series AMR, the outer cylinder of the nested structure is fixed, and the sinusoidal magnetic field is generated by the rotation of the inner cylinder. The magnetic field is composed of NdFeB permanent magnet. The system is controlled by programmable controller (PLC), which can change the rotation angle and speed of magnet. The power system is composed of servo motor and reducer, and the transmission system is gear drive. The circulation of heat transfer fluid is controlled by water pump and solenoid valve.

Currently, the composite magnetic refrigeration device can carry >100 L refrigeration space, a maximum refrigeration temperature difference 26 °C and minimum cold end temperature up to −6.5 °C. The cooling power of magnetic refrigerator reaches 1000 W, meaning that canned beer can be cooled from 25 to 5 °C like currently home-made refrigerator. In addition, the long-term operation effect is good.

7.4.2.2 Double-Ring Double-Set Magnetic Refrigerator

The double-loop two-group integrated magnetic refrigerator is composed of magnetic field system, cooler (including magnetic refrigeration material), cold and hot end heat exchanger, heat exchange fluid, driving mechanism and PLC controller, etc. [35]. The magnetic field of the refrigerator is composed of two sets of inner and outer nested annular cylindrical magnets; each set of external magnets is fixed. The inner magnets are driven and rotated by the motor, and the center of the inner magnet has a cylindrical working space. The internal magnets of the two sets of magnets keep the synchronous linkage through gear meshing, and ensure that the magnetic field phase produced by the two sets of magnets is staggered for half a week, that is, when the internal magnets are driven to rotate, the magnetic field of the two working spaces changes periodically and the phase difference is 180°.

7.4.2.3 Dual-Effect Magnetic Refrigerator

The double effect magnetic refrigerator has two magnetic working beds. When one magnetic working bed is always kept in the state of excitation exothermic, the other is in the state of demagnetization and endothermic. This double-effect can improve the refrigeration efficiency of reciprocating magnetic refrigerator. The refrigerator keeps the permanent magnet that provides the magnetization field stable, and the two magnetic working beds move in and out of the magnetic field alternatively. The refrigerator details can be found in [36].

7.4.2.4 Swing Magnetic Refrigerator

Reciprocating linear reciprocating motion, without the phenomenon of pipeline entanglement, simplifies the design of fluid pipeline circulation. At the same time, it has the disadvantages of low frequency and low refrigeration efficiency because reciprocating motion has the limitation of motion frequency. The swing motion mode meets the above requirements. The so-called swing is the reciprocating angle motion similar to the pendulum. Swing motion has the characteristics of rotating angular motion (which makes the magnetic working fluid locate quickly and increase the frequency). It also has the characteristics of reciprocating linear motion (no pipe entanglement, simplified structure) [36].

In a swing motion mode [36], the rotating shaft rotates 90° counterclockwise, the rotating shaft temporarily, the flow body, then the clockwise rotation 90°, the rotating shaft stops again, the flow body, so the circulation runs. The characteristic of the swing type is that there is no problem with pipe entanglement and winding. So the fluid pipeline can be as simple as reciprocating. The rapid positioning of the magnetic material box can be realized because of the angular motion.

7.4.3 Well-Studied Magnetic Refrigeration Prototype

The following is a brief introduction to some of the room temperature magnetic refrigeration prototypes. Most of the refrigeration materials used in magnetic refrigeration prototypes contain rare-earth elements such as Gd and La. In 1976, Brown [37] designed the world's first room temperature magnetic refrigeration prototype. A high-purity metal Gd was first used as a room temperature magnetic refrigerant. At a magnetic field intensity of 7 T provided by the superconductor, the temperature difference between the cold end and the hot end reaches 47 K after 50 cycles of reciprocating motion. In 1978, Steyert [38, 39] prepared the world's first rotating room temperature magnetic refrigeration prototype. Gd disk rotates in and out of the magnetic field to realize a temperature span 14 K at the frequency 1 Hz and refrigerating capacity 32 kW/L.

In 1997, the room temperature magnetic refrigeration prototype developed jointly by NASA and AMES Laboratory successfully operated for >5000 h [40]. The prototype is reciprocating using a superconductor magnetic field (50 kOe), Gd particles with a particle size of 0.15 and 0.30 mm. A maximum refrigeration power 600 W and maximum temperature difference 38 K (274–312 K) are obtained. The refrigeration efficiency reaches 60% of the Carnot cycle.

A rotating permanent magnet magnetic refrigeration machine was invented by Bohigas in 2000 [41]. The refrigerator uses 8 permanent magnets to assemble into two excitation sources; the magnetic working fluid is banded Gd, encapsulated in two cylindrical plastic plates with a diameter of 11 cm, thick 8 mm and a diameter 7 cm, thick 8 mm; Olive oil is injected into the plastic plate as a heat transfer medium. The magnetic field intensity was only 0.3 T. The temperature span was only 1.6 K in the initial experiment. After the magnetic field intensity reaches 0.9 T, the maximum temperature span is 5 K.

A magnetic refrigeration prototype was designed and manufactured by Zimm et al. [42]. The prototype uses an active cooler to cycle with water as heat transfer medium. The prototype uses a NdFeB permanent magnets (1.5 T), and Gd and $La(Fe_{1-x}Si_x)_{13}H_y$ ($x = 0.12, y = 1.0$) alloy powder. The prototype is stable, low noise and adjustable operating frequency (0.5–4 Hz). The results show that the refrigerating capacity decreases with the increase of temperature span. As the velocity increases, a maximum refrigerating capacity of 50 W is reached.

Lu et al. [43] developed room temperature magnetic refrigeration prototype using active cooler cycle mode. The prototype uses permanent magnets (fields up to 1.7 T), 9 × 18 × 12 mm work gap, two sources of heat (30 ml/source), 112 g Gd working material, mixed water and soft matter heat transfer medium. Pneumatic device drives working fluid reciprocating in and out of magnetic field. A maximum of 8 K refrigeration temperature span is achieved with about 10 W output power. A similar rotating room temperature magnetic refrigerator is designed by Sichuan University [44].

Zheng et al. [45] designed an active room temperature magnetic refrigeration system. The system consists of permanent magnet, active cooler, hydraulic heat transfer circuit and motion control unit. Flux density of the self-designed magnet can reach 1.5 T, and the amount of magnet is reduced by 40% after using finite element optimization design.

In a word, solid magnetic cooling technology has attracted the attention of researchers all over the world because of its advantages of high efficiency and energy saving, green environmental protection and so on. The application of magnetic working material at present is mainly rare-earth alloys and compounds. However, rare-earth alloys are expensive and partly toxic. It is not the most ideal magnetic cooling material. The magnetic shape memory alloy has excellent magnetic and thermal properties. Through reasonable structure design and optimization, it can greatly reduce the hysteresis of the alloy, widen the refrigeration temperature range and improve its mechanical stability. In addition, the phase transition temperature of the alloy can be feasibly adjusted to room temperature. So, these alloys have become ideal magnetic refrigeration materials at room temperature.

7.5 Conclusions

Magnetic shape memory alloys have been demonstrated as potential actuators working at high frequency. The high magnetic-field-induced strain renders MSMA highly attractive for precise position control, which has been proved possible by Ni–Mn–Ga single crystals and films. However, the application of MSMA for actuators and sensors needs further investigation, where the working stability and fatigue behavior are still rarely proved.

Room temperature magnetic refrigeration based on the magnetocaloric effect has been realized. New prototype has been designed using various working systems and magnetic field sources. Permanent magnets such as NdFeB can act as a magnetic source for such refrigerators. The working mode of the prototype has been studied but obviously needs further optimization. The prepared prototype has to be subjected to intensive working verification before it is widely accepted as an alternative to the conventional gas compressing technique.

References

1. Z.G. Shen, Study on the processing property of memory alloy materials and its application in spaceborne field. Machinery **56**, 8–11 (2018)
2. Q. Li, M.B. Zhu, Thermal deformation control of shape memory alloy in space-borne antenna. Comput. Eng. Des. **30**, 3698–3701 (2009)
3. F.C. Cai, X.H. Meng, Non-pyrotechnic device for joining and separating full. Spacecr. Recov. Remote Sens. **4**, 50–55 (2005)
4. E.R. Fosness, S.J. Buckley, W.F. Gammill, Development and release devices efforts at the air force research laboratory space vehicles directorate, in *AIAA Space 2001—Conference and Exposition*, Albuquerque, NM, 28–30 Aug 2001
5. M. Lucy, R. Harfdy, E. Kist et al., Report on alternative devices to pyrotechnic on space craft. NASA Langley Technical Report Server: NASA-96-10css-mhl
6. D. Dowen, S. Christiansen, Development of a reusable, low-shock clamp band separation system for small spacecraft release applications, in *15th Annual/USU Conference on Small Satellites*, pp. 1–10
7. K.R. Gall, M.S. Lake, J. Harvey et al., Development of a shockless thermally actuated release nut using elastic memory composite material, in *44th Structure, Structural Dynamics, and Materials Conference*, Norfolk, AIAA 2003-1582
8. R.T. Ruggeri, R.C. Bussom, D.J. Arbogast, Development of a 1/4-scale NiTinol actuator for reconfigurable structures. Proc. SPIE 69300L (2008)
9. J.H. Mabe, F.T. Calkins, M.B. Alkislar, Variable area jet nozzle using shape memory alloy actuators in an antagonistic design. Proc. SPIE 69300T (2008)
10. http://webdocs.cs.ualberta.ca/database/MEMS/sma_mems/sma.html
11. J.J. Mohd, M. Leary, A. Subic et al., A review of shape memory alloy research, applications and opportunities. Mater. Des. **56**, 1078–1113 (2014)
12. X. Zhao, H.Q. Zhang, Clinical application of shape memory alloy in orthopaedics. World's Lat. Med. Inf. Dig. **19**(11), 95, 279 (2019)
13. Y. Yu, X.Z. Li, Z.H. Zheng, Application of nickel-titanium shape memory alloys in orthopedics. J. Clin. Rehabil. Tissue Eng. Res. **14**(47), 8840–8842 (2010)
14. W.B. Zhao, Material characteristics and medical application of nickel-titanium shape-memory alloy. J. Clin. Rehabil. Tissue Eng. Res. **11**(22), 4376–4379 (2007)
15. Q. Li, J.C. Tang, Biocompatibility and orthopaedic applications of nickel-titanium shape memory alloy. J. Clin. Rehabil. Tissue Eng. Res. **14**, 4745–4750 (2010)
16. Q. Li, Y.Y. Xia, J.C. Tang, Biocompatibility of surface-modified nickel titanium alloys and the effects of their continuous compressive stress on fracture healing. J. Clin. Rehabil. Tissue Eng. Res. **13**(38), 7593–7596 (2009)
17. Z.H. Chen, L.G. Zhou, M.J. Yan, Research of thermostat with temperature-sensing unit of shape memory alloys. J. Tung Wah Univ. (Nat. Sci. Ed.) **28**(6), 92–95 (2002)
18. X.C. Zhou, Application of shape memory alloy, a magic fabric in fabric. Mod. Chem. Res. **3**, 135–137 (2019)
19. N. Zhang, Application of shape memory alloy on Alaskan Road viaduct replacement bridge. World Bridge **46**, 91 (2018)
20. H. Shi, B. He, Z. Wang, Magneto-mechanical behavior of magnetic shape memory alloy and its application in hydraulic valve actuator. J. Mech. Eng. **54**(20), 235–243 (2018)
21. S. Thomas, P. Emmanouel, M. Laufenberg, Magnetic shape memory actuators for fluidic applications, in *Proceedings of 9th International Fluid Power Conference* (2014)
22. S.J. Murray, M. Marioni, S.M. Allen et al., 6% magnetic-field-induced strain by twin-boundary motion in ferromagnetic Ni-Mn-Ga. Appl. Phys. Lett. **77**, 886–888 (2000)
23. M. Chmielus, X.X. Zhang, C. Witherspoon et al., Giant magnetic-field-induced strains in polycrystalline Ni-Mn-Ga foams. Nat. Mater. **8**, 863–866 (2009)
24. T. Schiepp, M. Maier, E. Pagounis et al., FEM-simulation of magnetic shape memory actuators. IEEE Trans. Magn. **50**, 989–992 (2014)

25. Magnetic Shape Memory Actuators. http://www.themsmnet.net/devices.html
26. S.A. Wilson, R.P.J. Jourdain, Q. Zhang et al., New materials for micro-scale sensors and actuators. Mater. Sci. Eng. R **56**, 1–129 (2007)
27. Boise State Micro-Pump Aids Neurological Research. https://news.boisestate.edu/update/2014/09/24/micro-pump-developed-boise-state-ids-neurological-research-england/ (2014)
28. W.M. Guo, *Orthopaedic Applications of Ferromagnetic Shape Memory Alloys* (Massachusetts Institute of Technology, 2008)
29. P.Y. Jin, J.X.H. Huang, H.W. Yan, A room-temperature reciprocal magnetic refrigerator adopting digital control technology. Chin. Rare Earth **33**, 90–93 (2012)
30. P.X. Hou, C.P. Liu, J.H. Wu, Numerical simulation on porous rotary active magnetic regenerator in room temperature. Acta Armamentarii **36**, 938–945 (2015)
31. H. Jin, X.N. Xu, D.W. Lu, A simple method of analytical calculation of cooling power for magnetic refrigerator with active magnetic regenerator. Cryogenics 37–40 (2013)
32. P. Li, G.Y. Yu, M.Q. Gong, Analysis of magnetic forces and magnetic power in an active magnetic regenerative refrigerator. J. Eng. Thermophys. **28**(S1), 53–56 (2007)
33. H. Zhang, X.N. He, J. Shen, Cooling performance of a new hybrid room-temperature magnetic refrigerator. J. Eng. Thermophys. **34**, 5–8 (2013)
34. Z.Z. Wang, *The Research of Refrigeration Performance for a Combined Type Permanent Magnet Room Temperature Magnetic Refrigerator* (Inner Mongolia University of Science and Technology, Baotou, 2015)
35. Z.J. Li, X.Y. Xia, Research on the integration of double ring double group magnetic refrigeration machine. Rare Earth Inf. **375**, 30–31 (2015)
36. Z. Bai, *Design and Study on the Room Temperature Magnetic Refrigerator* (South China University of Technology, Guangzhou, 2012)
37. G.V. Brown, Magnetic heat pumping near room temperature. J. Appl. Phys. **47**(8), 3673–3680 (1976)
38. W.A. Steyert, High temperature refrigerator. US Patent No. 4107935 (1978)
39. W.A. Steyert, Stirling-cycle rotating magnetic refrigerators and heat engines for use near room temperature. J. Appl. Phys. **49**(3), 1216–1226 (1978)
40. K.A. Gschneidner Jr., V.K. Pecharsky, Thirty years of near room temperature magnetic cooling: where we are today and future prospects. Int. J. Refrig. **31**(6), 945–961 (2008)
41. X. Bohigas, E. Molins, A. Roig et al., Room-temperature magnetic refrigerator using permanent magnets. IEEE Trans. Magn. **36**(3), 538–544 (2000)
42. C. Zimm, A. Boeder, J. Chell et al., Design and performance of a permanent-magnet rotary refrigerator. Int. J. Refrig. **29**(8), 1302–1306 (2006)
43. D.W. Lu, L. Yu, X. Jin, Study on a model of room temperature magnetic refrigeration with permanent magnet. Cryogenics **4**, 33–35 (2003)
44. Y.G. Chen, Y.B. Tang, B.M. Wang et al., *A Permanent Magnet Rotary Magnetic Refrigerator* (International Institute of Refrigeration, 2007), pp. 309–315
45. Z.G. Zheng, H.Y. Yu, X.C. Zhong et al., Design and performance study of the active magnetic refrigerator for room-temperature application. Int. J. Refrig. **32**(1), 78–86 (2009)

MIX
Papier aus verantwortungsvollen Quellen
Paper from responsible sources
FSC® C105338

If you have any concerns about our products,
you can contact us on
ProductSafety@springernature.com

In case Publisher is established outside the EU,
the EU authorized representative is:
Springer Nature Customer Service Center GmbH
Europaplatz 3, 69115 Heidelberg, Germany

Printed by Libri Plureos GmbH
in Hamburg, Germany